# MARINE BENTHIC DYNAMICS

The Belle W. Baruch Library in Marine Science

THE BELLE W. BARUCH LIBRARY IN MARINE SCIENCE NUMBER 11

# Marine Benthic Dynamics

Edited by Kenneth R. Tenore
Bruce C. Coull

Published for the Belle W. Baruch Institute for Marine Biology and

Coastal Research by the

UNIVERSITY OF SOUTH CAROLINA PRESS

**Library of Congress Cataloging in Publication Data**

Belle W. Baruch Symposium in Marine Science, 11th,
    Georgetown, S. C., 1979.
    Marine benthic dynamics.

    (The Belle W. Baruch library in marine science;
no. 11)
    Includes index.
    1.  Marine ecology--Congresses.  2.  Benthos--
Congresses.  I.  Tenore, Kenneth R.  II.  Coull,
Bruce C.  III.  Belle W. Baruch Institute for
Marine Biology and Coastal Research.  IV.  Title.
V.  Series:  Belle W. Baruch library in marine
science; no. 11.
QH541.5.S3B44   1979        574.5'2636        80-15941
ISBN 0-87249-401-2

# DEDICATION

"The beginning is the signal part of any work...for that is the time at which the character is being formed and most readily receives the desired impressions." Plato, <u>The Republic</u>.

Howard Sanders, more than any other single scientist, has contributed not only to our knowledge of the benthos but also, more essentially, to the character of the study itself. His search for ecological principles that regulate the structure of benthic communities has led to pioneer work in animal-sediment relations and species diversity. The very subject of this Symposium - secondary production of the benthos - received early attention in Howard's dissertation on the benthos of Long Island Sound. He continues to set the pace for the study of deep-sea benthos while providing a firm foundation upon which other benthic ecologists can focus and build. He truly deserves the respect and admiration of his fellow scientists.

Even beyond his scientific acumen, Howard's love of study and knowledge of the benthos has helped countless young (and old) scientists to discover and renew their own enthusiasm by helping to draw the thread of interest and scientific curiosity from their research. The esprit de corps of his team at Woods Hole attests to his dynamic leadership. At seminars, Howard's questions focus always, and elaborately, on the general principles illustrated by a piece of research rather than to belabor a mistake, an omission or the obvious.

We gladly take this opportunity to thank Howard for the concern and care which characterize his love and dedication to the study of benthos and to our profession.

# PARTICIPANTS

CONTRIBUTORS

ROBERT C. ALLER, Department of Geophysical Sciences, University of
Chicago, Chicago, IL 60637.

DANIEL M. ALONGI, Department of Invertebrate Zoology, Virginia Institute of Marine Science, Gloucester Point, VA 23062.

WOLF E.H.W. ARNTZ, Institut Für Meereskunde, Universität Kiel, 23
Kiel, Federal Republic of Germany.

SUSAN S. BELL, Department of Biology, University of South Florida,
Tampa, FL 33620.

I.I. CHERBADGI, Institute of Marine Biology, Akademi Nauk SSSR,
Vladivostok 69022, USSR.

BRUCE C. COULL, Baruch Institute, University of South Carolina,
Columbia, SC 29208.

PAUL K. DAYTON, Scripps Institution of Oceanography, La Jolla, CA
92093.

JACK W. FELL, Rosensteil School of Marine and Atmospheric Science,
University of Miami, Miami, FL 33149.

B.N. FURNOS, Graduate School of Oceanography, University of Rhode
Island, Kingston, RI 02881.

MICHEL GLÉMAREC, Laboratoire d'Oceanographie Biologique, Université
de Bretagne Occidentale, 29283 Brest Cedex, France.

S.S. HALE, Department of Natural Resources, 323 East 4th Ave.,
Anchorage, AK 99501.

ROGER B. HANSON, Skidaway Institute of Oceanography, Savannah, GA
31406.

BARRY HARGRAVE, Marine Ecology Laboratory, Bedford Institute of
Oceanography, Dartmouth, Canada BZY 4A2.

BRENDA J. HARRISON, Department of Oceanography, University of British Columbia, Vancouver, B.C. Canada V6T 1W5.

PAUL G. HARRISON, Department of Botany, University of British Columbia, Vancouver, B.C. Canada V6T 1W5.

CARLO H. HEIP, Laboratorium voor Morphologie et Systematiek, Rijksuniversiteit-Gent, Ledeganckstraat 3, B-9000 Gent, Belgium.

JOHN E. HOBBIE, Marine Biological Laboratory, Woods Hole, MA 02543.

J.R. KELLY, Graduate School of Oceanography, University of Rhode Island, Kingston, RI 02881.

CINDY LEE, Woods Hole Oceanographic Institution, Woods Hole, MA 02543.

JEFFREY S. LEVINTON, Department of Ecology and Evolution, State University of New York, Stony Brook, NY 11794.

N.V. LOOTZIK, Institute of Marine Biology, Akademi Nauk SSSR, Vladivostok 29022, USSR.

GLENN R. LOPEZ, Curriculum in Marine Science, University of North Carolina, Chapel Hill, Chapel Hill, NC 27514.

I.M. MASTER, Rosenstiel School of Marine and Atmospheric Science, University of Miami, Miami, FL 33149.

ALAIN MENESGUEN, Laboratoire d'Oceanographie Biologique, Université de Bretagne Occidentale, 29283 Brest Cedex, France.

ERIC L. MILLS, Department of Oceanography, Dalhousie University, Halifax, N.S., Canada B3H 4J1.

ROBERT J. NAIMAN, Woods Hole Oceanographic Institution, Woods Hole, MA 02543.

S.Y. NEWELL, Rosenstiel School of Marine and Atmospheric Science, University of Miami, Miami, FL 33149.

SCOTT W. NIXON, Graduate School of Oceanography, University of Rhode Island, Kingston, RI 02881.

JOHN S. OLIVER, Scripps Institution of Oceanography, La Jolla, CA 92093.

CANDACE A. OVIATT, Graduate School of Oceanography, University of Rhode Island, Kingston, RI 02881.

MARIO M. PAMATMAT, Tiburon Center for Environmental Studies, San Francisco State University, PO Box 855, Tiburon CA 94920.

MIKHAIL V. PROPP, Institute of Marine Biology, Akademie Nauk USSR, Vladivostok 69022, USSR.

DONALD C. RHOADS, Department of Geology and Geophysics, Yale University, New Haven, CT 06520.

DONALD L. RICE, Marine Science Coastal Carolina College, Conway, SC 29526.

JOHN R. SIBERT, Pacific Biological Station, Fisheries and Marine Service, Nanaimo, B.C. Canada.

V.G. TARASOFF, Institute of Marine Biology, Akademi Nauk SSSR, Vladivostok 69022, USSR.

KENNETH R. TENORE, Skidaway Institute of Oceanography, Savannah, GA 31406.

JOHN H. TIETJEN, Department of Biology, City College of New York, New York, NY 10031.

RICHARD M. WARWICK, NERC Institute Marine and Environmental Research, Prospect Place, The Hoe, Plymouth, U.K.

JOSEPHINE Y. YINGST, Department of Biology, Wayne State University, Detroit, MI 48202.

BERNT F. ZEITZSCHEL, Institut für Meereskunde, Universität Kiel, 23 Kiel, Federal Republic of Germany.

OTHER PARTICIPANTS

DENNIS M. ALLEN
University of South Carolina

LEON M. CAMMEN
Bedford Institute of Oceanography

ROBERT S. CARNEY
National Science Foundation

EDWARD J. CHESNEY
Skidaway Institute of Oceanography

BRUCE E. DORNSEIF
Skidaway Institute of Oceanography

BETTYE W. DUDLEY
University of South Carolina

S. RAGNAR ELMGREN
Universitet Stockholms

ROBERT J. FELLER
University of South Carolina

JOHN W. FLEEGER
Louisiana State University

STUART E.G. FINDLAY
University of Georgia

SEBASTIAN A. GERLACH
Institut für Meeresforschung

J. FREDERICK GRASSLE
Woods Hole Oceanographic Institution

ROGER H. GREEN
University of Western Ontario

OLAV W. GIERE
Universität Hamburg

JONATHAN GRANT
University of South Carolina

BRIAN E. LAPOINTE
Skidaway Institute of Oceanography

HELEN M. MCCAMMON
U.S. Department of Energy

DAVID W. MENZEL
Skidaway Institute of Oceanography

MARGARET A. PALMER
University of South Carolina

CHARLES H. PETERSON
University of North Carolina

HOWARD L. SANDERS
Woods Hole Oceanographic Institution

PHILLIP L. SHAFFER
University of South Carolina

F. JOHN VERNBERG
University of South Carolina

DAVID K. YOUNG
Naval Oceanographic Research
and Development Administration

# SCENES FROM THE MARINE BENTHIC
## DYNAMICS SYMPOSIUM

# PREFACE

In past years important advances have been made in studying
community structure of the benthos.  More recently, interest has
developed in understanding benthic processes that affect secondary
production.  The 11th Belle W. Baruch Symposium in Marine Science
allowed scientists to share their knowledge from laboratory and
field studies on the role of bioenergetics, populations dynamics,
animal-sediment relations and benthic-pelagic coupling on the production of benthos.

The great strides made in describing benthic communities have
partially obscured the fact that the benthos is a changing, dynamic
subsystem of marine environments.  For example, population dynamics
of benthos produce temporal changes (cycles?) in community structure.  The influence that competition and predation have on benthic
production is yet to be documented.  Difficulties in modeling marine
ecosystems indicate a need to better understand the effect on benthic production of changing energy and substrate transfer between
trophic levels.  Such knowledge is important if we are to understand
mechanisms affecting ground fisheries.  Detritus-based food chains
are important in marine coastal ecosystems, yet essential differences between their trophic dynamics and traditional "grazing" food
chains has yet to be defined.  The benthos are integral parts of any
marine ecosystem, and research in benthic-pelagic coupling, especially of the role of benthos and sediment processes in regulating
pelagic production, is exciting (and to a benthic ecologist -
satisfying!)

Because the number of invited participants was restricted, it was not possible to exhaustively cover all appropriate research . themes. The subjects covered at the Symposium resulted from advice solicited from a cross-section of scientists interested in benthos (and the prejudices of the conveners for meiofauna and detritus-based systems!). Omissions, such as any work from the deep sea, resulted from the inability of invited colleagues to attend the Symposium. However, the scientists involved had a range of the perspectives on bioenergetics, population dynamics, modeling, field experimentation, and microbial studies. Further, in publishing these proceedings we wished to present more than a collection of new research papers. Thus we asked Drs. Eric Mills, Richard Warwick, Bernt Zeitzschel, and Paul Dayton to present "state-of-the-art" contributions that helped us relate individual contributions to a general understanding of benthic dynamics. Other contributions were solicited to further develop these themes.

The Department of Energy, Division of Biomedical and Environmental Research provided funds for the publication of this volume. We extend a special thanks to Dr. Helen McCammon of the Department of Energy for her interest and support. Dr. F. John Vernberg, Director, Belle W. Baruch Institute for Marine Biology and Coastal Research, University of South Carolina and Dr. David Menzel, Director, Skidaway Institute of Oceanography, both provided institutional support and personal encouragement, understanding, and commitment to the symposium. The symposium and this volume would have not been possible without the unfailing help of Ms. Gertrude Brown, who typed and distributed symposium information, and Ms. Bettye W. Dudley and Dr. Dennis M. Allen, who planned the logistics of the meeting. Ms. Dudley also indexed this volume. In addition, the research assistants and students of the University of South Carolina and Skidaway Institute helped with numerous chores at the symposium and made it that much more enjoyable for all of us (especially the editors). A special thanks to Mr. John Van Dalen for production of this volume and to Ms. Lynn O'Malley, who assisted in redacting. And finally, we wish to thank the many reviewers for their careful critiques as well as the authors of this volume for their participation, patience, and promptness. We hope the readers of this volume will sense the vibrant ambiance that we felt during the Symposium.

Savannah
February 1980

# TABLE OF CONTENTS

## DETRITUS

# SECONDARY PRODUCTION

# Population Dynamics
# and Secondary Production of Benthos

R. M. Warwick

ABSTRACT: The contrasting population dynamics of macrofauna and meiofauna are discussed in relation to their reproductive biology and to the problems involved in calculating their production.

## INTRODUCTION

During the last decade considerable data has accumulated on the production[*] rates of marine benthic invertebrates, as Waters (1977) similarly points out for fresh waters. The measurement of production has recently been reviewed extensively by a number of authors, including Winberg (1971), Crisp (1971), Winberg *et al.*, (1971), Zaika (1973), Zelinka and Marvan (1976) and Waters (1977)

---

[*]Throughout this paper, by "production" I am referring to somatic production, Pg in IBP terminology (Petrusewicz, 1967), the component of production which is passed on to the next higher trophic level or the decomposer food chain. This does not include the production of gametes, Pr.

and will not be repeated here.  Rather, this paper concerns some
of the peculiar features of the population dynamics of marine
benthic invertebrates that may give rise to problems or inaccuracies in determining production in the field.  It discusses some of
the generalizations that can be drawn from such studies and evaluates the use of short-cut approximations such as annual P/B ratios
in estimating production.

### Population Dynamics of Macrofauna and Meiofauna

Certain features of the reproductive biology of marine benthic
invertebrates profoundly affect their population dynamics, which
in turn generate varying problems in the field assessment of production rates.  When considering the whole size-spectrum of metazoan benthos, a rather arbitrary dichotomy between macrofauna
and meiofauna has been erected, largely on the basis of size (see
e.g., McIntyre, 1969).  However, there is also a fundamental functional dichotomy between these two size-classes of animals in their
means of population maintenance.

Macrofauna species, except those from the high arctic and deep
sea, generally produce large numbers of gametes; fertilization is
external, and a planktonic larval phase ensures dispersal in the
majority of species (Thorson, 1950).  Spawning is usually seasonal
and more or less synchronized.  Because their life span is generally
longer than one year, the adult animal must tolerate the full seasonal range of environmental conditions, but their speed of dispersal enables them to exploit uncertain environments fully.  Their
large body size confers on them the ability to feed on a wide range
of sizes of food particles, which is advantageous in times of shortage (Calow, 1977).  This relatively unspecialized feeding behavior
contributes in part to the lower species diversity of macrofauna
compared with meiofauna.  The vicissitudes of larval life in the
plankton (Thorson, 1966), together with the frequently intense predation soon after settlement (e.g., Segerstråle, 1962, 1965) often
result in highly variable recruitment from year to year which,
accompanied by the possibility of dramatic adult mortalities in certain seasons, gives rise to wide and unpredictable fluctuations in
biomass and numbers from time to time.  Survivorship curves are
strongly concave and in extreme cases there may be no successful
settlement at all for a series of years, which may result in
species being represented in a population by a single year-class
(e.g., Birkett, 1959; Buchanan, 1966).

By contrast, a number of features of the reproductive behavior
of meiofauna illustrate their much more conservative strategy.  In
general, a small number of large gametes are produced, fertilization
is by copulation and development is direct, with no pelagic larval
phase (Remane, 1952).  Brood protection is common:  embryos are
frequently carried by the mother, or the eggs may develop in pro-
tective cocoons from which the young emerge in an advanced stage of
development (Swedmark, 1964).  Viviparity is frequent, and may
be a facultative capacity used under adverse conditions such as low
temperature (Gerlach and Schrage, 1971).  In some meiofaunal syllid
polychaetes, the smaller, young forms are viviparous, and the older

animals oviparous (Fauvel, 1923; Thorson, 1950). Parthenogenesis has been proven or implied for some marine nematodes (Hope, 1974) and gastrotrichs (Hummon, 1974), but not, as yet, for marine meiofaunal crustaceans (although the phenomenon is known in fresh water both for certain harpacticoid and ostracod genera). Hermaphroditism is also common, and is the rule in marine gastrotrichs (Hummon, 1974) and gnathostomulids (Sterrer, 1974). Some examples of this conservative behavior are illustrated in Figure 1. Dispersive capability is sacrificed by the meiofauna and must be met in other ways such as by adult locomotion, by suspension in the water column during storms, on floating materials, transfer by birds, etc. (Gerlach, 1977).

Reproduction in meiofauna is typically continuous but may be curtailed during certain seasons and may vary within a species over its geographic range or with habitat. For example, among harpacticoid copepods, *Tachidius discipes* does not reproduce below temperatures of 4 to 6°C (Heip, 1973). It breeds throughout the year on intertidal mudflats in southwest England where winter sea temperatures do not fall below this level (Teare, 1978), but only during the summer months in a similar habitat in the north of Scotland (Goodman, 1978). *Microarthridion littorale* breeds throughout the year subtidally in South Carolina (Coull and Vernberg, 1975), but intertidally ceases reproduction in winter in the same locality (Fleeger, 1979).

Some meiofaunal groups have the remarkable ability of rapidly altering their population sex-ratio in response to changes in population density, a capability made possible by their short generation time. Since a single male can usually fertilize several females, a sex ratio in favor of males is obviously advantageous when the population density is high, and a ratio in favor of males at low population densities will increase the chances of some females being fertilized if males are more active in seeking out the more passive females. Penter (1969) showed that there was an increasing proportion of males with decreasing population density for several estuarine harpacticoid copepod species, and Battaglia (1964) demonstrated similar sex-ratio switching in the copepod *Tisbe*, and discussed the genetic control of this mechanism. Such a mechanism may also explain the disappearance of males from high density monospecific cultures of some marine nematodes as reported by Hopper and Meyers (1966). Disproportionate sex ratios in favor of females have frequently been reported for nematodes both in the field and in culture (e.g., Tietjen, 1967), and the same is true of many natural harpacticoid copepod populations (Battaglia, 1964).

Meiofauna species may spend their non-reproductive periods at different life-stages, with the possibility of quiescence in a resistant stage in unfavorable conditions or seasons. Thus, in contrast to the macrofauna, some meiofauna species may appear to be absent from the substrate for certain periods of the year. Thiesen (1966), for example, found that three ostracod species in Nivå Bay, Denmark, overwintered as adults and four as resting eggs; in one of the latter species a true diapause was demonstrated.

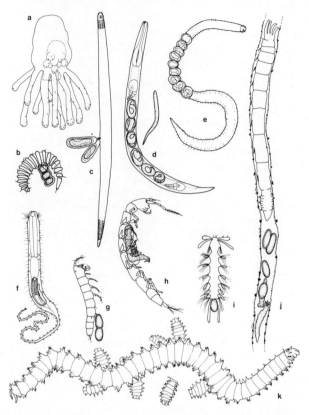

Figure 1.  Some examples of the conservative reproductive behavior of meiofauna:  (a) the coelenterate *Otohydra vagans* incubates
one or two embryos to an advanced stage in an ectodermal brood-pouch
(based on Swedmark and Teissier, 1958); (b) the nematode *Desmosco-
lex falcatus* carries a string of one to four developing eggs in a
packet clasped by a pair of specially elongated subventral setae
(original); (c) the nematode *Pselionema annulatum* var. *beauforti*
carries a pair of stalked egg capsules containing developing embryos
attached to the vulva of the female (sketched from photographs in
Hopper, 1973); (d) the nematode *Metachromadora vivipara* is ovovivi-
parous:  embryos developing in the uterus of the female hatch immediately after expulsion from the vulva (newly hatched larva drawn to
same scale as adult) (original); (e) the nematode *Desmodora (Croconema) ovigera* carries developing embryos externally in a row on the
sides of the body (redrawn from Ott, 1976); (f) a number of gastrotrichs produce only one egg at a time, and in *Urodasys viviparus*
the embryo develops to an advanced stage inside the mother (redrawn
from Wilke, 1954); (g) there is a tendency towards reduction in numbers of eggs produced by benthic harpacticoid copepods, especially
interstitial forms, the development taking place to the naupliar
stages in external ovisacs; *Psammastacus confluens* carries only two
eggs (redrawn from Nicholls, 1935); (h) the isopod *Angeliera phreati-
cola* with its single embryo in the brood-pouch of the female (redrawn from Delamare-Deboutteville, 1960); (i) the archiannelid *Neril-
lidium troglochaetoides* towing its single embryo (redrawn from Swedmark, 1964); (j) developing embryos of the sabellid polychaete
*Manayunkia aestuarina* brooded in the tube of the mother (original);
and (k) five-setiger juveniles of the syllid polychaete *Sphaerosyllis
bulbosa* attached by their tail ends in the ventral side of the mother:
one has just been released (original).

Some nematodes might overwinter as resting eggs (Hope, 1974), and this appears to be true of some turbellaria also (Pawlak, 1969). This may also apply to benthic copepods, which are often absent from populations in certain seasons (e.g., Heip, 1973; Coull and Vernberg, 1975), although in *Halicyclops magniceps* the adult females are known to hibernate in the deeper layers of sediment, probably in the burrows of *Nereis*, where they will escape predation (Heip, 1975). In some fresh water harpacticoids of the family Canthocamptidae, the adults are known to encyst (Dussart, 1967), although, so far, this has not been recorded in the marine representatives of this family. However, obligatory encystment for at least half of its lifetime is found in marine archiannelids of the genus *Dinophilus* (Jagersten, 1953). Physiological quiescence during periods of seasonal food shortage or during suboptimal environmental conditions for adult survival or reproduction may be regarded as part of an overall conservative strategy for population maintenance, but such a strategy can only be adopted in species which have generation times considerably shorter than one year.

These conservative features of meiofauna population maintenance tend to induce population stability. Thorson (1946) showed that in Danish waters macrofauna species with a long planktonic larval life have much more temporally unstable populations than those with direct non-pelagic development or with only a very short pelagic life. In a similar way, continually reproducing meiofauna species from relatively stable environments often have remarkably constant populations compared with their macrofaunal counterparts from the same habitat (e.g., Warwick and Buchanan, 1971, and Buchanan and Warwich, 1974; Kendall, 1979), while in seasonally reproducing species annual peaks of abundance tend to be very regular from year to year (e.g., Coull and Vernberg, 1975; Heip, 1980). Very few studies of meiofauna populations have continued for longer than a year, but in such studies temporal separation of species has been noted and attributed either to cyclical changes in the abundance of different food types in species with apparently different food requirements (Warwick, 1977), or as a method of temporal resource sharing in species with apparently similar food requirements (Coull and Vernberg, 1975). Meiofauna in general have rather narrowly specialized food-size and food-species requirements which, in conjunction with temporal partitioning of the environment, result in a higher species diversity as compared with the macrofauna (Tietjen and Lee, 1977).

Many of the features of the population biology of meiofauna correspond with those features which Pianka (1970) regards as typical of $K$-strategists, whereas the macrofauna have many characteristics of $r$-strategists. Of course, the absolute body size and length of life do not conform with this suggestion, but Christiansen and Fenchel (1977) have cautioned against comparisons based on animals of different body size, while Gadgil and Solbrig (1972) argue that $r$-strategists need not necessarily have a shorter lifespan. Pianka (1970) suggests that $r$-strategy leads to greater productivity and $K$-strategy to greater efficiency. It might be expected, then, that meiofauna production rates will be less than predicted by extrapolation from measured macrofauna lifetime turnover rates. One correlate of $K$-strategists is that they tend to have less concave

survivorship curves than $r$-strategists. We know little about the
shapes of survivorship curves for meiofauna, but undoubtedly they
are less concave than those for macrofauna with planktonic larvae.
Parallels can be drawn with insect populations in which the curves
are much less concave in species that protect their young (Itô,
1959).

Assessment of Macrofauna Production

The essential requirement for estimating production is com-
puting growth and recruitment/mortality curves. To do this from
the analysis of a time series of field samples, a prerequisite is
the ability to identify cohorts of similar age whose history in
terms of growth and changes in numbers can be followed. In view
of their more or less seasonal reproductive behavior and the fact
that growth slows down, stops or even becomes negative in the
winter, it is usually possible to separate and follow year-classes
of temperate-latitude macrobenthic species. Indeed, attempts to
estimate production of all the important species in a macrofauna
community have in all cases proved successful, with few 'problem'
species being encountered (Sanders, 1956; Buchanan and Warwick,
1974; Warwick and Price, 1975; Cederwall, 1977; Wolff and de Wolf,
1977; Warwick et al., 1978). These authors have further shown that
a large proportion of total community production can be attributed
to rather few species (Fig. 6), which makes estimates for whole
macrofauna communities a relatively practicable task.

Table 1 summarizes the data presently published on marine
macrobenthos production rates determined by field assessment.
Some additional Australian data (as yet not fully published) which
it is not presently possible to evaluate, are listed by Robertson (1979). The problems involved in determining these rates
have proved, in general, to be logistic. First, such studies
are very labor intensive (Waters, 1977), particularly when several
species are being studied simultaneously. Samples collected at
frequent intervals throughout the period of study must be divided
into year classes and the numbers and weight of each year class
determined separately. This involves a large amount of routine
sorting, identification, counting, measuring and weighing. Second-
ly, for sublittoral habitats, it is difficult to acquire (especially large) research vessels for short collecting trips repeated
at frequent intervals. Small wonder, then, that attempts to short-
cut these laborious procedures have been made, especially efforts
to establish generalizations about the ratio between annual production and mean annual biomass ($P/\bar{B}$) in order to predict production
from biomass data.

Zaika (1970), Parsons et al. (1977) and Robertson (1979) have
attempted to relate the annual population $P/\bar{B}$ to the maximum life
span. However, for the data in Table 1, this relationship is so
poor (Fig. 2) as to be almost useless for the prediction of $P/\bar{B}$.
Similar relationships drawn up on a log-log scale are deceptive.
For instance, in the example described by Robertson (1979), even
at its most accurate (at the mean lifespan of 2.9 yrs) one can
only ascertain with 95% confidence that the $P/\bar{B}$ lies between 1.0

Table 1. Annual production of macrobenthos (production in dry weight unless otherwise stated).

| SPECIES | PRODUCTION m$^{-2}$.y$^{-1}$ | P/$\bar{B}$ | MAX. AGE (YRS) | LOCALITY | REFERENCE |
|---|---|---|---|---|---|
| *Nephtys incisa* | 9.34 g | 2.16 | 3 | Long Island Sound, U.S.A. 4–30 m | Sanders, 1956 |
| *Cistenoides gouldii* | 1.70 g | 1.94 | 2 | " | " |
| *Yoldia limatula* | 3.21 g | 2.28 | 2 | " | " |
| *Pandora gouldiana* | 6.13 g | 1.99 | 2 | " | " |
| *Moira atropos* | 2.52 g | 0.70 | 6 | Biscayne Bay, Florida, U.S.A. 3m | Moore and Lopez, 1966 |
| *Tagelus divisus* | 21.0 g | 1.78 | 2 | Biscayne Bay, Florida, U.S.A. L.W.S. | Fraser, 1967 |
| *Ampharete acutifrons* | 0.719 g (wet) | 4.58 | 1 | Long Island Sound, U.S.A. 9–17 m | Richards and Riley, 1967 |
| *Neomysis americana* | 36.2 mg | 3.66 | 1? | " | " |
| *Crangon septemspinosa* | 0.519 g | 3.82 | 3 | " | " |
| *Asterias forbesi* | 4.52 g | 2.64 | 3 | " | " |
| *Tellina martinicensis* | 0.23 g | 2.4 | 2 | Biscayne Bay, Florida, U.S.A. 3m | Penzias, 1969 |
| *Chione cancellata* | 8.9 g | 0.42 | 7 | Biscayne Bay, Florida, U.S.A. M.L.W.S. | Moore and Lopez, 1969 |
| *Dosinia elegans* | 0.13 g | 1.25 | 2 | Biscayne Bay, Florida, U.S.A. 3m | Moore and Lopez, 1970 |
| *Pectinaria hyperborea* | 10.6 g | 4.6 | 2 | St. Margaret's Bay, Nova Scotia, 60 m | Peer, 1970 |
| *Scrobicularia plana* | 60 Kcal | 0.29 | 7? | North Wales, Lower shore | Hughes, 1970a |
| " | 13.3 Kcal | 0.67 | 4 | North Wales, Upper shore | " |
| *Anodontia alba* | 14.09 g | 1.43 | 4+(?) | Biscayne Bay, Florida, U.S.A. Low water | Moore and Lopez, 1972 |
| *Strongylocentrotus droebachiensis* | 401.0 Kcal | 0.80 | 6 | St. Margaret's Bay, Nova Scotia, Intertidal | Miller and Mann, 1973 |
| *Neanthes virens* | 45.2 Kcal | 1.62 | 3 | Thames estuary, U.K. Inter-tidal | Kay and Brafield, 1973 |
| *Ammotrypane aulogaster* | 359 mg | 2.08 | ? | Northumberland, U.K. 80 m | Buchanan and Warwick, 1974 |

Table 1 (continued)

| Species | Weight | | Location | Reference |
|---|---|---|---|---|
| *Heteromastus filiformis* | 297 mg | 2 | | " |
| *Spiophanes kroyeri* | 196 mg | 3 | | " |
| *Glycera rouxi* | 192 mg | 5 | | " |
| *Calocaris macandreae* | 142 mg | 9.5 | | " |
| *Abra nitida* | 118 mg | 3 | | " |
| *Lumbrineris fragilis* | 78 mg | 3 | | " |
| *Chaetozone setosa* | 50 mg | 3 | | " |
| *Brissopsis lyrifera* | 108 mg | 4 | | " |
| *Mya arenaria* | 11.6 g | 3 | Petpeswick Inlet, E. Canada, Intertidal | Burke and Mann, 1974 |
| *Macoma balthica* | 1.93 g | 3 | | " |
| *Littorina saxatilis* | 3.25 g | 1 | Ythan Estuary, Scotland, Intertidal | Chambers and Milne, 1975 |
| *Macoma balthica* | 10.07 g | 6 | | " |
| *Nephtys hombergi* | 7.34 g | 3 | Lynher Estuary, U.K. Inter-tidal | Warwick and Price, 1975 |
| *Ampharete acutifrons* | 2.32 g | 1 | | " |
| *Mya arenaria* | 2.66 g | 8 | | " |
| *Scrobicularia plana* | 0.48 g | 9 | | " |
| *Macoma balthica* | 0.31 g | 6 | | " |
| *Cerastoderma edule* | 0.21 g (wet) | 7 | | " |
| *Ampelisca brevicornis* | 4.26 g (wet) | 1.25 | Helgoland Bight, 28 m | Klein *et al.*, 1975 |
| | 2.43 g (wet) | 1.25 | | " |
| *Pectinaria californiensis* | 2.02 gC | 1.2 | Puget Sound, Washington, U.S.A. 34 m | Nichols, 1975 |
| " | 2.798 gC | 2.1 | Puget Sound, Washington, 203 m | " |
| " | 3.471 gC | 1.8 | Puget Sound, Washington, 254 m | " |
| " | 1.386 gC | 1.9 | Puget Sound, Washington, 207 m | " |
| " | 4.816 gC | 2.4 | Puget Sound, Washington, 71 m | " |
| *Cerastoderma edule* | 29.25 g | 5 | Southhampton Water, U.K. Intertidal | Hibbert, 1976 |
| " | 71.36 g | 5 | | " |
| " | 46.44 g | 5 | | " |
| *Mercenaria mercenaria* | 3.99 g | 8 | | " |
| " | 14.00 g | 8 | | " |
| *Venerupis aurea* | 6.19 g | 9 | | " |
| " | 0.70 g | 5 | | " |
| " | 1.25 g | 5 | | " |
| *Crassostrea virginica* | 3828 Kcal | ? | South Carolina, U.S.A. Inter-tidal | Dame, 1976 |

Table 1 (continued)

| | | | | | |
|---|---|---|---|---|---|
| *Littorina littorea* | 6.13 g | 0.61 | ? | Grevelingen Estuary, Netherlands, Intertidal | Wolff and de Wolf, 1977 |
| *Hydrobia ulvae* | 7.23 g | 1.78 | 1 | " | " |
| " | 8.80 g | 1.24 | 1 | " | " |
| " | 12.79 g | 1.36 | 2 | " | " |
| *Cardium edule* | 10.21 g | 0.69 | 3.5 | " | " |
| " | 119.82 g | 2.56 | 3.5 | " | " |
| " | 51.76 g | 1.13 | 3.5 | " | " |
| *Macoma balthica* | 3.40 g | 1.93 | 8.10 | " | " |
| " | 0.95 g | 1.00 | 8.10 | " | " |
| " | 0.07 g | 0.30 | 8.10 | " | " |
| " | -0.74 g | -0.25 | 8.10 | " | " |
| *Arenicola marina* | 3.79 g | 1.14 | 3 | " | " |
| " | 6.26 g | 0.72 | 3 | " | " |
| " | 3.32 g | 0.99 | 3 | " | " |
| *Pontoporeia affinis* | 3.17 g | 1.90 | 3 | North Baltic, 64 m | Cederwall, 1977 |
| *Pontoporeia femorata* | 3.03 g | 1.43 | 3 | " | " |
| *Harmothoe sarsi* | 0.23 g | 1.99 | 3 | " | " |
| *Pharus legumen* | 16.12 g | 0.56 | 6 | Carmarthen Bay, South Wales, 13.5 m | Warwick *et al.*, 1978 |
| *Spiophanes bombyx* | 3.35 g | 4.86 | ? | " | " |
| *Ensis siliqua* | 1.37 g | 0.27 | 10 | " | " |
| *Donax vittatus* | 0.72 g | 2.10 | 2.5 | " | " |
| *Magelona papillicornis* | 0.69 g | 1.10 | 3 | " | " |
| *Venus striatula* | 0.62 g | 0.41 | 10 | " | " |
| *Ophiura texturata* | 0.46 g | 0.68 | 3 | " | " |
| *Tellina fabula* | 0.29 g | 0.90 | 6 | " | " |
| *Glycera alba* | 0.28 g | 0.97 | 3 | " | " |
| *Sigalion mathildae* | 0.17 g | 0.44 | ? | " | " |
| *Tharyx marioni* | 0.015 g | 0.79 | 2 | " | " |
| *Astropecten irregularis* | 0.0004 g | 0.005 | ? | " | " |
| *Echinocardium cordatum* | -0.012 g | -0.02 | 3 | Gulf of Finland, 33-35 m | Sarvala (in press) |
| *Harmothoe sarsi* | 0.376 g | 2.4 | 2 | " | " |
| " | 0.401 g | 3.1 | 3 | " | " |

and 4.4.  When considering a number of species for the purposes
of estimating whole community production, the errors involved
in applying this relationship for individual species may balance
out, but for single species the estimate may be wildly wrong.
Data variability may arise partly through methodological errors,
which will vary from species to species (see below), but partly
also from variability in year-class strength alluded to earlier.

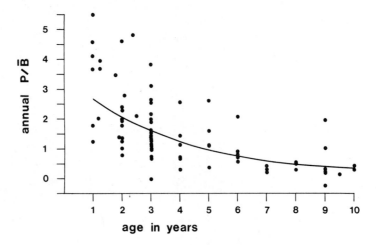

Figure 2.  The relationship between maximum
life span and annual P/B̄ derived from data in Table
1.  Regression line fitted from the equation log
P/B̄ = 0.538 - 0.112 age (yrs).  (r = -0.753).

Allen (1971) demonstrated the relationship between production
and mean biomass for single cohorts exhibiting various growth and
mortality patterns, while Waters (1969) showed empirically that
the P/B̄ of such a cohort over its life span was commonly between
2.5 and 5, with a mode of 3.5.  Thus, if a cohort lives longer than
one year, as most in Table 1 do, then the annual P/B̄ will be this
value divided by the life span in years.  Generally, however, single cohorts have not been followed throughout their life span in
these studies, but cohorts representing several years have been
studied simultaneously over a single annual cycle.  If the cohorts
were identical in numbers, growth and mortality then the annual
population P/B̄ would, of course, be the same as that of a single
cohort over its life span (Allen, 1971), but with dramatic irregularities in recruitment from year to year, this is rarely the case.
Waters (1977) concluded for fresh waters that the P/B̄ ratio is
not dependent on environmental factors, except indirectly as the
environment may affect voltinism.  He was able to draw on information from a number of multivoltine species, particularly insect
larvae, but from marine habitats no information on production is
available for multivoltine species.  For such species, a number of
generations are present during the span of a year, but each cohort
will have an essentially similar P/B̄ so that the annual population

P/$\bar{B}$ will simply be a function of voltinism. In long-lived species, however, several year classes are simultaneously present, each having a different P/$\bar{B}$, so that the annual P/$\bar{B}$ of the population as a whole will depend on the age distribution.

Because the annual P/$\bar{B}$ of a cohort usually decreases with age, it follows that populations dominated by older year classes will have a lower P/$\bar{B}$ than those comprising younger individuals. It is these environmentally induced irregularities in year-class strength that give rise to a great deal of the variability in the annual P/$\bar{B}$ ratio shown in Figure 2. The few studies in which the same species has been studied at different sites, or at the same site in different years, illustrate this point. Thus, Hughes (1970, 1970a) showed that for a lower shore population of *Scrobicularia plana*, dominated by animals greater than four years old, the P/$\bar{B}$ was 0.29, whereas on the upper shore the population consisted entirely of animals less than four years old and had a P/$\bar{B}$ of 0.67. Nichols (1975) also showed that the P/$\bar{B}$ of *Pectinaria californiensis* varied from 3.3 to 5.5 depending on the proportion of older, less productive animals in the population. The annual P/$\bar{B}$ of a population of *Nephtys hombergi* declined over a five-year period due to a series of less successful settlements resulting in an increase in the proportion of older animals in the population (Price and Warwick, in prep.). On the other hand, Sarvala (1971 and in press) studied a population of *Harmothoe sarsi* in which the growth rate did not decline with increasing age, and found that the P/$\bar{B}$ in 1967 when two year classes (0 and 1+) were present was 2.4, whereas the following year when three year classes were present the P/$\bar{B}$ was 3.1. In senescent populations where no recent recruitments have occurred, the P/$\bar{B}$ may be very low, and, in some cases, even negative, e.g., Wolff and de Wolf (1977) for *Macoma balthica* and Warwick *et al.* (1978) for *Echinocardium cordatum*. Thus, the only realistic measure of annual P/$\bar{B}$ for a population is the mean of the ratios for the individual year classes weighted by their mean biomasses (Allen, 1971), and shortcut approximations based on the life span are fraught with danger. Allen has suggested that the reciprocal of mean age might provide a useful approximation to the P/$\bar{B}$ ratio because it applies directly to some growth and mortality models. However, the mean age of a population is impossible to estimate unless cohorts can be separated and followed throughout the year, in which case conventional and more accurate estimates of production would be possible.

Hibbert (1976) showed that the annual P/$\bar{B}$ of individual year classes was remarkably constant in three species of bivalve mollusc. This opens the possibility that, if the mean biomass of each age class is known, the P/$\bar{B}$ of the population as a whole might be predicted more easily. In Figure 3 I have extended Hibbert's observations to cover all macrofauna species for which annual P/$\bar{B}$ of individual year classes can be determined from the literature. This exercise was made more difficult by terminological disagreement among authors regarding the designation of year classes. Some authors change the 0 group designation to 1+ when the animals are one year old or when the new year class appears

in the samples; others change from 0 to 1+ in their first winter
(i.e., when the calendar year changes in the northern hemisphere).
The latter terminology is commonly used in fisheries biology and
is most useful for invertebrates such as bivalve molluscs that lay
down winter growth checks on the shell which can be used for ageing
purposes.  Benthic ecologists would therefore be advised, for the
sake of uniformity, to adhere to this latter terminology.  Animals
spend a variable amount of time as 0 group depending on their date
of settlement, and therefore it is useless to attempt comparison
of annual P/$\bar{\text{B}}$ for this year class.  Furthermore, methodological
problems regarding sieve cut off (see below) often result in underestimating production of younger animals, which for small species
may still apply in the 1+ or older year classes.  In Figure 3 I have
been careful to omit data on year classes which still appear to be
recruiting to sieve size (i.e., are increasing in numbers).  Despite this, the variability in the data, although not as great as
for whole populations (Fig. 1), is still rather large.  Differences
in growth and mortality give rise to most of this variability; for
example, in species with asymptotic growth the 1+ group has a higher
P/$\bar{\text{B}}$ than the 2+, whereas in species tending towards exponential
growth this may be reversed.  Also, using this year-class terminology, the production of 1+ and subsequent year classes may depend
on the length of time the animals spent as 0 group.

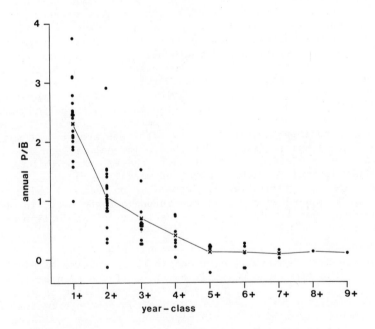

Figure 3.  Annual P/$\bar{\text{B}}$ of individual year classes.
Crosses joined by a line indicate mean values.

Aspects of the life cycle of macro-invertebrates that affect the annual P/B̄, such as growth and reproductive strategies, have clearly developed as a compromise to ensure optimum survival in the face of complex environmental pressures. In some cases, these strategies can be related to predation pressures (e.g., Seed and Brown, 1978) and, in others, to the availability (permanent or seasonal) of preferred foods (e.g., Thompson, 1964). The extent to which they are genetically determined puts limits on the range of environmental conditions under which a given species can survive. Certain species are very flexible. The surface deposit-feeding polychaete *Ampharete acutifrons*, for example, is an annual species on an intertidal mudflat with high surficial primary production that permits rapid growth; but predation by fish is severe, reducing the numbers from several thousands per $m^2$ in the spring to tens or only units per $m^2$ in autumn, when spawning occurs (Warwick and Price, 1975, Price and Warwick in prep.). In a sublittoral population, however, where there is no *in situ* primary production and predation is presumably less severe, the same species has a three-year life cycle (Warwick and George, in prep.) (Fig. 4). This latter population has a much lower annual P/B̄ than the former, and any notion that a species has a fixed P/B̄ cannot be supported. In addition, several marine groups, particularly polychaetes bridging the meiofauna/macrofauna dichotomy in size, can vary their mode of development (pelagic or non-pelagic) depending on environmental conditions (Thorson, 1950).

Even though most problems of estimating macrofauna production are logistic rather than practical, there are two practical problems that are often overlooked. The first, hinted at above, is estimating production prior to the stage when the animals reach a size at which they are retained on the macrofaunal ecologist's sieve (with usually 0.5 mm apertures), and the second is underestimating due to regeneration of tissue after cropping by predators.

Estimates of the true production or P/B̄ of a species over its lifespan should include production while the larvae are still in the plankton and production of newly settled spat less than 0.5 mm in size. Because survivorship curves for benthic macrofauna are initially strongly concave and growth is often rapid in early life, it follows that a large part of total production may be confined to this early period, especially in species with short life spans. For this reason, production for many of the species used in Figure 2 has been more or less underestimated. In some cases the error may be quite large. Buchanan and Warwick (1974), for example, found that the small polychaete *Heteromastus filiformis*, which has a maximum lifespan of 2.5 years, was only fully recruited to sieve size after 14 months. Obviously, to improve accuracy, smaller samples sieved more finely should be collected during the periods when juveniles are expected to be present in the sediment, and plankton samples taken at regular intervals when the larvae are expected in the plankton. However, this would require very precise timing and a great deal of effort.

Subsequent regeneration of cropped tissue, particularly bivalve siphons (e.g., Birkett, 1959; Trevallion, 1971; Warwick *et*

*al.*, 1978) or polychaete tails (e.g., Gibbs, 1968; Wolff and de
Wolf, 1977) will also result in underestimating production by methods
that follow the mean individual weights of members of a cohort (Wolff
and de Wolf; Warwick *et al.* ops. cit.).  This is not thought to be
a major source of error as, even in heavily exploited populations,
cropping amounts to a rather small proportion of total production
(Trevallion, 1971).  Wolff and de Wolf (1977) further point out
that size-selective mortality will affect the production estimate,
but this, again, would probably lead to rather small errors and has
never been quantified.

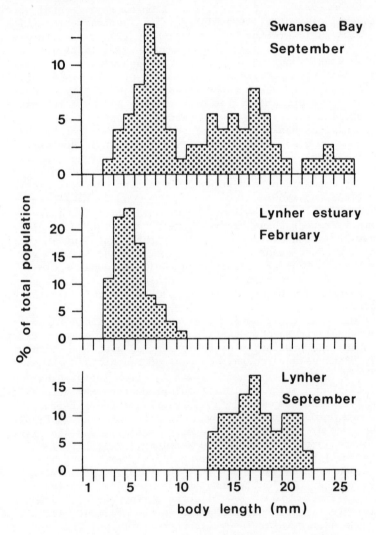

Figure 4.  Size frequency histograms of the
polychaete *Ampharete acutifrons* from a subtidal
(Swansea Bay) and intertidal (Lynher estuary) locality.

Accepting possible inaccuracies in determination, we can make
some generalizations about the partitioning of production among
trophic groups in macrofauna communities from the few studies
in which all the important producers have been investigated (Fig. 5).
As expected, carnivores contribute a uniformly low proportion of
the total production.  The production of deposit feeders shows a
steady decline with water depth from about 10 g dry wt. $m^{-2}$ in
shallow estuaries to 1.7 $g.m^{-2}$ at 80 m depth.   At greater depths
there is no primary production on the sediment surface and deposits
in general, contain less available food material.  Filter feeders,
however, have very varied levels of production in shallow water
but their production is non-existent, or virtually so, in the 46
and 80 m depth stations.  Comparison of production of filter
feeders and deposit feeders per unit area is, of course, an unfair
one because, in areas of strong currents, the former are able to
draw on a large primary-productive area relative to the area they
occupy (Crisp, 1975), and individual species may consequently have
a remarkably high production.

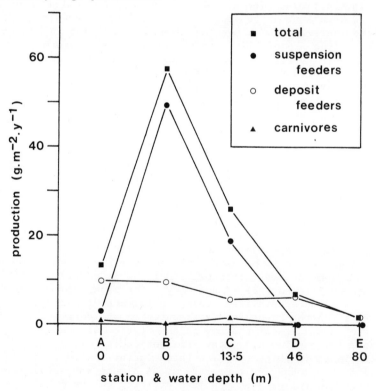

Figure 5.  Partitioning of macrobenthos production among trophic groups.  Station A, Lynher estuary,
U.K. (Warwick and Price, 1975); Station B, Grevelingen estuary, Netherlands (Wolff and de Wolf, 1977);
Station C, Carmarthen Bay, S. Wales (Warwick *et al.*,
1978); Station D, Baltic (Cederwall, 1977); Station E,
Northumberland, U.K. (Buchanan and Warwick, 1974).

Assessment of Meiofauna Production

In continuously reproducing populations it is not possible to follow the history of recognizable cohorts in a time-series of samples. Largely because of this methodological problem, very few life histories for marine meiofauna species are known, save for a few species with only one or very few generations per year in which generation overlap is not excessive (e.g., Thiesen, 1966; Lasker *et al.* 1970; Heip, 1976; Wieser and Kanwisher, 1960; Barnett, 1970; Skoolmun and Gerlach, 1971), and, indeed, not a single field estimate of production of a marine meiofaunal species has been published to date. Even in the examples listed above, the determination of life histories has often required very painstaking analysis of large numbers of samples collected at frequent time intervals, recruitment and growth of juveniles often being compressed into a short span of time.

In most cases, however, it will be necessary to determine the size-specific growth rate and apply this to measured standing stocks in the field. The growth of marked or otherwise identifiable individuals could be measured directly at intervals of time in the field, but this would pose problems of small mobile species in sediment. (The use of vital stains has not been explored in this context, their persistence and physiological effects being largely unknown). Despite the uncertainties involved in extrapolating from laboratory experiments, it seems that this will be inevitable for many meiofauna species. The application to the field situation of growth rates determined in laboratory culture is open to much criticism, largely because the more favorable food and temperature conditions will give artificially high growth rates and generation times. The number of generations per year for many meiofauna species is much lower than would be predicted from their generation time in culture (see table in Gerlach, 1971). Determining growth indirectly from laboratory determinations of respiration and growth efficiency ($K_2$), the so-called physiological method (Winberg, 1971), intuitively raises more objections: the $K_2$ may be much higher under laboratory conditions than in the field (e.g., Pasternak, 1977) and any proportion of the assimilated energy that is excreted is ignored. If such methods are to be used, one can only invoke the trite caution that conditions in the laboratory should be as nearly natural as possible. With meiofauna this is not easy. For example, we know so little about the specialized food requirements of individual species that providing the right food, presented in the correct way, is a major problem in itself and may be one of the main reasons why certain species, particularly many dominant nematodes, cannot as yet be established in culture. With these latter species, the problem of estimating production therefore seems almost intractable, at least until investigations now just starting on the precise ecological requirements of individual species in various meiofauna groups begin to reach fruition. Furthermore, partly as a result of this specialized feeding behavior, production in meiofaunal communities will tend to be partitioned more evenly among species (Fig. 6) so that investigating a few dominants will not give good estimates of the total community.

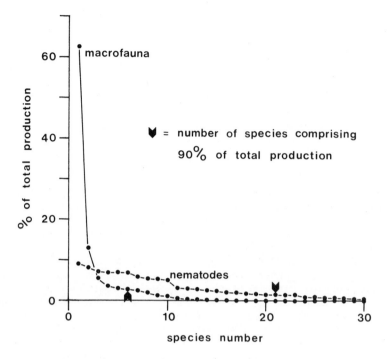

Figure 6. Partitioning of production among the thirty most important species in a macrofauna community and a community of meiofaunal nematodes with a similar number of species. Macrofauna from a *Venus* community based on data of Warwick *et al.* (1978). Meiofaunal nematodes from an estuarine mudflat based on data of Warwick and Price (in press) assuming that annual production is proportional to annual respiration.

Applying lifetime turnover ratios deduced from macrofauna studies offers one possibility for estimating meiofauna production if the number of generations expected each year can be determined. However, this has several weaknesses, the principal one of which is a fundamental difference in the growth pattern of meiofauna compared with macrofauna. Most meiofauna groups do not breed until they reach maximum size. Many (including the dominant nematodes and copepods) moult through a series of larval instars, no reproduction occurring until the final adult instar, after which Pg is zero and in females all available energy is diverted to the production of offspring (Pr), whereas in males energy is expended in seeking and copulating with the females. The situation is invariably different for macrofauna species, in which growth continues after the first breeding period and through subsequent spawnings so that the available energy is partitioned between Pg and Pr. This means that in meiofauna the adult population is completely non-productive in terms of the energy being made available

for higher trophic levels, and that only the juveniles are productive in this respect.  The biomass of juveniles in a population,
averaged over the year, is proportionally rather low.  Warwick and
Price (in press), for example, showed that juveniles comprised 33%
of the total nematode biomass and Teare (1978) obtained similar
values for copepods in a similar habitat.  These figures are
probably maximal, as other values in the literature assuming the
same numbers to weight ratio for adults and juveniles, are lower
(Warwick and Buchanan, 1971; Lorenzen, 1974; Juario, 1975).  This
implies that, in contrast to macrofauna populations, only a third
of the biomass of meiofauna populations is likely to be productive,
which should be borne in mind when considering the use of the $P/\bar{B}$
ratio.  The stage of the life cycle in which a species spends its
non-reproductive period will also affect the annual $P/\bar{B}$.  Consider
two species which pass through the same number of generations and
have the same total production during their reproductive season,
but one spends the non-reproductive period as low-biomass resting
eggs and the other as high biomass adults.  Although production is
the same, the annual $P/\bar{B}$ will be higher in the former than the
latter.  Obviously, then, one should find out as much as possible
about the natural history of the species in question (whether its
reproduction is continuous or seasonal, how it spends non-reproductive periods, and, if possible, how many generations it passes through
annually) before predictions based on turnover ratios can be made.
Far too little is known about the natural history of meiofauna
species, but it is a discipline that will require a period of painstaking research before any empirical relationships between production and life-cycle phenomena can be established, and such
research will need to lean heavily on the experimental studies mentioned above to investigate development under natural conditions in
the laboratory.

      Meiofauna workers, perhaps more than macrofauna workers and
perhaps out of necessity, have tended to assume that a particular
taxon, be it an individual species, phylum or even "meiofauna" as a
whole, has a fixed annual $P/\bar{B}$.  However, it is most likely that
many species will have flexible life cycles adapted to local regimes
of food availability, predation pressure, etc., in the same way as
I have indicated for macrofauna with the *Ampharete* example.  It would
be very difficult to reproduce these regimes in the laboratory.  Longer generation times and lower $P/\bar{B}$ might be expected in deeper water
because of reduced food availability, lower temperatures, etc.

      In the deep sea, the macrofauna worker faces problems similar
to those of the shallow water meiofauna worker.  In any case, here
the macrofauna/meiofauna distinction breaks down both in terms of
body size and the strategy for population maintenance, because
"macrofauna" species in the deep sea tend to be very small with
continuous reproduction and direct non-pelagic development.  However,
the problems of maintaining animals under natural conditions in the
laboratory to determine growth rates are perhaps more daunting than
for shallow water meiofauna.

REFERENCES

Allen, K. R. 1971. Relation between production and biomass. J. Fish. Res. Bd. Can. 28: 1573-1581.

Barnett, P. R. O. 1970. The life cycles of two species of *Platy-chelipus* Brady (Harpacticoidea) on an intertidal mudflat. Int. Rev. ges. Hydrobiol. Hydrogr. 55: 169-195.

Battaglia, B. 1964. Advances and problems of ecological genetics in marine animals, pp. 451-461. *In* Genetics Today. Proceedings of the XI International Congress of Genetics, The Hague, The Netherlands, September, 1963. Pergamon Press.

Birkett, L. 1959. Production in benthic populations. Cons. L. Int. Explor. Mer. (Unpubl. Rep. C.M., No. 42).

Buchanan, J. B. 1966. The biology of *Echinocardium cordatum* (Echinodermata: Spatangoidea) from different habitats. J. mar. biol. Ass. U.K. 46: 97-114.

Buchanan, J. B. and R. M. Warwick. 1974. An estimate of benthic macrofaunal production in the offshore mud of the Northumberland coast. J. mar. biol. Ass. U.K. 54: 197-222.

Burke, M. V. and K. H. Mann. 1974. Productivity and production: biomass ratios of bivalve and gastropod populations in an eastern Canadian estuary. J. Fish. Res. Bd. Can. 31: 167-177.

Calow, P. 1977. Ecology, evolution and energetics: a study in metabolic adaptation. Adv. ecol. Res. 10: 1-62.

Cederwall, H. 1977. Annual macrofauna production of a soft bottom in the northern Baltic proper, pp. 155-164. *In* B. F. Keegan, P. Ó. Céidigh and P. J. S. Boaden (eds.), Biology of Benthic Organisms.

Chambers, M. R. and H. Milne, 1975. The production of *Macoma balthica* (L.) in the Ythan estuary. Est. coastl mar. Sci. 3: 443-455.

Christiansen, F. B. and T. M. Fenchel. 1977. Theories of populations in biological communities. Springer-Verlag.

Coull, B. C. and W. B. Vernberg. 1975. Reproductive periodicity of meiobenthic copepods: seasonal or continuous? Mar. Biol. 32: 289-293.

Crisp, D. J. 1971. Energy flow measurements, pp. 197-279. *In* N. A. Holme and A. D. McIntyre (eds.), Methods for the study of marine benthos. IBP Handbook No. 16. Blackwell.

Crisp, D. J. 1975. Secondary productivity in the sea, pp. 71-89. *In* Productivity of World Ecosystems. National Academy of Sciences, Washington, D.C.

Dame, R. F. 1976. Energy flow in an intertidal oyster population. Est. coastl. mar. Sci. 4: 243-253.

Delamare-Deboutteville, C. 1960. Biologie des eaux souterraines littorales et continentales. Hermann, Paris.

Dussart, B. 1967. Les copépodes des eaux continentales d'Europe occidentale. Tome 1: Calanoides et Harpacticoides. N. Boubée et Cie, Paris.

Fleeger, J. W. 1979. Population dynamics of three estuarine meiobenthic harpacticoids (Copepoda) in South Carolina. Mar. Biol. 52: 147-156.

Fraser, T. H. 1967. Contributions to the biology of *Tagelus divisus* (Tellinacea: Pelecypoda) in Biscayne Bay, Florida. Bull. mar. Sci. 17: 111-132.

Gadgil, M. and O. T. Solbrig. 1972. The concept of *r*- and *K*- selection: evidence from wild flowers and some theoretical considerations. Am. Nat. 106: 14-31.

Gerlach, S. A. 1971. On the importance of marine meiofauna for
benthos communities. Oecologia, 6: 176-190.

Gerlach, S. A. 1977. Means of meiofaunal dispersal. Mikrofauna
Meeresboden, 61: 89-103.

Gerlach, S. A. and M. Schrage. 1971. Life cycles in marine meiobenthos. Experiments at various temperatures with *Monhystera
disjuncta* and *Theristus pertenuis* (Nematoda). Mar. Biol. 9:
274-280.

Gibbs, P. E. 1968. Observations on the population of *Scoloplos
armiger* at Whitstable. J. mar. biol. Ass. U.K. 48: 225-254.

Goodman, K. 1978. The ecology of the meiofauna of the Ythan estuary, Aberdeenshire. Unpublished Ph.D. thesis, University of
Aberdeen.

Heip, C. 1973. Partitioning of a brackish water habitat by copepod species. Hydrobiologia, 41: 189-198.

Heip, C. 1975. Hibernation in the copepod *Halicyclops magniceps*
(Lilljeborg, 1853). Crustaceana 28: 311-313.

Heip, C. 1976. The life-cycle of *Cyprideis torosa* (Crustecea,
Ostracoda). Oecologia 24: 229-245.

Heip, C. 1980. The influence of competition, predation, environmental stability on population dynamics and production of harpacticoid copepods. K. R. Tenore and B. C. Coull (eds.). Marine
Benthic Dynamics. Univ. of South Carolina Press, Columbia.
Carolina Press, Columbia.

Hibbert, C. J. 1976. Biomass and production of a bivalve community on an intertidal mud-flat. J. exp. mar. Biol. Ecol.
25: 249-261.

Hope, W. D. 1974. Nematoda. Chapter 8. *In* Reproduction of marine
invertebrates. Vol. 1. Acoelomate and pseudocoelamate metazoans.
Academic Press.

Hopper, B. E. 1973. Free-living marine nematodes from Biscayne Bay,
Florida. VI. *Ceramonematidae*: Systematics of *Pselionema annulatum* var. *beauforti* Chitwood, 1936, and a note on the production
and transport of an egg capsule. Proc. Helm. Soc. Wash. 40: 265-
272.

Hopper, B. E. and S. P. Meyers. 1966. Aspects of the life cycle of
marine nematodes. Helgoländer wiss. Meeresunters. 13: 444-449.

Hughes, R. N. 1970. Population dynamics of the bivalve *Scrobicularia plana* (da Costa) on an intertidal mud-flat in North Wales. J.
anim. Ecol. 39: 333-356.

Hughes, R. N. 1970a. An energy budget for a tidal flat population
of the bivalve *Scrobicularia plana* (da Costa). J. anim. Ecol.
39: 357-381.

Hummon, W. D. 1974. Gastrotricha. Chapter 10. *In* Reproduction of
marine invertebrates Vol. 1. Acoelomate and pseudocoelomate
metazoans. Academic Press.

Itô, Y. 1959. A comparative study on survivorship curves for
natural insect populations. Japanese J. Ecol. 9: 107-115.

Jagersten, G. 1953. Life cycle of *Dinophilus*, with special reference to the encystment and its dependence on temperature. Oikos, 3: 143-165.

Juario, J. V. 1975. Nematode species composition and seasonal fluctuations of a sublittoral meiofauna community in the German Bight. Veröff. Inst. Meeresforsch. Bremerh. 15: 283-337.

Kay, D. G. and A. E. Barfield. 1973. The energy relations of the polychaete *Neanthes* (=*Nereis*) *virens* (Sars). J. anim. Ecol. 42: 673-692.

Kendall, M. A. 1979. The stability of the deposit feeding community of a mudflat in the River Tees. Est. coastl. mar. Sci. 8: 15-22.

Klein, G., E. Rachor and S. A. Gerlach. 1975. Dynamics and productivity of two populations of the benthic tube-dwelling amphipod *Ampelisca brevicornis* (Costa) in Helgoland Bight. Ophelia 14: 139-159.

Lasker, R., J. B. J. Wells and A. D. McIntyre. 1970. Growth, respiration, reproduction and carbon utilization of the sand-dwelling harpacticoid copepod *Asellopsis intermedia*. J. mar. biol. Ass. U.K. 50: 147-160.

Lorenzen, S. 1974. Die Nematoden fuana der sublitoralen Region der Deutschen Bucht, insbesondere im Titan - Abwassergebeit bei Helgoland. Veröff. Inst. Meeresforsch. Bremerh. 14: 305-327.

McIntyre, A. D. 1969. Ecology of marine meiobenthos. Biol. Rev. 44: 245-290.

Miller, R. J. and K. H. Mann. 1973. Ecological energetics of the seaweed zone in a marine bay on the Atlantic coast of Canada. III. Energy transformations by sea urchins. Mar. Biol. 18: 99-114.

Moore, H. B. and N. N. López. 1966. The ecology and productivity of *Moira atropos* (Lamarck). Bull. mar. Sci. 16: 648-667.

Moore, H. B. and N. N. López. 1969. The ecology of *Chione cancellata*. Bull. mar. Sci. 19: 131-148.

Moore, H. B. and N. N. López. 1970. A contribution to the ecology of the lamellibranch *Dosinia elegans*. Bull. mar. Sci. 20: 980-986.

Moore, H. B. and N. N. López. 1972. A contribution to the ecology of the lamellibranch *Anodontia alba*. Bull. mar. Sci. 22: 381-390.

Nicholls, A. G. 1935. Copepods from the interstitial fauna of a sandy beach. J. mar. biol. Ass. U.K. 20: 379-403.

Nichols, F. H. 1975. Dynamics and energetics of three deposit-feeding benthic invertebrate populations in Puget Sound, Washington. Ecol. Monogr. 45: 57-82.

Ott, J. A. 1976. Brood protection in a marine free-living nematode; with the description of *Desmodora* (*Croconema*) *ovigera* n.sp. Zool. Anz. 196: 175-181.

Parsons, T. R., M. Takahashi and B. Hargrave. 1977. Biological Oceanographic Processes (2nd ed.) Pergamon.

Pasternak, A. F. 1977. Changes in some energy transformation indices during the growth of the mysid *Neomysis mirabilis* (Czernjavsky). Oceanology, 17: 470-473.

Pawlak, R. 1969. Zur Systematik und Okologie (Lebenszyklen, Populations-dynamik) der Turbellarien-Gattung *Paromalostomum*.

Helgolander wiss. Meeresunters. 19: 417-454.

Peer, D. L. 1970. Relation between biomass, productivity, and loss to predators in a population of a marine benthic polychaete, *Pectinaria hyperborea*. J. Fish. Res. Bd. Can. 27: 2143-2153.

Penter, D. M. 1969. The ecology of intertidal harpacticoids in the Swale, North Kent. Unpublished Ph.D. thesis, University of London.

Penzias, L. P. 1969. *Tellina martinicensis* (Mollusca: Bivalvia): Biology and productivity. Bull. mar. Sci. 19: 568-579.

Petrusewicz, K. 1967. Suggested list of more important concepts in productivity studies (definitions and symbols), pp. 51-82. *In* K. Petrusewicz (ed.), Secondary Productivity of Terrestrial Ecosystems, V. 1. Warsaw and Cracow.

Pianka, E. R. 1970. On *r*- and *K*- selection. Am. Nat. 104: 592-597.

Remane, A. 1952. Die Besiedlung des Sandbodens im Meere und die Bedeutung der Lebensformtypen für die Ökologie. Verh. dtsch. zool. Ges. (1951): 327-339.

Richards, S. W. and G. A. Riley. 1967. The benthic epifauna of Long Island Sound. Bull. Bingham Oceanogr. Coll. 19: 89-135.

Robertson, A. I. 1979. The relationship between annual production: biomass ratios and life spans for marine macrobenthos. Oecologia 38: 193-202.

Sanders, H. L. 1956. Oceanography of Long Island Sound, 1952-1954. X. The biology of marine bottom communities. Bull. Bingham Oceanogr. Coll. 15: 345-414.

Sarvala, J. 1971. Ecology of *Harmothoe sarsi* (Malmgren) (Polychaeta, Polynoidae) in the Northern Baltic area. Ann. Zool. Fennici, 8: 231-309.

Sarvala, J. (in press). Production of *Harmothoe sarsi* (Polychaeta) at a soft bottom locality near Tvärminne, southern Finland. Pr. morsk. Inst. ryb. Gdyni.

Seed, R. and R. A. Brown. 1978. Growth as a strategy for survival in two marine bivalves, *Cerastoderma edule* and *Modiolus modiolus*. J. anim. Ecol. 47: 283-292.

Segerstråle, S. G. 1962. Investigations on Baltic populations of the bivalve *Macoma balthica* (L.). Part II. What are the reasons for the periodic failure of recruitment and the scarcity of *Macoma* in the deeper waters of the inner Baltic? Commentat. biol. 24 (7): 1-26.

Segerstråle, S. G. 1965. Biotic factors affecting the vertical distribution and abundance of the bivalve, *Macoma balthica* (L.) in the Baltic sea. Bot. Gothoburg. 3: 195-204.

Skoolmun, P. and S. A. Gerlach. 1971. Jahreszeitliche Fluktuationen der Nematodenfauna im Gezeitenberich des Weser-Astuars (Deutsche Bucht). Veroff. Inst. Meeresforsch. Bremerh. 13: 119-138.

Sterrer, W. 1974. Gnathostomulida. Chapter 6. *In* Reproduction of marine invertebrates, Vol. 1. Acoelomate and pseudocoelomate metazoans. Academic Press.

Swedmark, B. 1964. The interstitial fauna of marine sand. Biol. Revs. 39: 1-42.

Swedmark, B. and G. Teissier. 1958. *Otohydra vagans* n.g. n.sp.,

Hydrozoaire des sables apparenté aux Halammohydridées. C.R. Acad. Sci. Paris, 247: 238-240.

Teare, M. J. 1978. An energy budget for *Tachidius discipes* (Copepoda, Harpacticoida) from an estuarine mud-flat. Unpublished Ph.D. thesis, University of Exeter.

Thiesen, F. 1966. The life history of seven species of Ostracoda from a Danish brackish water locality. Meddr Danm. Fisk.-og Havunders. N.S.4: 215-270.

Thompson, T. E. 1964. Grazing and the life cycles of British nudibranchs. Brit. ecol. Soc. Symp. 4: 275-297.

Thorson, G. 1946. Reproduction and larval development of Danish marine bottom invertebrates. Medd. Komm. Danm. Fisk.-og Havunders., ser Plankton, 4: 1-523.

Thorson, G. 1950. Reproductive and larvae ecology of marine bottom invertebrates. Biol. Rev. 25: 1-45.

Thorson, G. 1966. Some factors influencing the recruitment and establishment of marine benthic communities. Neth. J. Sea Res. 32: 267-293.

Tietjen, J. H. 1967. Observations on the ecology of the marine nematode *Monhystera filicaudata* Allgén, 1929. Trans. Amer. Microsc. Soc. 86: 304-306.

Tietjen, J. H. 1969. The ecology of shallow water meiofauna in two New England estuaries. Oecologia, 2: 251-291.

Tietjen, J. H. and J. J. Lee. 1977. Feeding behavior of marine nematodes, pp. 21-36. *In* B.C. Coull (ed.), Ecology of marine benthos. University of South Carolina.

Trevallion, A. 1971. Studies on *Tellina tenuis* (da Costa). III. Aspects of general biology and energy flow. J. exp. mar. Biol. Ecol. 7: 95-122.

Warwick, R. M. 1977. The structure and seasonal fluctuations of phytal marine nematode associations on the Isles of Scilly. pp. 577-585. *In* B. F. Keegan, P. Ó Céidigh and P. J. S. Boaden (eds.), Biology of Benthic Organisms. Pergamon.

Warwick, R. M. and J. B. Buchanan. 1971. The meiofauna off the coast of Northumberland. II. Seasonal stability of the nematode population. J. mar. biol. Ass. U.K. 51: 355-362.

Warwick, R. M., C. L. George, and J. R. Davies. 1978. Annual macrofauna production in a *Venus* community. Est. coastl. mar. Sci. 7: 215-241.

Warwick, R. M. and R. Price. 1975. Macrofauna production in an estuarine mud-flat. J. mar. biol. Ass. U.K. 55: 1-18.

Warwick, R. M. and R. Price. 1979. Ecological and metabolic studies on free-living nematodes from an estuarine mud-flat. Est. coastl. mar. Sci. 9: 257-271.

Waters, T. F. 1969. The turnover ratio in production ecology of freshwater invertebrates. Am. Nat. 103: 173-185.

Waters, T. F. 1977. Secondary production in inland waters. Adv. ecol. Res. 10: 91-164.

Wieser, W. and J. W. Kanwisher. 1960. Growth and metabolism in a marine nematode *Enoplus communis* Bastian. Z. vergl. Physiol. 43: 29-36.

Wilke, U. 1954. Mediterrane Gastrotrichen. Zool. Jb. (Syst.), 82: 497-654.

Winberg, G. G.   1971.   Methods for the estimation of production of
    aquatic animals.   Academic Press.
Winberg, G. G., K. Patalas, J. C. Wright, A. Hillbricht-Ilkowska, W.
    E. Cooper and K. H. Mann.   1971.   Methods for calculating pro-
    ductivity pp. 296-317. *In* W. T. Edmonson and G. G. Winberg (eds.)
    A manual of methods for the assessment of secondary productivity
    in fresh water.   IBP Handbook No. 17.   Blackwell.
Wolff, W. J. and L. de Wolf.   1977.   Biomass and production of
    zoobenthos in the Grevelingen estuary, the Netherlands.   Est.
    coastl. mar. Sci.   5:   1-24.
Zaika, V. E.   1973.   Specific production of aquatic invertebrates.
    Wiley.
Zaika, V. E.   1970.   Productivity of marine molluscs as dependent on
    their lifetime.   Oceanology, 10:   702-708.
Zelinka, M. and P. Marvan.   1976.   Notes to methods for estimating
    production of zoobenthos.   Folia Fac. Sci. Nat. Univ. Purkynianae
    Brunensis, Biol. 17 (10):   5-54.

# The Structure and Dynamics of Shelf and Slope Ecosystems off the North East Coast of North America

Eric L. Mills

ABSTRACT: Two marine ecosystems off the coast
of Nova Scotia, Canada are used to illustrate
the differences that may result from spatial
variations in primary production and depth.
Demersal fish are the main product of a shelf
ecosystem centered at 90 m, while pelagic fish
are produced in a system centered at about
320 m where primary production is increased due
to stabilization and nutrient enrichment at an
oceanic front. In comparison with the North
Sea, primary production is higher in the Nova
Scotian systems but zooplankton and fish production are considerably lower.

A major difficulty in delineating plausible quantitative food webs off Nova Scotia is
the problem of providing sufficient energy
to the pelagic components if the benthic components have realistic values. This is the
result of a series of uncertainties about the
amount of primary production, the importance
of microzooplankton and gelatinous herbivores,
the turnover of macrozooplankton, the food

requirements of the fish, and the proper transfer efficiencies.  On the benthic side there
are problems in estimating the amount and kind
of food reaching the bottom; in assessing the
importance of bacteria as food, in determining
relationships of macrofauna, meiofauna, non-
bacterial microfauna, and bacteria; and in
assigning turnover rates to all components.

Comparing the Nova Scotian fish production systems to those off the east coast of
the United States, there are indications that
less primary production is available to the
non-fish elements of the Nova Scotian systems
than farther south.  If so, food webs are
likely to be less complex off Nova Scotia than
in the more southern ecosystems, or transfer
efficiencies may be higher.


INTRODUCTION

In the temperate zones there is evidence from the similarity
of fish catches all over the world that coastal ecosystems may be
similarly structured (Dickie, 1972; Mills, 1975).  However, in
coastal waters there are physical and biological inhomogeneities on
scales from kilometers to hundreds of kilometers that may affect
production and also the components of marine ecosystems.  This
paper explores the structure of ecosystems off the east coast of
Canada based on the hypothesis that seemingly uniform ecosystems
in different areas (on both large and small geographic scales)
may be differently structured and that these differences in structure may be accompanied by striking differences in secondary production.

Some of the forces that produce inhomogeneities in nearshore
marine areas are the following:

1) Historical factors:  There is intriguing evidence that the
outflow of the St. Lawrence River is linked with fish and invertebrate catches on the Scotian Shelf and in the Gulf of Maine up
to several years later (Sutcliffe *et al.*, 1976; 1977).  The
mechanism is likely to be related to the amount of primary production and to changes in the survival of larvae resulting from
increased nutrient supply to the euphotic zone.  The whole system
may be linked to estuarine entrainment in the Gulf of St. Lawrence and ultimately to river hydrology and climatic cycles.
Cause and effect are hard to prove in these cases, but this does
not mean that cyclic and secular climatic changes can be ignored
as a structuring force in the marine environment.

Another historically-related factor affecting marine ecosystems on a broad scale is evident in the different between the
glaciated and unglaciated areas of the east coast of North America.
Comparing regions northeast and southwest of Long Island (roughly
the location of the terminal moraine of the Wisconsin ice sheet),
it is clear that the northern sector is primarily a drowned con-

tinental borderland, most of which is still sinking.  Large estuaries are few and commercial production is centered on species
like cod, haddock, and flounder that live in cold, rocky, or coarse
sand environments.  South of Cape Cod, especially south of Long
Island, the coastal plain is broad and low, the shelf smoother
and temperate, estuaries large and many.  The commercial catch
of marine animals is composed of far more pelagic-estuarine species
depending on terrigenous input to the sea.  These broad but profound differences may easily be seen in compilations of fishery
statistics (e.g., ICNAF, 1977) or in charts and diagrams of the
distributions of commercially-important fishes and invertebrates
(e.g., Hare, 1977).

2) Depth of the water column and nutrient regeneration:
Hargrave (1973) has shown how the depth of the water column and
levels of primary production may be related in marine or freshwater ecosystems.  For any given level of production, of course,
more surface production will reach the bottom if the water column
is short, for example, close to coasts, than if it is longer.  For
water columns of equal depth, those with high primary productivity
will have more of the production consumed in the water column, whereas
in those with low productivity, more of the primary production will be
consumed on the bottom.  Thus, in terms of the fraction of material
reaching the consumers, high surface production and high pelagic production are linked, but there is a complex, not direct, relationship
between surface production and the production of the benthos (Mills,
1975).

3) Geographically-localized sources of nutrients:  The
Scotian Shelf off eastern Canada (Fig. 1) shows at least two patterns of geographical change in levels of production.  In the
central area of the shelf (for example, southeast of Halifax),
most fishery production is of demersal species; pelagic species
like mackerel and herring are a minor component of the catch.
Farther to the southwest, near the entrance to the Bay of Fundy,
there is a major herring fishery.  This and other kinds of evidence (visible plankton patches, sea bird and whale aggregations)
show that a major planktonic production system is in action,
probably linked to turbulent advection of nutrients into the
euphotic zone.  If we follow an inshore-offshore transect, rather
than one directed the length of the shelf, a shifting zone of
high primary production may be detected just beyond the edge of
the continental shelf at an average depth of about 800 m.  There,
at what is often evidently a prograde front between slope water
and continental shelf waters (Bowman and Esaias, 1978; Fournier,
1978; Fournier *et al.*, 1977), chlorophyll a and primary production are higher than closer to shore or farther seaward.  Sea-
bird numbers are higher there than elsewhere, especially during
winter (Brown *et al.*, 1975), whaling was centered there (Sutcliffe and Brody, 1977b), and foreign fishing fleets concentrate a pelagic fishery at the middle depths of the slope close
to the front (Grant and Rygh, 1973; Fig. 2).  Fronts of this kind
are far from rate (see, for example, the 1978 survey by Legeckis),
and it seems very likely that they may have a profound effect in
modifying the structure and levels of production of marine ecosystems on relatively small scales.

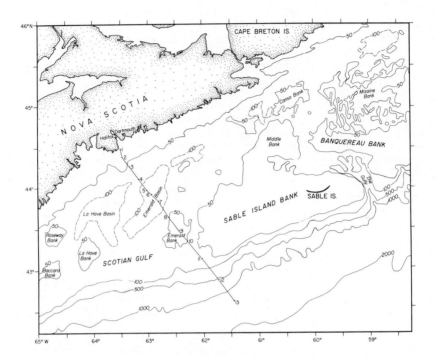

Figure 1.  The Scotian Shelf and slope, showing bathymetry
(in meters) and the transect mentioned in the text.  Brown's
Bank and the entrance to the Bay of Fundy are off the diagram
at lower left. (from Mills and Fournier, 1979).

A reasonable amount of information is now available on fish
catches, zooplankton and benthos standing crops, and phytoplankton production across the Scotian Shelf into deeper water of the
continental slope, as well as from regions to the south, such as
the Gulf of Maine, Georges Bank, and the middle Atlantic Bight of
the U. S. east coast.

I have used some of this information to examine the structure
of marine ecosystems at shelf and slope depths, rather than concentrating solely on the benthic dynamics of shelf and slope ecosystems.  In doing this, I shall focus on several serious problems
that remain unsolved in understanding benthic production, pelagic
production, and the relations between the two, mediated by transfer
processes in the water column.

In doing so, I make the assumption, justifiable or not, that
nature may reasonably be represented by food webs with at least
partially distinct trophic levels.  Alternative formulations are
certainly possible (e.g., Isaacs, 1973), and eventually may even
prove more heuristically useful, but I have chosen to work within
a familiar and conventional framework.

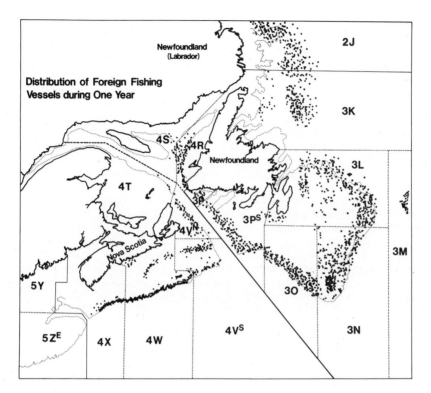

Figure 2. The distribution of foreign fishing vessels during 1972 off the east coast of Canada. Note the close coincidence of the shelf edge and intense fishing effort (from Grant and Rygh, 1973). Areas with numbers and letters are ICNAF divisions.

THE STUDY OF THE SCOTIAN SHELF AND SLOPE ECOSYSTEMS

The region off Nova Scotia (ICNAF Subareas 4Vn through 4X) has supported a major fishery for more than 200 years. In recent years the catch from this region has been about $0.5 \times 10^6$ metric tons, about two-thirds the catch taken between the Gulf of Maine and southern Cape Cod, and about half that taken in the middle Atlantic Bight region of the U.S. East coast (ICNAF, 1977; Hare, 1977). The fishery of the Scotian Shelf and slope region is a major economic factor in the Atlantic Provinces of Canada. Throughout the 1960's and early 1970's, it was increasingly exploited by European fishing nations and other countries such as Japan. Despite this, until a few years ago we did not even have a rough estimate of the annual primary production of this cold water region, let alone estimates of secondary production (apart from the fishery).

The Scotian Shelf (Fig. 1) is topographically rough and dissected, consisting of fishing banks with an average depth of 90 m extending from Banquereau in the northeast to Brown's Bank (near

the entrance to the Bay of Fundy) in the southwest, basins as deep
as 300 m, and eroded or reworked glacial deposits (King and MacLean,
1976). Its total area to the 180 m contour is 62,000 km$^2$. Where
depth allows, the water column has a three-layered structure, consisting of a surface mixed layer having seasonal temperature changes,
an intermediate layer from 50 to 150 m with year-round temperatures
less than 5$^o$C, and a bottom layer derived from slope water with
high temperature and salinity (Houghton *et al.*, 1978; Smith *et al.*,
1978). The general circulation is toward the southwest at about 9
cm sec$^{-1}$ (Sutcliffe *et al.*, 1976).

In 1974, my colleague R. O. Fournier and his associates began
to study the factors controlling primary production on the Scotian
Shelf and adjacent slope by occupying stations along a transect
extending 270 km SE of Halifax (see Fig. 1) across the shelf-
slope front. Their program, which included hydrography, nutrient
chemistry, estimates of primary production, measurement of phyto-
and zooplankton standing stocks and the determination of zooplankton excretion and respiration, is still continuing. Early results
have been summarized by Fournier *et al.* (1977). In 1976, I began
work on the standing crop of benthos on the transect out to 1700 m.
By 1978, Fournier and I believed that a synthesis of our results
would be helpful as a summary, and to allow us to formulate new
hypotheses about the ecosystems of the Scotian Shelf and slope (Mills
and Fournier, 1979). In doing this, we based our conclusions on work
that had been conducted as follows:

The distribution of chlorophyll, especially in the front region
beyond the shelf edge, was determined by continuous profiling with
a fluorometer. Primary production throughout the year was determined by $^{14}$C incubations in deck incubators, simulating conditions
*in situ*, although the cruises have not yet coincided exactly with
the spring bloom. Figures for annual primary production were calculated for shelf (0-180 m) and slope (180-730 m) by multiplying
average daily carbon fixation in each region by 365. We converted
carbon fixation to kilocalories using Platt and Irwin's (1973) relation, 1gC = 11.4 kcal.

Zooplankton biomass, as mg dry weight m$^{-2}$, was derived from
oblique tows using a 0.75 m net with 239 μm mesh. We estimated
annual production by assuming an annual P/B ratio of 7, assuming,
as Steele (1974) did before us, that the standing crop of zooplankton (mainly herbivorous copepods) represents 30-50% of the production of each generation and that there are three generations
per year. For zooplankton, we used the relation 1 g dry weight =
3 kcal (Wiebe *et al.*, 1975).

The macrobenthos from the bank tops (about 90 m) to 1700 m on
the slope was collected using a 0.25 m$^2$ box corer (Hessler and Jumars, 1974). The relation between macrobenthic standing crop and
depth proves to be $\log_{10}B = 1.37 - 0.69Z$, where B is preserved
weight in g and Z = depth of collection in km (Fig. 3). At the
shallower depths (shelf and upper slope) this gives values close
to those obtained by Haedrich and Rowe (1977) southeast of Massachusetts. For purposes of our model, Fournier and I calculated
weights in specific depth ranges from the regression equation and
corrected them to fresh, unpreserved weights, based on data from

my laboratory that the samples decrease 15% in weight after preservation.  Using the corrected values and the relation 1 g wet
weight = 0.6 kcal (Brawn *et al.*, 1968; Thayer *et al.*, 1973; Tyler,
1973), we then calculated annual production using P/B ratios of 2.0–
2.5, assuming that polychaetes and small Crustacea are the main
producers and that the values will fall somewhere between those
obtained for benthos in West Greenland (Curtis, 1977) and in Long
Island (Sanders, 1956).  We obtained data on fish catch and standing crop from Kohler (1968) and Scott (1971), who present information on the stocks gathered at almost the same time during the
mid-1960s.  Scott's information on standing crops, based on trawl
surveys in five depth intervals from 0–730 m, is particularly valuable, for we have found nothing to match it in the more recent literature.  Assuming that the natural mortality for demersal fish (cod,
haddock, flatfish, etc.) was 20% and for pelagic species (some hake,
argentines, redfish, etc.) 50%, we corrected Scott's figures for
mortality, then converted to kcal using the relation 1 g wet
weight = 1.25 kcal (Brawn *et al.*, 1968; Thayer, 1973).  To calculate annual production we estimated P/B ratios of 0.3 for demersal
fish and 0.6 for the pelagics using ratios calculated for a number
of species from the data provided by Scott (1971) and Kohler (1968).
In our calculations, large piscivorous fish were responsible for
the natural mortality of the other fish and the conversion was
assigned a transfer efficiency (defined as { (production at a trophic level) x 100} {production at the preceding level}$^{-1}$) of 10%.

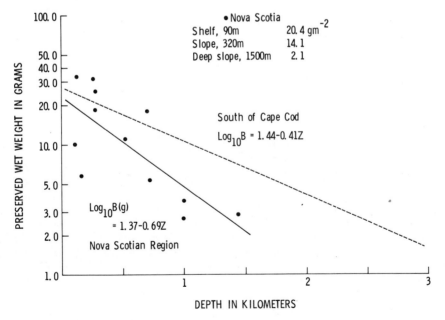

Figure 3.  The distribution of benthic wet weight biomass
with depth along a transect SE of Halifax, Nova Scotia, Canada
(solid line) and SE of Cape Cod, Massachusetts (latter from
Haedrich and Rowe, 1977).

Other elements of these shelf-slope ecosystems are far more difficult to deal with and we were forced to use indirect means to estimate the production of a number of benthic and pelagic components, often by following or modifying the methods used by Steele (1974). For example, we have assumed that a large fraction of the material produced by phytoplankton reaches the bottom either as zooplankton feces or directly as settling cells. There it is consumed directly or indirectly by the macro- and meiobenthos. Gerlach (1978) has presented evidence that macrofaunal and meiofaunal production are nearly equal in shallow subtidal ecosystems. This gives us a hypothetical figure for meiofaunal production. By assuming that Gerlach is correct in saying that bacteria are the main food of benthic organisms ranging in size from ciliates to macrofauna, we estimated bacterial production using a transfer efficiency of 20% to the combined meio-macrofauna. The transfer efficiency between primary producers and bacteria is likely to be 35-40% (Gosselink and Kirby, 1974; Payne, 1970; John Robinson, personal communication). We applied the higher figure to the shallow waters of the shelf and the lower to the slope to estimate the input of phytoplankton cells and zooplankton feces to the bottom.

On the pelagic side of the shelf and slope food webs, we filled the gap between herbivorous zooplankton and pelagic fish by using the same relation Steele (1974) found useful, that pelagic fish require 12 X their own production. Two other components are far more difficult to deal with. Johansen's (1976) work and Fournier's unpublished results suggest that microzooplankton, which must consume significant amounts of phytoplankton, are very abundant at some times in some places. Heinbokel and Beers (1979) have indicated that ciliates may be a powerful grazing force off the California coast. We cannot assign microzooplankton either a biomass or a turnover rate, nor assess the effect of their grazing. In addition, gelatinous zooplankton like salps (probably also appendicularians) are sometimes very abundant, particularly in the front region near the shelf edge. They, too, are of unknown significance, though because of their filtering and turnover rates, pelagic tunicates probably affect the water column and may well contribute large amounts of organic material to the bottom as feces and decaying remains.

The outcome of our first formulation of the ecosystems important on the Scotian Shelf and slope (Mills and Fournier, 1979) is shown in Table 1 and Figure 4. Although we have tabulated the data using the depth intervals from Scott's (1971) survey of fish standing crop, it is evident that the most significant difference is between the continental shelf and the region of the front over the continental slope. Primary production is distinctly higher in the front region, as is total fish production, a result that originally led us to look for unique biological structure seaward of the shelf break. Note that demersal fish production is twice as high on the shelf as on the slope and that pelagic fish far outproduce the demersals on the slope. Scaling the results relative to primary production shows clearly the overriding importance of pelagic fish production in the slope system (Table 2). Demersal and pelagic fish production show oppositely-directed gradients along our inshore-offshore transect. Averaging the fish production across the whole transect obscures the importance of the front region because the shelf has a much greater area than does the front region.

Our results may be compared with the North Sea ecosystem as described by Steele (1974) (Table 2). The catch of demersal fish off Nova Scotia is nearly identical to that in the North Sea, but overall fish catch is 47% lower in Nova Scotian waters because

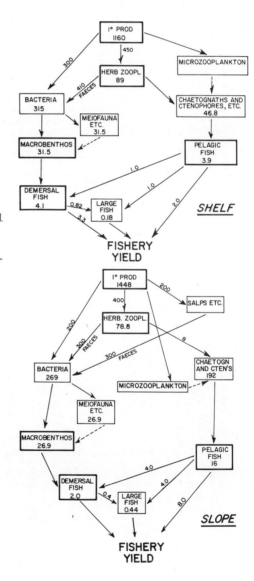

Figure 4. Hypothetical food webs centered at 90 m (shelf) and 320 m (slope) SE of Nova Scotia, Canada. Figures are in kcal $m^{-2}yr^{-1}$. Those in heavy boxes are based on direct measurements or fishery statistics and those on lines represent yield from one trophic level to the next if it could be inferred. Figures in light boxes have been calculated or assigned as described in the text. (from Mills and Fournier, 1979).

Table 1. The major components of production on the Scotian Shelf and slope based on original data plus fishery surveys and catches. Except for primary production, the uppermost number in each category is biomass per m² while the second is production in kcal $m^{-2}yr^{-1}$. P/B ratios are 7 for zooplankton, 0.3 for demersal fish, 0.6 for pelagic fish, 2.0 for the 0-90 m benthos and 2.5 for the remaining benthos. (From Mills and Fournier, 1979)

| Depth range (m) | Shelf | | | Slope | | | |
|---|---|---|---|---|---|---|---|
| | 0-90 | 90-180 | 0-180 | 180-275 | 275-360 | 360-730 | 180-730 |
| Area ($10^9$m²) | 17.68 | 24.91 | 42.59 | 5.17 | 0.83 | 0.81 | 6.81 |
| Primary production | | | | | | | |
| g Carbon | 66.1 | 127.6 | 102.1 | 127.6 | 127.6 | 127.6 | 127.6 |
| kcal | 753.5 | 1448 | 1159.6 | 1448 | 1448 | 1448 | 1448 |
| Zooplankton | | | | | | | |
| g Dry wgt. | 1.9 | 5.9 | 4.2 | 4.0 | 4.0 | 2.0* | 3.8 |
| kcal | 39.9 | 123.9 | 89.0 | 84.0 | 84.0 | 42.0* | 78.8 |
| Benthos | | | | | | | |
| g Wet wgt. | 24.0* | 22.1 | 22.9 | 19.1 | 16.6 | 11.8 | 17.9 |
| kcal | 28.8 | 33.1 | 31.5 | 28.6 | 24.9 | 17.6 | 26.9 |
| Demersal fish | | | | | | | |
| g wet wgt. | 8.0 | 9.1 | 8.6 | 4.7 | 2.7 | 1.1 | 4.0 |
| kcal | 3.8 | 4.3 | 4.1 | 2.3 | 1.3 | 0.5 | 1.9 |
| Pelagic fish | | | | | | | |
| g wet wgt. | 0 | 4.4 | 2.6 | 10.3 | 15.4 | 7.3 | 10.6 |
| kcal | 0 | 6.6 | 3.9 | 15.6 | 23.0 | 11.0 | 15.9 |
| Total fish | | | | | | | |
| kcal | 3.8 | 10.9 | 8.0 | 17.9 | 24.3 | 11.5 | 17.9 |

*Estimated value

Table 2. Comparison of production between the Scotian Shelf and slope (our data) and the North Sea (Steele, 1974). All units are kcal m$^{-2}$yr$^{-1}$. Figures in parentheses are from Steele (1974), while asterisked (*) figures have been calculated from his data using our conversion factors and/or P/B values. Note that Steele's figures for benthic production are based on P/B = 5, ours on 2 to 2.5. If P/B = 2 applied in the North Sea, macrobenthic production would be 8 to 20 kcal m$^{-2}$yr$^{-1}$. The division of fish catch between the Scotian Shelf and slope (†) has been estimated by assigning species catches to shelf and slope; no direct information where catches occurred was available. Nova Scotian pelagic catch does not include argentines and grenadiers, for which we had no data, but because these species are taken in a small area, this is unlikely to increase the figure greatly. Figures in brackets are production scaled per unit primary production, allowing the comparison of regions and components of the food webs on a common basis.

| | Shelf | Slope | Average N.S. | North Sea |
|---|---|---|---|---|
| Primary production | 1160 | 1148 | 1199 | (900) 1026 |
| Zooplankton | 89 [0.077] | 78.8 [0.054] | 88 [0.073] | (170) 128* [0.125] |
| Macrobenthos | 35 [0.027] | 26.9 [0.019] | 30.7 [0.026] | 20-50 [0.019-0.049] |
| Demersal fish | 4.1 [0.004] | 1.96 [0.001] | 3.8 [0.003] | (2.6) 3.3* [0.003] |
| Pelagic fish | 3.9 [0.003] | 15.95 [0.011] | 5.6 [0.005] | (8) 10* [0.009] |
| Total fish production | 8.0 [0.007] | 17.9 [0.012] | 9.4 [0.008] | (10.6) 13.3*[0.013] |
| Fish catch, total | 2.8† [0.002] | 11.8† [0.008] | 3.9 [0.003] | (6.0) 7.4* [0.007] |
| pelagic | | | 1.8† [0.002] | 5.1 [0.005] |
| demersal | | | 2.2† [0.002] | 2.3 [0.002] |

the production of pelagic fish occurs in a small area rather than being spread across the continental shelf. Despite the fact that primary production on the Scotian Shelf and slope is on average 17% higher than in the North Sea, zooplankton production is 31% lower. As the scaled values in Table 2 show, this is a major difference between the two systems along with the production of pelagic fish. Other features are quite similar. Once again we find differences in food webs and their commercial output that appear to be governed by processes in the water column.

Steele has been able to provide a plausible scheme that provides sufficient energy to the pelagic side of the North Sea food web but not the benthic side. In my analysis of the Scotian Shelfslope ecosystems, as long as the benthic side is in reasonable balance and follows the conventional lines:

primary producers → bacteria → { meiofauna and macrofauna } → demersal fish;

the pelagic side will be unbalanced. The most dramatic problem is that the production of herbivorous zooplankton is far too low to support the links:

primary producers → herbivorous zooplankton → pelagic invertebrate predators → pelagic fish

with conventional transfer efficiencies. Something is wrong: perhaps in our estimates; perhaps in the transfer efficiencies we have used; or perhaps in our delineation of the food webs. If the pelagic components should be re-examined, perhaps so, too, should our seemingly clearcut results with the benthic side so that critical questions can be asked about all the components of our models. By doing so we can formulate new, falsifiable hypotheses as we continue to study the secondary production of these complex marine systems.

CRITIQUE OF THE SCOTIAN SHELF-SLOPE MODELS

Taking the elements of our models, as illustrated in Figure 4, step by step, I believe that we should now be concentrating on answering questions about general levels of primary production, about the transfer of material to the bottom and its utilization, and about the feeding links and transfer efficiencies of pelagic animals.

The major uncertainties may be listed as follows:

1) Estimating primary production: It would ease our problems considerably if primary production were higher than 102–128 gC m$^{-2}$ yr$^{-1}$ (1160–1448 kcal m$^{-2}$yr$^{-1}$) calculated from $^{14}$C uptake. Our estimates are probably 25% too low because we have missed the spring bloom (if we extrapolate the results described by Platt and Conover, 1975); in addition, Fournier *et al.*, (1979) have shown that winter production at the front may be significant when transitory stability keeps phytoplankton in the euphotic zone. Only a few

events of this kind each winter might increase production in the
front region by 35%. However, even with these additions, it is
unlikely that annual production across the Scotian Shelf is higher
than 200 gC m$^{-2}$yr$^{-1}$, a figure far below the startlingly high value
of 400-500 gC m$^{-2}$yr$^{-1}$ suggested by Cohen and Wright (1978) for
Georges Bank, not far to the south of the Scotian Shelf. If our
formulation of the problems is correct, it would require an order
of magnitude correction in estimates of primary production to have
much effect. I believe this is very unlikely to occur.

2) The benthic components:

a) We are not sure how much phytoplankton production reaches
the bottom, either in absolute amounts or the relative amounts
of zooplankton feces, directly settling cells and phytoplankton
debris. There may be significant differences in these components
on very small scales, as shown in Table 3, for at 800 m on the
slope off Nova Scotia (close to the average position of the front),
the amount of fresh chlorophyll a in the sediment rises dramatically. Curiously enough, I cannot find evidence that macrofaunal
biomass increases, although the sediments are full of the fecal
pellets of benthic animals, unlike locations just above and below
800 m. Walsh *et al.* (1978) have made the point that "benthic-
pelagic coupling" is far greater in the New York Bight than in
the Peru Current upwelling region, but our results show that
benthic-pelagic coupling may vary greatly on the scale of kilometers, given sharp changes in surface production and enough time
and stability for separate ecosystems to develop.

Table 3. The amount of chlorophyll a, pheopigment, and chl a/
pheopigment ratio in sediments along a transect SE from Halifax, Nova
Scotia. The asterisk marks a depth corresponding to the location of
a front between slope and continental shelf water where high chlorophyll
values and $^{14}$C uptake occur. Data courtesy of B. T. Hargrave, Marine
Ecology Laboratory, Dartmouth, Nova Scotia.

| Depth, m | Chlorophyll a, µg g$^{-1}$ dry sediment | Pheopigment, µg g$^{-1}$ dry sediment | Chlorophyll a as % of total pigment |
|---|---|---|---|
| 80 | 0.03 | 0.36 | 7.7 |
| 86 | 0.17 | 0.85 | 17 |
| 400 | 0.44 | 2.34 | 16 |
| 720 | 0.46 | 3.77 | 11 |
|  | 0.34 | 2.23 | 13 |
| *800 | 12.83 | 19.03 | 40 |
|  | 8.43 | 14.30 | 37 |
| 1700 | 0.22 | 1.97 | 10 |
|  | 0.14 | 1.94 | 6.7 |

b)  The role of bacteria is problematical:  Gerlach (1978) may well be correct that most benthic organisms subsist on bacteria, and that the role of the meiofauna is critical in gardening bacteria, that is, releasing their production.  On the other hand, Cammen (1978) has shown that an omnivorous polychaete cannot balance its carbon budget by consuming bacteria alone and that it must and does assimilate non-living particulate organic matter (see also Sibert, Tenore and Rice, this volume, on the role of detritus in benthic nutrition).  If this is more broadly applicable, our estimates of bacterial production may be grossly in error and the link between phytoplankton and benthic animals will become far more difficult to quantify, even if we are able to measure the input of zooplankton feces.  No one knows yet whether benthic bacteria at shelf-slope depths are mainly physiologically-quiescent (as they appear to be in the deep-sea), or if they are significant producers.  Perhaps a transition occurs from active microbial activity on the shelf to much lower rates or quiescence on the slope, but we have no way of knowing if this occurs or where it might happen.

c)  What turnover rates do macrobenthic animals really have?

The P/B values in our model were little better than "guesstimate" interpolations from limited work in very different areas. Very few studies of macrobenthic production have been done in the Scotian Shelf region (one exemplary study is by Peer, 1970).  We desperately need the kind of information tabulated recently by Warwick *et al.* (1978) for a nearshore environment of the British Isles.

d)   How are meiofauna, macrofauna and other elements of the benthos related?

Warwich (1980) has pointed out the difficulties of estimating meiofaunal production.  Gerlach (1978) calculated roughly equivalent production by the macrofauna and meiofauna of shallow North European waters.  To have this relationship, the P/B ratios of the meiofauna had to be in the range 4-10.  If the equivalence of macrofaunal production is correct near shore, is it also true at greater depths?  Coull *et al.* (1977) have documented the decrease of meiofauna in increasingly deeper water toward abyssal depths, but at the very greatest depths in the most food-poor oceans, nearly all the organisms are of meiofaunal size (Thiel, 1975).  What occurs between 90 and 1700 m along our transect?  We do not know yet, although studies in progress in my laboratory should give us a partial answer soon.  The answer may be very important in interpreting the relation between pelagic production and the sediments, for if there is really no increase in macrofaunal biomass or production where fresh chlorophyll in abundance reaches the bottom at 800 m, there is likely to be increased production of meiofauna or bacteria, perhaps both, and greatly decreased transfer efficiencies to demersal fish, a result we seem to see beyond the shelf edge.

Another unknown aspect of the benthic situation is the significance of microfauna like ciliates and foraminifera.  I find large numbers of benthic foraminifera across the shelf into deep water, many of them apparently living, but I cannot assess their importance as consumers or in energy transfer.  They may well form a link

between particulate and dissolved organic matter and the meiobenthos or selective deposit-feeding macrobenthos; these are merely
speculations until research shows us what to expect.

   3)  The pelagic components:

   Our main problem (see Table 1 and Figure 4) is that even in
the shelf ecosystem, pelagic fish production is too high for the
production of herbivorous zooplankton we can calculate. This problem is particularly acute in the slope ecosystem. Some of the
topics discussed below increase our difficulties but some may al-
leviate them.

   a)  The role of microzooplankton and of gelatinous herbivores

   We cannot yet estimate the abundance of microzooplankton organisms (e.g., tintinnids, other ciliates, rotifers, invertebrate
larvae, etc.), or their turnover. Almost certainly microzooplankton add an extra step between the phytoplankton and the "herbivorous" zooplankton (which is likely to be partially carnivorous) and
thus reduce the amount of energy transferred to the larger herbivores and on to invertebrate predators like chaetognaths and cteno-
phores. Heinbokel and Beers (1979) suggest that there may not be
a large loss because the growth efficiencies of tintinnids and
other ciliates are very high and they are readily eaten by other
zooplankton. This problem urgently needs resolution. I believe
it is the most important gap or inaccuracy in our formulation of
the pelagic food web. Salps and appendicularians, if they are
consistent members of the offshore ecosystem, as casual observations suggest, should have a similar effect by diverting primary
production from pelagic consumers and transferring it as partially-
assimilated cells in fecal material to the benthos. Despite this,
it is the slope ecosystem, where these gelatinous herbivores seem
to be most abundant, that has the most productive pelagic food web.

   b)  The abundance and production of macrozooplankton

   We may have underestimated the abundance of macrozooplankton
by failing to catch its larger, more mobile members. The margin
of error is quite great; for example, Brodie *et al.*, 1978 calculated macrozooplankton abundance of 0.1 g m$^{-3}$ (wet weight) near the
edge of the Scotian Shelf, yet at least 17.5 g m$^{-3}$ should be present to support the whale stocks that are present. They resolved
this anomaly by showing through sonar and high speed net tows that
the euphausiid *Meganyctiphanes norvegica* was present in layers
5-20 m thick at depths of 100-150 m and that its abundance there
lay somewhere between 8 and 26 g m$^{-3}$. Our sampling almost certainly
missed this impressive biomass of large herbivores, thus our macrozooplankton values are too low. However, when integrated over the
euphotic zone to give production per m$^2$, the absent euphausiids are
unlikely to more than double the production rate we have calculated.
In the pelagic ecosystem, even doubling herbivore production will
only bring it to the same level as chaetognath and ctenophore production, calculated from the food requirements of the fish (see
part d).

   Another possible source of error is the P/B value of 7 we chose
for herbivorous zooplankton. Overall, if we were to include the
microzooplankton species ( a problem in themselves - see preceding)
a far higher value would be more suitable. But for copepods alone,

which are the most evident herbivorous zooplankton along the northeast coast of North America, a P/B value of 7 falls nicely into the
range of values tabulated by Conover (1979) for a range of species
in several temperate continental shelf areas. We must look elsewhere for the solution of the zooplankton-pelagic fish conundrum.

     c)  The production of fish

     In our original models we estimated P/B = 0.3 for demersal fish
and 0.6 for pelagic ones, values that we considered reasonable based
on the realized production and estimated biomass of commercially-
important stocks on the Scotian Shelf. Might these values be in
error, especially for pelagic fish, reducing our estimate of pelagic production if we have overestimated P/B for those species? Some
information that may be relevant is provided by Grosslein *et al.*,
1978, who calculated annual P/B ratios for several fish species
based on an energy balance equation and stock assessments by virtual
population analysis. Their averaged values were 0.29 for herring,
0.60 for cod (important off Nova Scotia), 0.34 for mackerel, 0.69
for silver hake (a heavily exploited species off Nova Scotia), 0.63
for yellowtail and 0.41 for haddock (a major component of the demersal catch off Nova Scotia). As with the P/B values we chose for
the macrobenthos, it is difficult to know how to select values.
Should we use values from metabolic models, or should we select values
based on catches and estimated biomasses (both of which fall far short
of full production and total standing stock of all age classes)?
There is sufficient agreement between our estimates (particularly for
hake and haddock) that, without completely brushing the discrepancies
under the carpet, I believe we have used figures that are reasonable
given our present state of knowledge.

     d)  Food requirements of the fish

     In our original formulation we calculated the food requirements
of the fish as 12 X their annual production. This was a figure used
by Steele (1974) in his calculation of the food budget for North Sea
pelagic fish, derived from Lasker's (1970) estimates for Pacific
sardine. Recently, new information has become available that allows
us to re-examine the original factor. Green (1978) calculated the
food requirements of Gulf of Maine herring using a metabolic model.
Averaged for all year classes, Gulf of Maine herring may require only
4.9 X their annual production in food. Although herring are not a
significant component of the ecosystems in our study, we can use this
food requirement to calculate a lower limit to the production occurring at the invertebrate predator level. In the shelf ecosystem, 3.9
kcal $m^{-2}yr^{-1}$ of pelagic fish production would require 19.1 kcal rather than the 46.8 we originally calculated, implying a transfer
efficiency of about 20%. This could be provided by 191 kcal of herbivorous zooplankton production via the invertebrate predator link,
a value far higher than we could justify, or directly by the 89 kcal
we calculated if the pelagic fish fed directly on copepods. In the
latter case, the transfer efficiency would be only 4.4%. Considering the pelagic ecosystem, 16 kcal $m^{-2}yr^{-1}$ of pelagic fish production
would require only 78.4 kcal rather than the 192 kcal we calculated
originally, with a transfer efficiency of about 20%, as in the shelf
ecosystem. This 78.4 kcal could readily be provided directly at equilibrium production by 78.8 kcal $m^{-2}yr^{-1}$ of copepods, leaving a little

over for invertebrate predators, or leaving considerably more if
our estimates of herbivore abundance are too low, as discussed earlier.

It is clear that we can significantly reduce the problem presented by pelagic fish in our models by reducing their food requirements or by leaving out the invertebrate predator link and allowing
the pelagic fish to feed directly on the herbivores, as herring do.
However, not all pelagic fish are herring, especially in our area,
and the correct formulation probably lies in a balance of the two
approaches. Establishing the balance plausibly depends on our ability to determine the feeding habits of the fish and the production
of all the major components of the zooplankton.

Before I become too sanguine about the prospects of reducing
the food requirements of the pelagic fish, I must point out that
Sherman *et al.* (1978) tabulate food requirements (based on a metabolic model) that differ considerably from the value I have just
considered. They estimate the food requirements of pelagic species
like herring, silver hake, argentines, redfish, mackerel, etc. as,
on average, 8.3 X annual production (range 7-10). So our original
estimate, while too high, was not a gross error. We must consider
ways of determining the true food relations and transfer efficiencies of pelagic fish as well as the effective production of herbivorous zooplankton and invertebrate predators. Band-aid solutions
will not be adequate.

4) Discrepancies between our results and other estimates of
marine production:

Sheldon, Sutcliffe and Paranjape (1977) have provided an original method of estimating production at any trophic level based
on the observation that there are roughly equal standing stocks of
organisms in logarithmic size intervals from the smallest phytoplankton cells to the largest consumers. Given values for the
standing stocks and turnover rates of phytoplankton and fish plus
estimates of their relative size ranges and the fish production,
they calculated primary production in the Gulf of Maine and concluded that it was 260 X fish production. In theory their method
should be applicable to any pair of trophic levels provided the
size range of the organisms, their doubling times, and the production of one component are known. Sherman *et al.* (1978) reversed
this method to estimate fish production from primary production
along the east coast of the United States from the Gulf of Maine
to the middle Atlantic Bight. According to their calculations by
this method, the ratio of phytoplankton production to fish production should range from about 500 in the middle Atlantic Bight
to 180 off southern New England. They used a factor of 260 in
the Gulf of Maine and on Georges Bank, following the analysis by
Sheldon *et al.* (1977). The actual fish production, based on
catches, is considerably lower than the calculated values, suggesting that there is surplus primary production available to other
components of the ecosystem in the areas they studied. The values
are shown in Table 4. I have included values from the Scotian
Shelf area for comparison, including the fish production that
should be realized ("theoretical fish production") if the value
of 260 calculated by Sheldon *et al.* applies in the Nova Scotian

area as it does in the closely-similar Gulf of Maine.

This analysis indicates that more energy from primary production is available to non-fish components of the ecosystems off
the east coast of the United States than on the Scotian Shelf.
This is consistent with the steep decline of benthic biomass with
increasing depth off Nova Scotia (Fig. 3).  It leaves us with the
problem that in the Nova Scotian ecosystems (especially on the
slope) more fish production is being realized than might be expected using classical food chain formulations and transfer efficiencies.  Food chains may be shorter off Nova Scotia, or transfer
efficiencies may be higher, especially on the slope, than they
are farther south.  It is certainly a testable hypothesis that the
food webs of Nova Scotian ecosystems are simpler than those farther
south (differences in species diversity would be a clue to this),
but it will be difficult, given the state of the art, to attempt
to falsify the hypothesis that transfer efficiencies differ significantly from ecosystem to ecosystem.

Table 4.  Primary production, fish production (both in g C m$^{-2}$yr$^{-1}$)
and the amount of primary production available to non-fish components of
ecosystems in five areas off the North American east coast, based largely
on the work of Sherman *et al.* (1978) plus estimates from the Scotian
Shelf (this paper).  Figures in row 2 are derived from Sheldon *et al.*
(1977) and Sherman *et al.* (1978).  Theoretical fish production (3a) was
calculated from primary production as outlined in the text.  Actual fish
production (3b) is based on stock assessments and P/B values; it represents
stocks present in the 1960's, before the overfishing of the 1970's.  Primary production needed for actual fish production (4) is 2 x 3b.  Surplus
primary production (5) is the measured primary production not needed for
actual production, i.e., [(1 -4)÷4]100.  In the first column (Nova Scotia),
values on the left apply to the shelf ecosystem, those on the right to
the slope ecosystem.

|   |   | Nova Scotia | Gulf of Maine | Georges Bank | Southern New England | Mid Atlantic Bight |
|---|---|---|---|---|---|---|
| 1. | Measured primary production | 102–128 | 150–200 | 450 | 150 | 150–200 |
| 2. | Estimated ratio of primary pro- duction to fish production | 260 | 260 | 260 | 180 | 500 |
| 3. | Fish production | | | | | |
|   | a. theoretical | 0.39–0.49 | 0.77 | | | |
|   | b. actual | 0.38–0.86 | 0.42 | 1.05 | 0.55 | 0.25 |
| 4. | Primary production needed for actual fish production | 99.8–223 | 109 | 273 | 99 | 123 |
| 5. | Surplus primary pro- duction, % | 2.2–0 | 38–83 | 65 | 52 | 22–63 |

## CONCLUDING DISCUSSION

This paper seems to have strayed rather far from the major
concerns of a symposium on benthic dynamics, but I believe this is
true only in a superficial way. Marine ecosystems do show underlying unity - almost certainly several patterns of unification -
that to a greater or lesser extent involve the benthos  as well as
the pelagic components. Just how different coastal marine ecosystems may in this respect be is evident by comparing Georges
Bank, which has very high primary production and high demersal fish
production (a function of high production by the invertebrate benthos), with the productive front region off the continental shelf
region of Nova Scotia, where high levels of primary production find
expression in pelagic fish production. The most significant difference between these two areas appears to be depth, a factor that
governs the speed and routing of phytoplankton production to the
bottom.

Between these two contrasting situations lies a variety of
different near-shore production systems with intermediate characteristics, or with lower overall levels of production due to reduced nutrient supply or low *in situ* regeneration of nutrients. No doubt
only a few of these systems require really detailed exploration, for
not all have high commercial importance or are likely to give us
new scientific insights. However, even where they are of
economic and scientific interest, for example along the urbanized,
resource-rich east coast of North America, it has proved distress-
ingly difficult to provide firm answers about how these ecosystems
with commercial fisheries at their apex really function. Even if
we have formulated the problems in the right way by delineating
trophic levels and food web linkages and by applying concepts like
transfer efficiencies, it is clear that the apparently straight-
forward linkages, for example from the phytoplankton to macrobenthic animals, still provide more profound problems than intellec-
tually satisfying answers. My analysis has aimed at showing that
the study of internal complexities and the external factors that
constrain them is really far more interesting at the moment than
a broad systems approach to marine ecosystems. Benthic ecologists,
having a tradition of attention to detail, should find that conclusion a congenial one.

## ACKNOWLEDGMENTS

I am particularly grateful to my colleague, R. O. Fournier of
Dalhousie University for much discussion, the use of data and his
sparkplug effect in initiating and continuing the study of the
Scotian Shelf region. Kathy Aldous, Sheila Byers, Tom Clair, John
Kearney, Chris Majka and Allison Mitchell worked valiantly to provide information on benthic biomass across the Scotian Shelf. B. T.
Hargrave, Marine Ecology Laboratory, Dartmouth, Nova Scotia, allowed
me to use his data on pigments in sediments. I am also grateful
to E. B. Cohen, J. R. Green, M. D. Grosslein and Kenneth Sherman,
all of the U.S. National Marine Fisheries Service, for allowing me to
cite and use information from their papers presented at the 1978

ICES meetings.  This research was supported by grants from the
Natural Sciences and Engineering Research Council of Canada.

REFERENCES

Bowman, M. J. and W. E. Esaias (eds.).  1978.  Oceanic fronts in
    coastal processes.  Springer-Verlag.
Brawn, V. M., D. L. Peer and R. J. Bentley.  1968.  Caloric content
    of the standing crop of benthic and epibenthic invertebrates of
    St. Margaret's Bay, Nova Scotia.  J. Fish. Res. Bd. Can. 25: 1803-
    1811.
Brodie, P. F., D. D. Sameota and R. W. Sheldon.  1978.  Population
    densities of euphausiids off Nova Scotia as indicated by net
    samples, whale stomach contents, and sonar.  Limnol. Oceanogr.
    23: 1264-1267.
Brown, R. G. B., D. N. Nettleship, P. Germain, C. E. Tull and T.
    Davis.  1975.  Atlas of eastern Canadian seabirds.  Ottawa:
    Canadian Wildlife Service.
Cammen, L.  1978.  The significance of microbial carbon in the
    nutrition of the polychaete *Nereis succinea* and other deposit
    feeders and detritivores.  Ph.D. thesis, North Carolina State
    University, Raleigh.
Cohen, E. B. and W. R. Wright.  1978.  Changes in the plankton on
    Georges Bank in relation to the physical and chemical environ-
    ment during 1975-76.  I.C.E.S., Biological Oceanography Commit-
    tee, C.M. 1978/L27, 13 pp (mimeographed).
Conover, R. J.  1979.  Secondary production as an ecological phe-
    nomenon.  *In* S. van der Spoel and A. C. Pierrot-Bults (eds.).
    Zoogeography and Distribution of Plankton.  Bohm, Scheltema
    Holkema (in press).
Coull, B. C., R. L. Ellison, J. W. Fleeger, R. P. Higgins, W. D.
    Hope, W. D. Hummon, R. M. Rieger, W. E. Sterrer, H. Thiel and J.
    H. Tietjen.  1977.  Quantitative estimates of the meiofauna from
    the deep sea off North Carolina, U.S.A.  Mar. Biol. 39:  233-240.
Curtis, M. A.  1975.  Life cycles and population dynamics of marine
    benthic polychaetes from the Disko Bay area of West Greenland.
    Ophelia 16:  9-58.
Dickie, L. M.  1972.  Food chains and fish production.  ICNAF
    Spec. Publ. No. 8:  201-221.
Fournier, R. O.  1978.  Biological aspects of the Nova Scotia
    shelf-break fronts.  *In* M. J. Bowman and W. E. Esaias (eds.),
    Oceanic Fronts In Coastal Processes, pp. 69-77.  Springer Ver-
    lag.
Fournier, R. O., J. Marra, R. Bohrer and M. van Det.  1977.  Plank-
    ton dynamics and nutrient enrichment of the Scotian Shelf.  J.
    Fish. Res. Bd. Can. 34:  1004-1018.
Fournier, R. O., M. van Det, J. S. Wilson and N. B. Hargreaves.
    1979.  The influence of the shelf-break front off Nova Scotia
    on phytoplankton standing stock in winter.  J. Fish. Res. Bd.
    Can. (in press).
Gerlach, S. A.  1978.  Food-chain relationships in subtidal silty
    sand marine sediments and the role of meiofauna in stimulating
    bacterial productivity.  Oecologia 33:  55-69.

Gosselink, J. G. and C. J. Kirby. 1974. Decomposition of salt marsh grass, *Spartina alterniflora* Loisel. Limnol. Oceanogr. 19: 825–832.

Grant, D. A. and P. R. Rygh. 1973. Surveillance of fishing activities in the ICNAF areas of Canada's east coast. Maritime Command Operations Research Branch, Halifax, N.S. Rep. 3/73.

Green, J. R. 1978. Theoretical food rations of Gulf of Maine and Georges Bank herring stocks. I.C.E.S. Pelagic Fish. Committee, C.M. 1978/H:39, 14 99 (mimeographed).

Grosslein, M. D., R. W. Langton and M. P. Sissenwine. 1978. Recent fluctuations in pelagic fish stocks of the Northwest Atlantic, Georges Bank region, in relationship to species interactions. I.C.E.S. Biological Oceanography Committee, C.M. 1978/L.24, 52 pp (mimeographed).

Haedrich, R. L. and G. T. Rowe. 1977. Megafaunal biomass in the deep sea. Nature 269: 141–142.

Hare, G. M. 1977. Atlas of the major Atlantic coast fish and invertebrate resources adjacent to the Canada-United States border area. Environment Canada, Fish. Mar. Serv. Tech. Rpt. No. 681, 97 pp.

Hargrave, B. T. 1973. Coupling carbon flow through some pelagic and benthic communities. J. Fish. Res. Bd. Can. 30: 1317–1326.

Heinbokel, J. F. and J. R. Beers. 1979. Studies of the functional role of tintinnids in the southern California bight. III. Grazing impact of natural assemblages. Mar. Biol. 52: 23–32.

Hessler, R. R. and P. A. Jumars. 1974. Abyssal community analysis from replicate box cores in the central North Pacific. Deep-sea Res. 21: 185–209.

Houghton, R. W., P. C. Smith and R. O. Fournier. 1978. A simple model for cross-shelf mixing on the Scotian Shelf. J. Fish. Res. Bd. Can. 35: 414–421.

International Commission for the Northwest Atlantic Fisheries (ICNAF). 1977. Statistical bulletin for the year 1976. Vo. 26, 236 pp.

Isaacs, J. D. 1973. Potential trophic biomasses and trace substance concentrations in unstructured marine food webs. Mar. Biol. 22: 97–104.

Johansen, P. L. 1976. A study of tintinnids and other Protozoa in eastern Canadian waters, with special reference to tintinnid feeding, nitrogen excretion and reproduction rates. Ph.D. thesis, Dalhousie University. 156 pp.

King, L. H. and B. Maclean. 1976. Geology of the Scotian Shelf. Envir. Can., Fish. Mar. Serv., Marine Sci. Paper 7, 31 pp and charts.

Kohler, A. C. 1968. Fish stocks of the Nova Scotia Banks and the Gulf of St. Lawrence. Fish. Res. Bd. Can., Tech. Rpt. No. 80, 8 pp and table, figures.

Lasker, R. 1970. Utilization of zooplankton energy by a Pacific sardine population in the California current. *In* J. H. Steele (ed.), Marine food chains, pp. 265–284. University of California Press.

Legeckis, R. 1978. A survey of worldwide sea surface temperature fronts detected by environmental satellites. J. Geophys. Res. 83 (C9): 4501–4522.

Mills, E. L.  1975.  Benthic organisms and the structure of marine
    ecosystems.  J. Fish. Res. Bd. Can. 32:  1657-1663.
Mills, E. L. and R. O. Fournier.  1979.  Fish production and the
    marine ecosystems of the Scotian Shelf, eastern Canada.  Mar.
    Biol. (in press).
Payne, W. J.  1970.  Energy yields and growth of heterotrophs.  Ann.
    Rev. Microbiol. 24:  17-52.
Peer, D. L.  1970.  Relation between biomass, productivity, and loss
    to predators in a population of a marine benthic polychaete,
    *Pectinaria hyperborea*.  J. Fish. Res. Bd. Can. 27:  2143-2153.
Platt, T. and R. J. Conover.  1975.  The ecology of St. Margaret's
    Bay and other inlets on the Atlantic coast of Nova Scotia.  *In*
    T. W. M. Cameron and L. W. Billingsley (eds.), Energy Flow - Its
    Biological Dimensions, pp. 249-259.  Ottawa, Royal Society of
    Canada.
Platt, T. and B. Irwin.  1973.  Caloric content of phytoplankton.
    Limnol. Oceanogr. 18:  306-310.
Sanders, H. L.  1956.  Oceanography of Long Island Sound. X.  The
    biology of marine bottom communities.  Bull. Bingham Oceanogr.
    Coll. 15:  345-414.
Scott, J. S.  1971.  Abundance of ground fishes on the Scotian
    Shelf.  Fish. Res. Bd. Can., Tech. Rpt. No.260, 8 pp and figures,
    tables.
Sheldon, R. W., W. H. Sutcliffe, Jr. and M. A. Paranjape.  1977.
    Structure of pelagic food chain and relationship between plank-
    ton and fish production.  J. Fish. Res. Bd. Can. 34:  2344-2353.
Sherman, K., E. Cohen, M. Sissenwine, M. Grosslein, R. Langton and
    J. Green.  1978.  Food requirements of fish stocks of the Gulf
    of Maine, Georges Bank, and adjacent waters.  I.C.E.S. Biologi-
    cal Oceanography Committee C.M. 1978/Gen.:8  (Symp.) 14 pp
    (mimeographed).
Sibert, J. R., and R. J. Naiman.  1980.  The role of detritus and
    the nature of estuarine ecosystems.  *In* K. R. Tenore and B. C.
    Coull (eds.), Marine Benthic Dynamics, Univ. of South Carolina
    Press, Columbia.
Smith, P. C., B. Petrie and C. R. Mann.  1978.  Circulation, varia-
    bility, and dynamics of the Scotian Shelf and slope.  J. Fish.
    Res. Bd. Can. 35:  1067-1083.
Steele, J. H.  1974.  The structure of marine ecosystems.  Harvard
    University Press.
Sutcliffe, W. H. Jr., K. Drinkwater and B. S. Muir.  1977A.  Cor-
    relations of fish catch and environmental factors in the Gulf
    of Maine.  J. Fish. Res. Bd. Can. 34:  19-30.
Sutcliffe, W. H., Jr. and P. F. Brodie.  1977B.  Whale distribu-
    tions in Nova Scotia waters.  Fisheries and Environment Canada,
    Fish. Mar. Serv., Tech. Rpt. No. 722, 83 pp.
Sutcliffe, W. H., Jr., R. H . Loucks and K. F. Drinkwater.  1976.
    Coastal circulation and physical oceanography of the Scotian
    Shelf and Gulf of Maine.  J. Fish. Res. Bd. Can. 33:  98-115.
Tenore, K. R., and D. L. Rice.  1980.  A review of trophic factors
    affecting secondary production of deposit-feeders.  *In* K. R.
    Tenore and B. C. Coull (eds.), Marine Benthic Dynamics, Univ.
    of South Carolina Press, Columbia.

Thayer, G. W., W. E. Schaaf, J. W. Angelovic and M. W. LaCroix. 1973. Caloric measurements of some estuarine organisms. Fish. Bull. 71: 289-296.

Thiel, H. 1975. The size structure of the deep-sea benthos. Int. Rev. ges. Hydrobiol. 60: 575-606.

Tyler, A. V. 1973. Caloric values of some North Atlantic invertebrates. Mar. Biol. 19: 258-261.

Walsh, J. J., T. E. Whitledge, F. W. Barvenik, C. D. Wirick, S. O. Howe, W. E. Esaias and J. T. Scott. 1978. Wind events and food chain dynamics within the New York Bight. Limnol. Oceanogr. 23: 659-683.

Warwick, R. M., C. L. George and J. R. Davies. 1978. Annual macrofauna production in a *Venus* community. Est. coast. mar. Sci. 7: 215-241.

Warwick, R. M. 1980. Population dynamics and secondary production of benthos. *In* K. R. Tenore and B. C. Coull (eds.), Marine Benthic Dynamics, Univ. of South Carolina Press, Columbia.

Wiebe, P. H., S. Boyd and J. L. Cox. 1975. Relationships between zooplankton displacement volume, wet weight, dry weight, and carbon. Fishery Bull. 73: 777-786.

# Functioning of a Muddy Sand Ecosystem: Seasonal Fluctuations of Different Trophic Levels and Difficulties in Estimating Production of the Dominant Macrofauna Species

Michel Glémarec
Alain Menesguen

ABSTRACT: In temperate environments there are important seasonal fluctuations at different trophic levels of the benthos. This has been studied in a muddy-sand community in the bay of Concarneau (South Brittany).

After a spring microphytobenthic maximum, difficult to estimate in 1978 because of severe climatic conditions, nematodes appeared in great numbers one month later. The same situation occurred in autumn following a maximum of phaeophytin a and a period of optimal sedimentary stability.

Interpretation of trophic relationships between microphytobenthos and meiobenthos is difficult. The macrofauna biomass oscillated between 10 and 30 g/m$^2$ during one year. In order to illustrate the macrobenthic variations, three populations have been studied in detail with the help of automatized histogram analysis: *Abra alba* (Bivalvia), *Ampelisca spinipes* (Amphipoda) and *Amphiura filiformis* (Amphiurid). Two different demographic strategies appeared that correspond to different behavior during settlement in a new area.

The evolution of the successive cohorts may
be very different from year to year.  Fluctuations
in the densities of *Amphiura filiformis*  studied
over seven years show that changes in recruitment are difficult to link with any abiotic factor, but that intraspecific competition exists.
This approach emphasizes the difficulties in estimating secondary production.  Production estimates are made for the three species.


INTRODUCTION

For several years our laboratory has studied the muddy sand
ecosystem of Concarneau Bay (Glémarec, 1978).  Among the several
stations studied, only one is to be considered in this paper.  We
have tried to evaluate the importance of each trophic level and at
the same time tried to determine the range of seasonal fluctuations
in this temperate region.  A detailed study of the populations of
three species of macrofauna emphasizes the problems which have
arisen in evaluating secondary production.

Site Description

The "Mousterlin" station is located on the west side of Concarneau Bay (Fig. 1), at a depth of 17 m, sheltered from western pre-
vailing swells but exposed to south and southeasterly winds.  During
strong storms, the surface of the sediment can be resuspended.  From
a climatic point of view, this station is situated at the boundary
of the infralittoral and circalittoral "zones", i.e., the range of
thermal fluctuations on the sea bed is less than 10ºC.
The sediment consists of muddy sand, with a mean percentage of
silt + clay of 14%.  This content varies seasonally, especially in
the superficial layer (Fig. 2A).  The maximum percentage of silt +
clay (33%) appears in the summer period when the sea is calm; silt +
clay is minimal (about 4%) in March.  The water content in this
superficial layer varies in the same manner.  These two parameters
have been studied from June 1977 until April 1978, as well as biological factors at the different trophic levels.

Microphytobenthos

Sediment pigment content and production exhibited seasonal variations similar to those found in the phytoplankton cycle (Fig. 2B).
Both chlorophyll $\underline{a}$ and Phaeophytin $\underline{a}$ were measured.  Because we did
not sample during the April - May 1977 period, we include data from
Boucher (1975) for April - May 1972.  Due to winter storms the sediment was unstable, water turbidity was at its maximum, and chlorophyll pigments in the sediments were at their minimum.  Resuspension
resulted in regeneration of nutrients accumulated during the summer.
In April, a certain hydrodynamic and sedimentary stability occurred.
With the increase in temperature and light, the microphytobenthic
pigments increased again in April - May.  A second less important
maximum may appear in September.

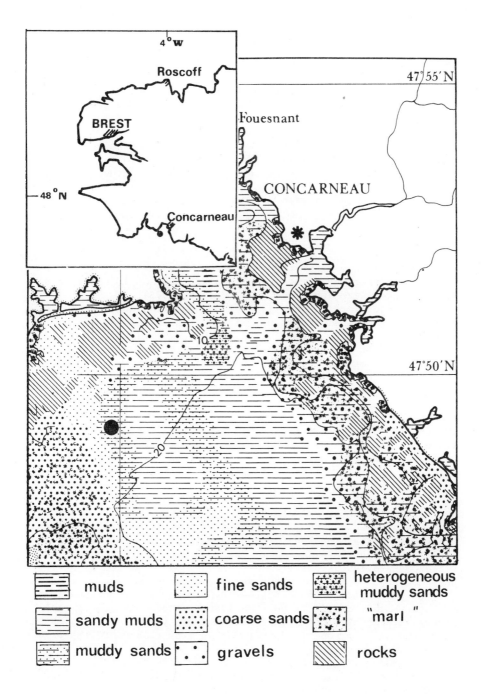

Figure 1. Location of the "Mousterlin" Station located by dot.

Chlorophyll a content varied from 0.2 $\mu g \cdot g^{-1}$ in March to 35 $\mu g \cdot g^{-1}$ in April, and Phaeophytin a from 2 $\mu g \cdot g^{-1}$ to 22 $\mu g \cdot g^{-1}$. The estimated primary production was 1.6 $\mu g \cdot g C^{-1} \cdot day^{-1}$ in 1972 (Boucher, 1975). The spring maximum hardly appeared in 1978. It may have been "fugitive", i.e., patchy and/or short duration, thus escaping measurement, or possibly not developed due to bad climatic conditions encountered in spring 1978 ($9^{\circ}$ in April 1978 versus $11^{\circ}$ in April 1972) (Fig. 3). Light is the main factor limiting microphytobenthos production at this station (Boucher, pers. comm.), but its effect on production is concealed by biomass fluctuations strongly linked to sediment stability. The latter is maximal in August (maximal water and silt contents), as is the percentage of Phaeophytin a, which is an index of the amount of detrital substances on the superficial sediment surface.


Meiobenthos

The total number of the meiobenthos organisms living in the upper centimeter varied between 500 and 3,000 $ind \cdot 10 \ cm^{-2}$; these numbers agree with other authors (Gerlach, 1978).

Nematodes form the dominant group, comprising 82-96% of the total number of individuals of meiofauna in this muddy sediment type. The meiobenthos were not evenly distributed vertically throughout the sediment. Most individuals were found in the upper few centimeters, and a sharp decrease in numbers occurred at the level of the Redox Potential Discontinuity. This vertical distribution correlated with changes in oxygen content, which quickly becomes the limiting factor for most organisms. Only the nematodes seemed to be distributed throughout the sediment. Information about stability conditions at the water-sediment interface is inferred from the fluctuations in the number of meiobenthic organisms in the top centimeter (Fig. 2C).

The density of nematodes was minimal in March (300 per 10 $cm^2$), which may be explained by the resuspension of the superficial layer. Nematodes disappear or find shelter at greater depths (Levenez, pers. comm.). After the microphytobenthos bloom in spring (as shown by the increase in sediment Phaeophytin a content), nematodes appeared in great numbers in May (3,200 per 10 $cm^2$). Their number decreased in summer, then again peaked in September (3,000 per 10 $cm^2$) one month after a Phaeophytin a maximum. This corresponded to the optimal period of sediment stability. As a consequence of the first storms that occurred in autumn, the high densities decreased.

Harpacticoid copepods represented only 0.4-1.8% of the meiobenthos. Muddy sand is a poor habitat compared with sands with higher interstitial circulation. The density of harpacticoids still shows two maxima: one in July that could be related to the spring increase in microphytobenthos; the other in October, after the nematode maximum. It is difficult to further interpret these trophic relationships. In spite of any differences between groups, and because nematodes dominate the meiobenthos, total meiofauna density is significantly correlated with the water content of the sediment, an index of the space available to these organisms.

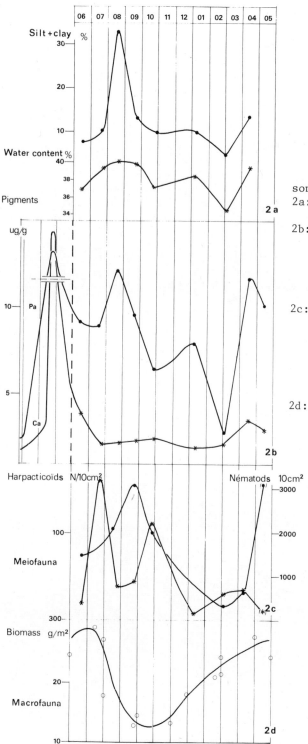

Figure 2. Seasonal fluctuations of:
2a: Silt + clay amount and water content
2b: Phaeophytin a (·) and Chlorophyll a (*). Data from Spring 1972 are adjoined on the left.
2c: Densities of meiobenthic organisms: nematodes (·) and Harpacticoid Copepods (*). Estimated values in brackets.
2d: Biomass of macrofauna.

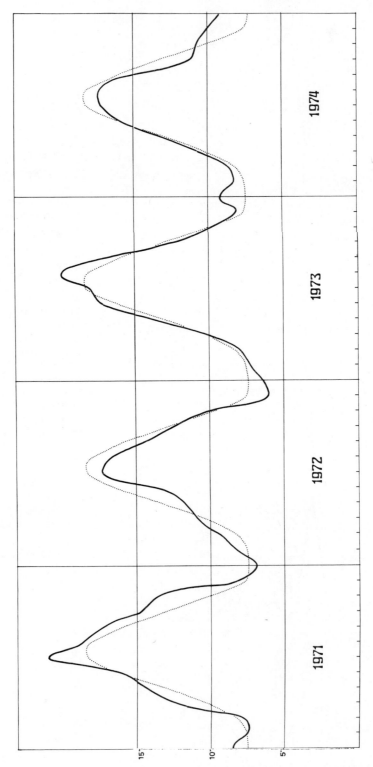

Figure 3. Air temperature fluctuations recorded near the station during nine years. Mean values on ten years are represented by the dotted line.

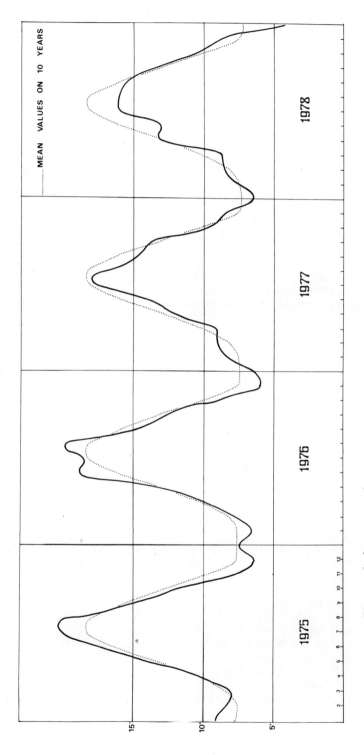

Figure 3 (continued)

Macrobenthos

Spatial and temporal changes of macrofaunal populations have been followed over seven years at three stations in the Bay of Concarneau. Multivariate analyses using inertia techniques elucidated the principal sources of variation in time and space (Chardy and Glémarec, 1977). After severe erosion of the sediment caused by a gale, a pioneer community first recolonized the area and then the community matured. An increasingly complicated trophic structure developed. The interactions between members of the same trophic level are superimposed on the elementary relations among trophic levels (Glémarec, 1978). The normal course of ecological succession is therefore under the control of cumulative effects of periodic and aperiodic climatic stresses (Glémarec, 1979).

The total macrobenthic biomass of the *Amphiura filiformis* community characteristic of this station varied between 10 g dry wt/m$^2$ in autumn to 30 g dry wt/m$^2$ in summer (Fig. 2D). These variations are due to: gonadal maturity (minimal in autumn and increasing from the end of winter); recruitment of young (principally at the end of winter and summer); and mortalities (occurring mainly in autumn). This cycle of variation in biomass was far more regular from one year to the next than the densities show. Densities fluctuate widely because a great number of young may suddenly settle, but these do not represent a significant biomass, and will often have disappeared a few months later.

In order to illustrate more precisely the quantitative variations of these macrobenthic organisms, macrofauna were sampled every six weeks with five to eight Smith-MacIntyre grabs (0.1 m$^2$) from November 1974 until March 1976. Three species were studied in detail: 1) the brittle star *Amphiura filiformis*, the main community dominant both in biomass and numbers; this species is characteristic of muddy sands; 2) the amphipod *Ampelisca spinipes*, the dominant peracaridan species both in number as well as biomass; and 3) the bivalve *Abra alba*, more ubiquitous than the two preceding species, and particularly well represented in this community.

The first step was to determine the population structure for each sampling date. There were no visible criteria for determining age classes of *Amphiura* and *Ampelisca* and insufficient information was obtained from the one or two growth striae found for *Abra*. Determination of the different cohorts was made by modal analysis of histograms of length. The lengths measured were: the disc diameter of *Amphiura*; the cephalic length of *Ampelisca*; and the maximal shell length of *Abra*.

The modal analysis of these successive histograms which is a discrimination of overlapping Gaussian components, was made in a semi-automatic way with a program using successively:

1) an original recurrent, smoothing-out method applied to the histogram. At each iteration, each class of the resulting histogram was divided into two adjacent and equal classes of half-amplitude and half-number of individual; a mobile-averaging process (limited to three adjacent classes) was then applied to this new histogram. Thus, the surface under the curve remained unchanged and the modal values were not greatly modified.

2) the logarithmic differences method of Bhattacharya (1967) applied to the smoothed-out curve, provided an initial estimation for the arithmetic average and standard deviation of each respective Gaussian component.

3) the maximum-likelihood method of Hasselblad (1966) using as initial values the preceding estimations. The program used was the version written in basic by Gros, Conan and Gonzalez (in Gros and Cochard, 1978) from the Fortran program "Normsep" written by Tomlinson (1970).

Figure 4 is a complete example of modal analysis performed with this program. The theoretical curve adjusted to the original histogram is marked by crosses. The life history of the population could then be summarized both by a growth curve and a mortality curve for each of the cohorts (cf. Fig. 6).

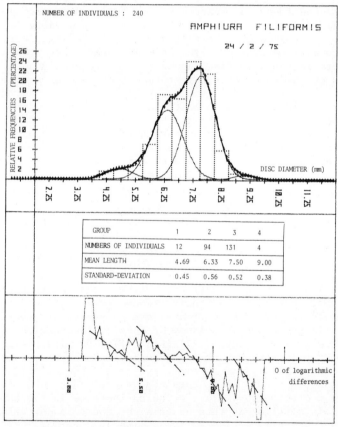

Figure 4. Illustration of the modal analysis technique applied to the *Amphiura's* population. Top section: histogram indicated by dotted lines, smoothed curve and Gaussian components by solid lines. Result of the fitting procedure is indicated by crosses. Bottom section: Bhattacharya's method applied to the smoothed curve.

The three species showed different responses (Figs. 5-10).
Whereas *Abra alba* showed an almost continuous recruitment, one well-
marked cohort for each season, *Ampelisca spinipes* and *Amphiura
filiformis* showed two marked recruitments in a year:  one in spring
(P-cohort), the other in autumn (A-cohort).   This is in agreement
with the life history established by Menesguen (1979) for two other
*Ampelisca* species, *A. brevicornis* and *A. spinimana*, in the same bay.
The maximum life span of *Abra alba* is approximately 14 months, while
the life spans of *Ampelisca* and *Amphiura* are usually two years.
Therefore, there are two demographic strategies corresponding to two
different behaviors during settlement in a new area.   Owing to its
widespread recruitment and to the fast renewal of its cohorts,
*Abra alba*, known as a pioneer species (Glémarec, 1978), seems to be
better adapted than the other species to recolonizing a perturbed
area.

A second important feature of the dynamics of such populations
was the interruption of growth during winter and early spring, especially for the adult cohorts that have high energetic requirements
due to gonadal maturation.   The growth of the young immature individuals born in the preceding autumn seemed frequently not to be
affected by the cold water temperature (see A-cohorts of *Ampelisca
spinipes* or *Amphiura filiformis*).

A third characteristic of these population dynamics was marked
difference from year to year of the successive cohorts.   The year
1975 seems to have been very unfavorable for all three populations
and especially for *Amphiura*:   the two cohorts that were recruited
in 1974 disappeared after only one year, and the spring cohort
seemed to have completely vanished after nine months.

During this year, the mortality of young recruits was very
high for all three species, but the reasons are not obvious.   Water
temperature was not exceptionally cold.   This is a remarkable example
of slight fluctuations which may affect the population dynamics and
destabilize their cycles from one year to the next.   If the important meteorological anomalies simultaneously perturb all the species,
there are at the same time less obvious factors which may only affect
some species.   A good illustration of these annual irregularities in
macrofaunal abundance, and their frequent correlation with variation
around mean temperature (estimated over a period of several years; see
Fig. 3), is the seven year-study of density fluctuations of *Amphiura
filiformis* from 1970 until 1977 (Fig. 11).   Despite the fact that
the demographic strategy of this species was characterized by two
recruitment periods, one in early spring and the other in autumn,
variations did occur.   For instance, in the years 1970 and 1971 there
were two recruitments and a continuing increase in the number of
individuals settling in this environment due to an exceptional 1971
autumn and an early spring in 1972.   The 1972 spring recruitment was
very important, but the density of individuals decreased very rapidly from the following autumn, probably because of the abnormally
cold 1972 summer (Fig. 3).   It is likely that summer reproductive
activity continued throughout the autumn, because this was observed
during the same period for bivalves from a station near the Glenan
archipelago (Glémarec and Bouron, 1978).   This unusually long reproductive period was immediately followed by a modest 1973 spring recruitment.   Such a rapid renewal of sexual activity does not allow

the animals to restore their reserves in autumn; therefore, the 1973 reproductive period was bound to be reduced and the expected 1973 autumnal recruitment will not appear.

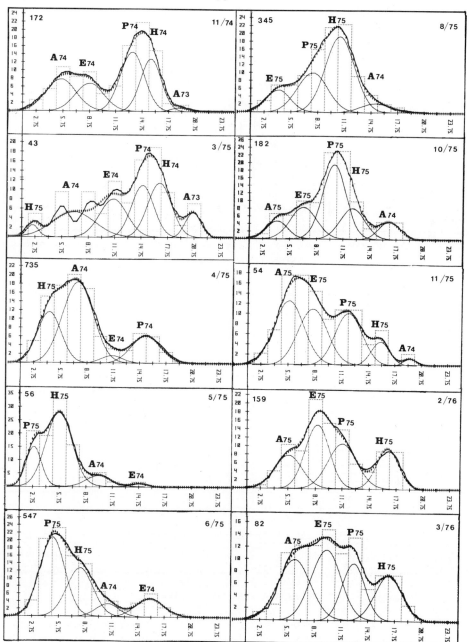

Figure 5. Size distributions of *Abra alba*. Total number of individuals is in the left corner, sample's date is in the right corner. Cohorts have been identified by their recruitment season: A = autumn; H = winter; P = spring; and E = summer.

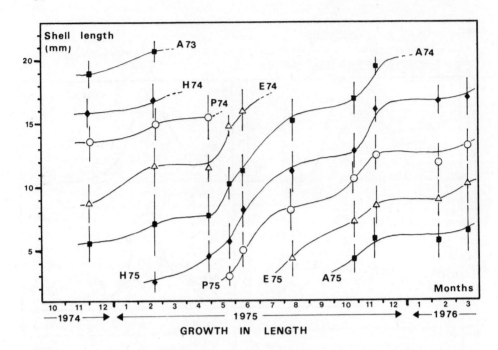

GROWTH IN LENGTH

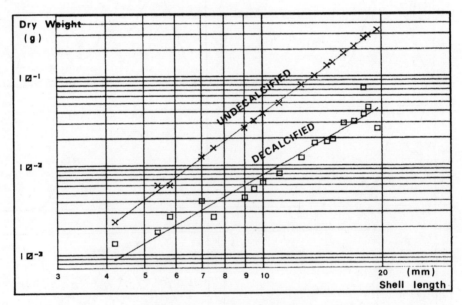

WEIGHT LENGTH CURVES

Figure 6.  6a: *Abra alba's* growth in length.  Symbols are as
follows: ◆ winter cohort, ○ spring cohort, △ summer cohort,
■ autumn cohort.  Standard deviation indicated by lines.  6b: weight-
length curves of *Abra Alba*.

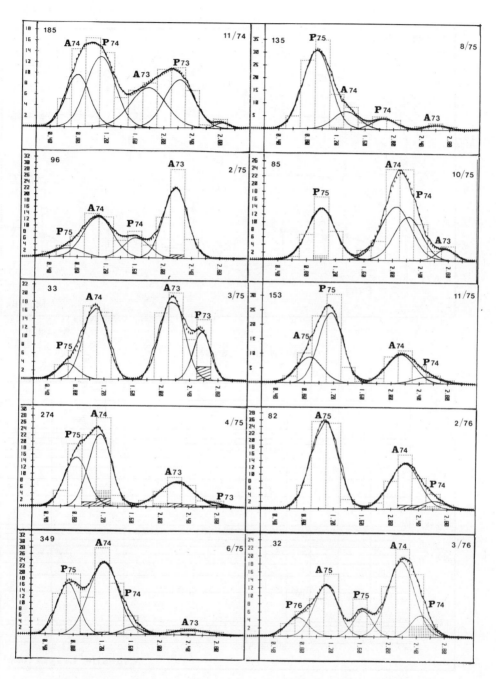

Figure 7.   Size distributions of *Ampelisca spinipes*.   Dotted areas = ovigerous females; hatched areas = males.   For legend see Figure 5.

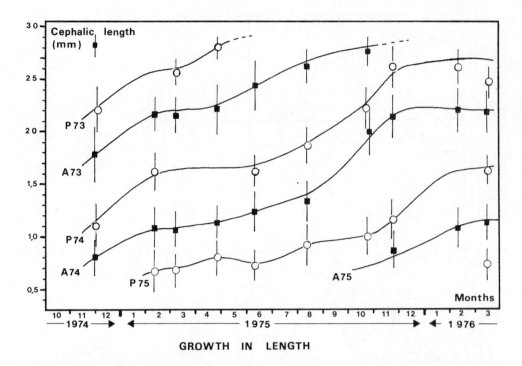

**GROWTH IN LENGTH**

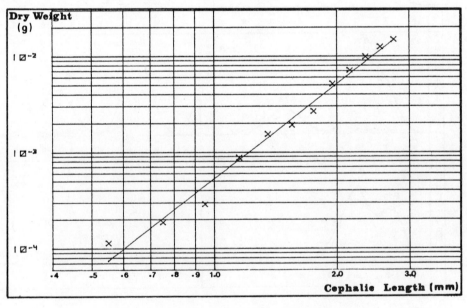

**WEIGHT - LENGTH CURVE**

Figure 8.   Growth in length (8a) and weight-length curve (8b) of *Ampelisca spinipes*.   For legend see Figure 6.

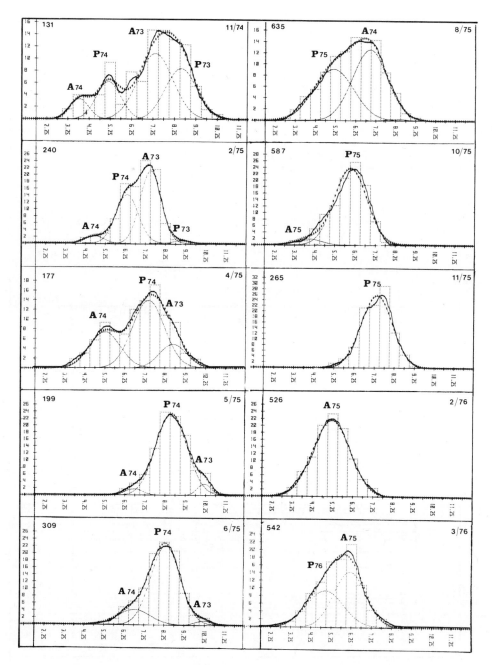

Figure 9. Size distributions of *Amphiura filiformis*. For legend see Figure 5.

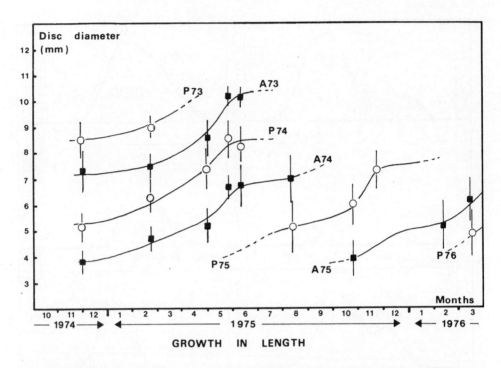

**GROWTH  IN  LENGTH**

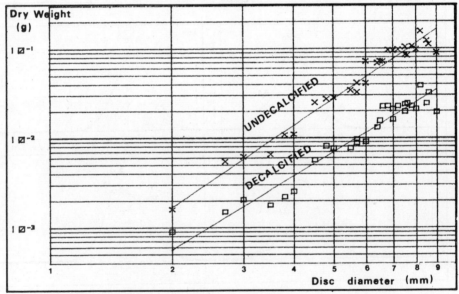

**WEIGHT   LENGTH   CURVES**

Figure 10.  Growth in length (10a) and weight–length curves (10b)
of *Amphiura filiformis*.  For legend see Figure 6.

These fluctuations are much more pronounced in 1975 and 1977 than for the first years, when the population is increasing. The population on the whole seems to be rather unstable. As a matter of fact, the smaller the population is in spring when the number of individuals are minimal, the more abundant the recruits will be. When populations are too large, animals are too numerous to allow the settled young to have a high chance of surviving. Intraspecific competition is so strong that most of them will either die or migrate. To ensure the stability of the *Amphiura filiformis* population itself, and inside the whole community, we estimated that an upper limit of 400 ind per m$^2$ must be exceeded.

We have estimated production of these three populations for the period November 1974 through November 1975. In order to obtain a realistic estimation of densities at the "Mousterlin" station, the percentages represented by the different cohorts were not deduced from the preceding histograms, which united all the measurements made from the whole of the bay. By using mean and standard deviation estimations, one can assign each individual of a grab to a particular cohort. Then, at each date a mean number of individuals per cohort could be estimated, and used to develop Allen curves (Fig. 12). It must be admitted that the standard deviation around the mean density of a cohort is always particularly high and may often be equal to the mean because of the over-dispersion of animals. In spite of the strong limitations of this method, production has been estimated for each cohort for which satisfactory figures could be calculated. For the period November 1974 through November 1975, the results are:

| SPECIES | PRODUCTION (g/m$^2$/year) | BIOMASS (g/m$^2$) | P/$\overline{B}$ ratio |
|---|---|---|---|
| *Abra alba* | 1.45 | 0.72 | 2.0 |
| *Ampelisca spinipes* | 0.18 | 0.07 | 2.4 |
| *Amphiura filiformis* | 15.70 | 5.57 | 2.8 |

The P/$\overline{B}$ ratios are not very high, especially for *Ampelisca spinipes*, whose turnover rate in this environment seems to be lower than that of *Ampelisca brevicornis* (P/$\overline{B}$ = 4, reported by Klein, Rachor, Gerlach, 1975 in the German Bight) and *Ampelisca tenuicornis* (P/$\overline{B}$ = 3.4, reported by Sheader, 1977 in the North Sea).

CONCLUSION

This study has clearly shown how important the seasonal fluctuations are at every level of the trophic structure. For macrofauna especially, significant anomalies about the temperature mean of a season can produce modifications in the maturation of sexual products, in their release, and in the duration of the planktonic stages. All these reasons can give rise to important irregularities in the density of the benthic organisms, irregularities which can destabilize the whole community from one year to the next, both in numbers and in composition. Inter- and intra-specific competition may be very strong

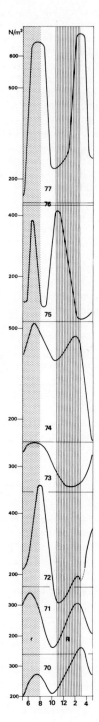

Figure 11.   Fluctuations
of *Amphiura's* density per square meter
from 1970 to 1977.  Dotted area = period
of spring recruitment (r); hatched area =
period of autumn recruitment (R).

in an unstable environment. One weak recruitment of *Amphiura* will facilitate the next because of the available room. At the same time, a competing species may have taken advantage of this situation and have increased its settlement in the environment, leading to a progressive modification of community structure. In spite of such a development of interspecific competition the "climax" concept seems not to apply to this periodically strongly perturbed environment.

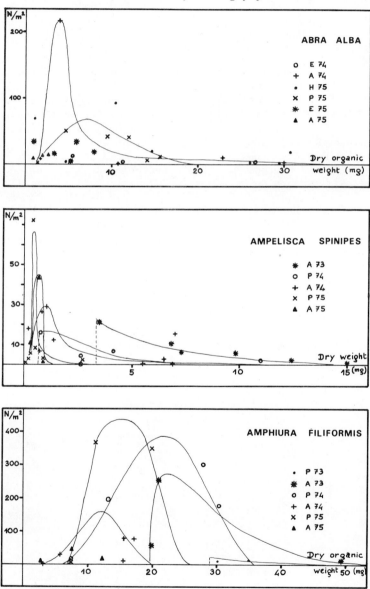

Figure 12. Allen's curves for production estimation.

REFERENCES

Bhattacharya, C. G.  1967.  A simple method of resolution of a distri-
  bution into Gaussian components.  Biometrics. 23 (1):  115-135.
Boucher, D.  1975.  Production primaire saisonniére du microphytobenthos des sables envasés en baie de Concarneau.  Thèse 3e cycle.
  Université de Bretagne Occidentale.  113 pp.
Chardy, P. and Glémarec, M.  1977.  Evolution dans les temps des peuplements des sables envasés en baie de Concarneau (Bretagne).  Biology
  of benthic organisms.  Keegan, B. F. (ed.), O'Ceidigh, P. and
  Boaden, P. J. S.  Pergamon.  165-172.
Gerlach, S. A.  1978.  Food-chain relationship in subtidal silty sand
  and marine sediments and the role of meiofauna in stimulating bacterial productivity.  Oecologia (Berl.) 33:  55-69.
Glémarec, M.  1978.  Problèmes d'écologie dynamique et de succession
  en bai de Concarneau.  Vie et Milieu.  28 (I, AB):  1-20.
Glémarec, M.  1979.  Les fluctuations temporelles des peuplements
  benthiques liées aux fluctuations climatiques.  Oceanologica Acta.
  2(3):  365-371.
Glémarec, M and Bouron, D.  1978.  Evolution de la maturité sexuelle
  chez six espèces de Bivalves des Glenan.  Haliotis 9 (1):  45-48.
Gros, P. and Cochard, J. C.  1978.  Biologie de *Nyctiphanes couchii*
  (Crustacea, Euphausiacea) dan le secteur nord du Golfe de Gascogne.
  Ann. Inst. Océanogr. Paris.  54 (I):  25-46.
Klein, G., E. Rachor, and S. A. Gerlach.  1975.  Dynamics and produc-
  tivity of two populations of the benthic tube-dwelling amphipod
  *Ampelisca brevicornis* (Costa) in Helgoland Bight.
Menesguen, A.  1979.  La macrofaune benthique de la baie de Concar-
  neau: Peuplements, dynamique de populations et prédation exercée
  par les poissons.  Thèse 3e cycle.  Université de Bretagne Occidentale (in press).
Sheader, M.  1977.  Production and population dynamics of *Ampelisca
  tenuicornis* (Amphipoda) with notes on the biology of its parasite,
  *Sphaeronella longipes* (Copepoda).  J. mar. biol. Ass. U.K. 57:
  955-968.

# Facultative Anaerobiosis of Benthos

Mario M. Pamatmat

ABSTRACT: Anaerobic heat production rates of
the mussels, *Mytilus edulis* and *Modiolus demissus*, average 95% below their aerobic level
of metabolism. The actual decrease varies with
the relative proportion of resting and active
metabolism during each measurement period, metabolic history, and nutritional state of the individual. Anaerobic metabolism is always less
than, but is a variable percentage of, aerial
metabolism. Anaerobic metabolism, like aerobic,
is characterized by alternating periods of resting (steady) and active (fluctuating) metabolism.
Aerobic resting metabolism is higher than anaerobic resting metabolism and fluctuations in
aerobic active metabolism are greater than fluctuations in anaerobic active metabolism. During
extended anaerobiosis, *Arctica islandica* shows
variations in average daily heat production rate
as a result of variations in the relative proportion of resting and active metabolism. The
level of resting metabolism drops slowly each day
while heat bursts from physical activity also
diminish with time. The relationship between
anaerobic heat production rate at $20°C$ and body

size in seven species that show a common regression line (two warm-temperate, *Polymesoda caroliniana* and *Modiolus demissus*; five boreo-arctic, *Mytilus edulis*, *Arctica islandica*, *Astarte borealis*, *Cardium edule*, and *Arenicola marina*) is described by a linear regression equation ($r^2$ = 0.83, 29 DF), $Y = 0.089 \times 10^{-4} + 1.58 \times 10^{-4} X$, where X is the g tissue dry weight and Y is heat production rate in watts. The average weight-specific heat production rate for the seven species equals $1.77 \times 10^{-4}$ W $g^{-1}$ (SE = $0.14 \times 10^{-4}$ W $g^{-1}$). Although ATP yield during anaerobiosis is much less than that during aerobic metabolism, the greater reduction in metabolic energy demand than in ATP yield may gain an energy saving for the facultative anaerobes that undergo the "succinate" pathway of anaerobic metabolism. The situation may be analogous to the energetic advantage that vertically migrating zooplankton are supposed to derive from feeding in warm surface waters and descending to cooler depths afterwards.

## INTRODUCTION

Facultative anaerobiosis appears to be widespread among the benthos. Its occurrence and importance among intertidal animals has been reviewed by Newell (1972). During prolonged tidal exposure, intertidal barnacles, *Chthamalus stellatus* and *Balanus balanoides*, become anaerobic as they close their opercular valves to prevent desiccation (Barnes *et al.*, 1963). *Modiolus demissus*, the most adapted of the bivalves to intertidal life (Lent, 1969), may keep their valves open during tidal exposure, not only to take up oxygen, but also for the purpose of evaporative cooling (Lent, 1968), but their tissues lying deep below exposed surfaces still become partly anaerobic (Moon and Pritchard, 1970; Booth and Mangum, 1978; Widdows *et al.*, 1979). The same is true of *Mytilus californianus* (Bayne *et al.*, 1976b) and *Mytilus edulis* (Coleman, 1973; Widdows *et al.*, 1979). On the other hand, *Uca pugnax* becomes anaerobic during high water when oxygen tension in its burrow declines below a critical level (Teal and Carey, 1967). *Nereis virens* is aerobic as long as it is pumping water through its burrow but it stops ventilating in rhythmic fashion and then becomes anaerobic (Scott, 1976). Rhythmic shell closing and pH decline, both indicative of anaerobiosis, have been observed in submerged *Mercenaria mercenaria* in aerated water (Gordon and Carriker, 1978). *Cirolana borealis*, an isopod, is a scavenger that eats its way into the flesh of dead fish where it becomes anaerobic during feeding (de Zwaan and Skjoldal, 1979). *Ctenodiscus crispatus*, an infaunal Asteroid, shows more resistance to hypoxia and hydrogen sulfide than epifaunal Asteroids and could be anaerobic in its deeper tissues even while absorbing oxygen with its epiproctal cone (Shick, 1976). *Arctica islandica* volun-

tarily goes anaerobic by burrowing deeper into the sediment and remains buried for periods up to 24 days before moving up to the surface again (Taylor, 1976).  There is no apparent reason for the self-
induced anaerobiosis.  Various benthic organisms that are subject to
the incidence of low oxygen tension in their natural habitat exhibit
"aerobic shutdown" (Mangum, 1970; Kushins and Mangum, 1971).  Below
a critical level of oxygen tension the animals stop taking up oxygen
and presumably become anaerobic.  *Uca pugnax* and *Littorina irrorata*
never stopped consuming oxygen as oxygen tension dropped to low
levels, but became partly and mostly anaerobic below their respective
critical oxygen tension (Pamatmat, 1978).  All of these observations
indicate that facultative anaerobiosis is a natural way of life
among the benthos.

Our present knowledge of respiration in these animals is still
largely based on measurements of oxygen consumption.  We know little
about their metabolic rates during their periods of anaerobiosis.
Such data are clearly important for accurate energy budget estimations.
Total metabolism of benthos, including both aerobic and anaerobic,
can now be determined by direct calorimetry.

Direct calorimetry has provided some information on the effects
of body size, temperature, and duration of anoxia on anaerobic metabolic rates of two bivalves, *Modiolus demissus* and *Polymesoda caroliniana* (Pamatmat, 1979).  Additional data will be presented in
this paper for five other species of bivales and the polychaete,
*Arenicola marina*, in an attempt to generalize present knowledge of
anaerobic metabolism and its adaptive significance to the benthos.

METHODS AND MATERIALS

The experimental animals from the subtidal (*Arctica islandica*,
*Mytilus edulis* and *Astarte borealis*) were collected from Kiel Bight
in the western Baltic Sea and stored in aquaria in slightly warmer
water but of the same salinity as in Kiel Bight (ca. 15°C and 17 °/oo
S).  The intertidal animals (*Cardium edule*, *Mya arenaria* and *Arenicola Marina*) were collected from a sandflat on the North Sea coast
near Busum and Wilhelmshaven, FRG, and similarly stored in aquaria
at 15°C and 17 °/oo S.

The shell of each bivalve was thoroughly scrubbed with a wire
brush underwater, rinsed with tap water, dried, and wrapped with
double layers of latex rubber balloon, then submerged in Dow Corning Methyl Silicone (Type 350, viscosity 200 centipoise) in a metal
cannister.  The cannister was placed in a rubber bag that was immersed in the calorimeter's external bath for temperature equilibra-
tion prior to placing inside the calorimeter.  The polychaetes were
either contained in a close-fitting glass vial or tubing, or buried
in natural sediment in a bottle, and all of these containers were
submerged in Silicone oil in a metal cannister.  In the case of
the experimental individual buried in sediment, an equal amount
of sediment in a separate bottle was used as a control.

The calorimeter used was a modified version of the heat-flow
twin calorimeter described by Pamatmat (1978).  The modification
consisted of the addition of a second twin chamber within the same
heat sink.  One twin chamber was used to measure the heat flux from

an experimental animal while the other served to monitor, simultaneously and continuously, the baseline for any drift or fluctuation that might occur. The instrument calibration constant is 2.16 x $10^{-5}$ W $\mu v^{-1}$ (= 5.16 x $10^{-6}$ cal $sec^{-1}$; 1 W = 1 J $sec^{-1}$ = 0.239 cal $sec^{-1}$).

When signals are changing slowly, thermopile voltages are directly proportional to heat production rates, but not during periods of rapid signal fluctuations, as shown by normal thermograms of macrofauna, because of instrument lag. The area under the curve over a time period represents the total amount of heat produced during that period and area divided by time represents the average heat production rate during that period. Areas were integrated by planimetry. Rates of heat production are expressed in watt. The tissues of experimental animals were dried to constant weight at 95°C.

RESULTS

Thermograms and Changes in Metabolic Activity with Time

One individual of *Arctica islandica* was kept anaerobic inside the calorimeter continuously for 10 days (Fig. 1). There are instantaneous changes and a changing rhythmic pattern in its metabolic heat production. The daily thermograms are characterized by a slowly decreasing basal level depicted by the dashed lines (top three panels, Fig. 1), and upward fluctuations from the basal level. The basal level represents resting or basal metabolism of the clam while the fluctuations show the increased heat production arising from the clam's physical activity. The hatched areas above the basal level denote the total amount of heat resulting from active metabolism during that time period.

During the first day there was a slow decline in basal heat production from 3.24 x $10^{-4}$ W to 1.62 x $10^{-4}$ W with relatively slight active metabolism. During the second day there was a more or less steady basal metabolism of 1.55 x $10^{-4}$ watt with one slight physical activity and one large producing a heat burst of 1.67 x $10^{-3}$ W-h. The basal level continued to decrease slightly each day to 1.29 x $10^{-4}$ W during the third, 1.15 x $10^{-4}$ W during the fourth, 1.17 x $10^{-4}$ W during the fifth and 1.13 x $10^{-4}$ W during the sixth day. After the sixth day, physical activity appeared to diminish in intensity but to occur more frequently as indicated by the thermogram's persistent elevation above basal level.

The average daily total heat production rate ranged from a high of 2.29 x $10^{-4}$ W during the second day to 1.44 x $10^{-4}$ W during the sixth day, being greatly influenced by physical activity. The peaks numbered 1 to 6 are believed to have arisen from one type of physical activity. The areas of these successive fluctuations diminish with time, suggesting that prolonged anoxia has a limiting effect on the animal's capacity for burst of activity.

Total anaerobic metabolism is partitioned between resting and active metabolism (Table 1), showing that the latter amounted to 6% of total metabolism during the first day, fluctuated daily but averaged 31% (SD = 9%) during the next nine days.

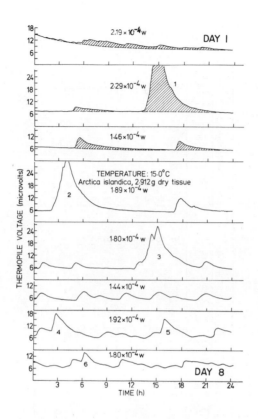

Figure 1. Ten-day calorimetry (first eight days only
are shown) on an anaerobic *Arctica islandica* at 15°C. The
thermogram on the first day starts after the calorimeter
has already stabilized (4-5 h after placing the sample inside). The values of average heat production rates are
shown for each day. Anaerobic metabolism, like aerobic
metabolism, is characterized by alternating periods of
resting and active metabolism. Resting or basal metabolic
rate is represented by broken lines. The hatched areas
above basal metabolic rate depict the amounts of heat evolved
as a result of physical activity. Resting metabolism decreases during the first day, and then more slowly afterwards. Small, but persistent, activity elevates the thermogram above basal level during the first day. Peaks numbered 1-6 are believed to arise from the same physical
activity, perhaps as a result of the clam's attempt to
emerge from its "buried" state; the decreasing areas of the
six successive peaks signify diminishing physical exertion,
perhaps as a result of energy store limitation. After five
days, the clam appears to show more frequent, but smaller
bursts of activity.

Table 1. Daily anaerobic heat output and output rates by *Arctica islandica* (= 2.912 g dry tissue weight) at 15°C for 10 days.

| 24-h Period | Total Heat Output (W-h) | Average Rate (W) | Resting Metabolism (W-h) | % of total | Active Metabolism (W-h) | % of total |
|---|---|---|---|---|---|---|
| 1 | $5.26 \times 10^{-3}$ | $2.19 \times 10^{-4}$ | $4.92 \times 10^{-3}$ | 94 | $3.36 \times 10^{-4}$ | 6 |
| 2 | 5.50 | 2.29 | 3.72 | 68 | 17.8 | 32 |
| 3 | 3.50 | 1.46 | 3.10 | 89 | 4.10 | 11 |
| 4 | 4.54 | 1.89 | 2.76 | 61 | 17.7 | 39 |
| 5 | 4.32 | 1.80 | 2.81 | 65 | 15.2 | 35 |
| 6 | 3.46 | 1.44 | 2.71 | 78 | 7.53 | 22 |
| 7 | 4.61 | 1.92 | 3.02 | 66 | 15.9 | 34 |
| 8 | 4.32 | 1.80 | 3.00 | 69 | 13.1 | 31 |
| 9 | 4.78 | 1.99 | 3.00 | 63 | 17.9 | 37 |
| 10 (= 17.6 h) | 3.63 | 2.06 | 2.20 | 61 | 14.4 | 39 |

$1 \text{ W} = 1 \text{ J sec}^{-1} = .239 \text{ cal sec}^{-1}$

$1 \text{ W-h} = 860 \text{ cal}$

After prolonged anoxia lasting several days to more than one week, there is invariably a strong odor of fermentation. When placed in oxygenated water it takes the clam (and mussels) some 10 to 30 min to open its shell. There is considerable variability among individuals. A milky-looking substance, possibly $CaCO_3$, sometimes oozes out of the valves. The valves then open widely. Once open, tapping on the shell or prodding the tissues will gen-- erally fail to elicit a response. Some might slowly close their valves but never do so completely. After several hours of a refrac- tory period of recovery, the individual normally becomes irritable again and shuts its valves readily upon the slightest tap on its shell. After only one or two days of anoxia, the experimental clam or mussel, soon after opening in oxygenated water, will read- ily close its shells when disturbed, although it may take somewhat vigorous prodding rather than just a slight tap on the shell to make some of them shut their valves.

Aerobic vs. Anaerobic Metabolism

One individual of *Arctica islandica*, kept anaerobic inside the calorimeter for one day, was more quiescent than the other individual shown in Figure 1, producing a steady rate of 1.68 x $10^{-4}$ W throughout the day (Fig. 2). It was taken out of the calorimeter, placed in oxygenated water, and returned aerobic into the calorimeter. The basal level of aerobic heat production is at 22 µv (= 4.86 x $10^{-4}$ W) and there are bursts of activity re- sulting in peaks of thermogram fluctuation reaching up to 54 µv (as compared with 30 to 33 µv for the much larger anaerobic indi- vidual in Figure 1) and amounting to 31% of total heat production during that period when it occurred. After a few heat bursts, the clam quieted down and then showed a slow increase in heat pro- duction to a level equaling the previous intermittent heat bursts. The clam was removed and the water returned to the calorimeter, showing no detectable bacterial activity in the water. The clam was then placed in another container with freshly oxygenated water, and returned aerobic once more into the calorimeter. The thermo- gram bottomed out at 20 µv (= 4.32 x $10^{-4}$ W), drifted slowly up- ward, then leveled off at 40 µv (= 8.64 x $10^{-4}$ W). The steady anaerobic metabolism during the first day equals 23% and 25% of the average aerobic metabolism during the second and third days, re- spectively. In comparison with the highest levels of active metab- olism during the first and second days, however, anaerobic metabo- lism during the first day amounted to 16 and 19%, respectively.

A small specimen of *Mytilus edulis* was first placed anaero- bic inside the calorimeter (Fig. 3). The thermogram shows con- tinuous change, fluctuating by as much as ± 0.6 µv with an average heat flux for the day of 7.56 x $10^{-6}$ W. The lowest thermogram level of 0.2 µv suggests resting metabolism of 4.32 x $10^{-6}$ W. The mussel was taken out of the calorimeter, transferred into oxygen- ated water and returned aerobic into the calorimeter. The thermo- gram shows a rapid drop at about the third hour, upward fluctuations after that on the order of 5 µv (10 times the magnitude of fluctu- ations during anaerobiosis), a protracted rest period lasting about

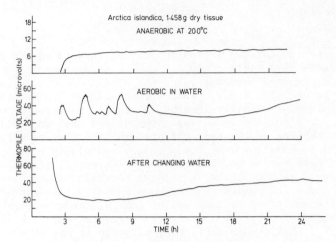

Figure 2.   Three-day calorimetry on an *Arctica islandica*:   anaerobic the first day, aerobic in water during two subsequent days.   The clam was placed inside the calorimeter at the zero time shown each day.   The first 4 h show rapidly changing thermopile voltage as the calorimeter regains thermal equilibrium.   As shown by the rising thermogram on the first day, the container was slightly cooler when placed inside than the calorimeter.   Anaerobic metabolism of this individual is more stable than that of the other specimen shown in Figure 1, and its bursts of aerobic physical activity produce higher peaks than the anaerobic heat bursts by the much larger clam in Figure 1.   After the initial aerobic heat bursts, this clam appears to show a steadily increasing level of physical activity without further heat bursts.

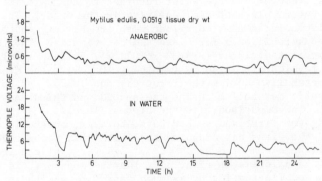

Figure 3.   Two-day calorimetry of *Mytilus edulis* at 20°C, first anaerobic, then aerobic in water.   Resting aerobic metabolism (2.59 x $10^{-5}$ W) is six times the level of resting anaerobic metabolism (4.32 x $10^{-6}$ W).   Thermogram fluctuations during aerobiosis are about ten times those during anaerobic fluctuation.

3 h from the 15th to the 18th h (resting metabolism = $2.59 \times 10^{-5}$ W), then more upward fluctuations beyond this period. This mussel's anaerobic metabolism during the first day amounts to 7% of its subsequent aerobic metabolism during the second day while its resting anaerobic metabolism equals 17% of its resting aerobic metabolism.

The metabolism of a second specimen of *Mytilus edulis* was measured four times, first in air, in oxygenated water twice, and finally anaerobic (Fig. 4). During the first day, in air the mussel showed a slowly decreasing level of heat production from 2 μv to 1.4 μv, with a slight drop during the 13th to 17th h period to a slightly lower level of 1.0 μv (= $2.16 \times 10^{-5}$ W). The average total heat production rate equals $3.39 \times 10^{-5}$ W during the entire period in air.

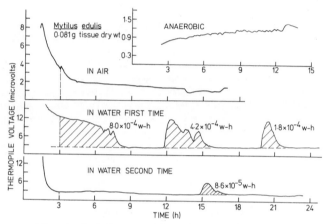

Figure 4. Four-day calorimetry of *Mytilus edulis* at 20°C, first in air, then in aerated water for two days, and then anaerobic on the fourth day. Average heat production rate in air ($3.39 \times 10^{-5}$ W) is 30 to 63% of average aerobic heat production rate in water. Anaerobic heat production rate during the fourth day is 76% of aerial metabolism, 23% of aerobic aquatic metabolism during the second day, and 48% during the third day. There is a distinct difference between this individual and the mussel shown in Figure 3.

When transferred into oxygenated water during the second day, the mussel exhibited a high thermogram level of about 12 μv (= $2.59 \times 10^{-4}$ W) decreasing to and staying at 2.7 μV (= $5.83 \times 10^{-5}$ W, its resting metabolism) for 4 to 5 h. The hatched areas bounded by the thermogram and the dashed line represent heat production arising from physical activity, possibly from ventilating or pumping. During the third day, in renewed oxygenated water, the mussel shows even more protracted periods of quiescence and a shorter period of activity. The heat energy liberated during four successive periods of physical activity in two days decreased by about 50% each time, from $8.0 \times 10^{-4}$ W-h the first time to $8.6 \times 10^{-5}$ W-h the last time. On the fourth day, when the mussel was made anaerobic, the thermogram shows a slight upward drift, some ripple, a slight upward fluctuation, generally leveling off at about 1.2 μV. Anaerobic metabolism during the fourth day amounts to 76% of aerial metabolism during the first day, 23% of aerobic metabolism during the second, and 48% during the third day. Metabolic rate in air is only a little higher than anaerobic metabolism and amounts to 30% of average heat production rate underwater (11.3 x $10^{-5}$ W). The rates of aerobic and anaerobic heat production in *Mytilus edulis*, *Cardium edule* and *Arctica islandica* are summarized in Table 2. In the experiment with *Mytilus edulis* weighing 0.081 g (Fig. 4), after measuring anaerobic heat production, the mussel was taken out, its mantle cavity fluid drained by prying the shell slightly, and the specimen immediately returned to the calorimeter. The second measurement in air (4.87 x $10^{-5}$ W) is 30% higher than the first (3.39 x $10^{-5}$ W), the difference probably being due to repayment of an oxygen debt during the anaerobic period just before the second measurement in air.

### Anaerobic Metabolism vs. Body Size

Experimental bivalves either show fairly steady average heat production during the first day or start with a relatively high initial rate which decreases to a more or less stable average level after several hours. The metabolic rates used in showing a correlation with body size are the average values for at least 3 h (Pamatmat, 1979), and often for much longer periods of up to more than 24 h after the thermogram has begun to show a uniform average level.

For the five boreo-arctic species studied in summer plus the two warm-temperate species (*Polymesoda caroliniana* and *Modiolus demissus*) studied in winter, all at 20°C, the relationship between anaerobic heat production and body size (Fig. 5) is described by the common linear regression equation ($r^2$ = 0.83, 29 DF),

$$Y = 0.089 \text{ x } 10^{-4} + 1.58 \text{ x } 10^{-4} \text{ X}$$

where X is tissue dry weight in grams and Y is heat production rate in watt. The values for each species are scattered about the common regression line, indicating no significant difference between them. The pooled data indicate that anaerobic metabolism is directly proportional to body weight and weight-specific anaerobic rate is independent of body size. The average weight-specific anaerobic rate is 1.77 x $10^{-4}$ W $g^{-1}$ (standard error of the mean = 0.14 x $10^{-4}$ W $g^{-1}$).

Four individuals of *Mya arenaria* showed rates of anaerobic heat production that are significantly higher than the foregoing group (t statistic = 11.1, 33 DF). The average weight-specific rate is 6.94 x $10^{-4}$ W $g^{-1}$ (SD = 1.75 x $10^{-4}$ W $g^{-1}$). A specimen of *Arctica islandica* that was fitted with a platinum wire electrode in its anterior adductor muscle (and whose myogram showed a perfect correlation of volleys of muscle action potential with short bursts of heat in the thermogram) also had a significantly higher heat production rate than the other bivalves.

Table 2. Relationship between aerobic and anaerobic heat production in *Mytilus edulis* and *Cardium edule* at 20°C and *Arctica islandica* at 15°C. The arrows indicate the sequence of treatment.

| Species | Size | Aerobic (in water) | Aerobic (in air) | Anaerobic | Anaerobic/Aerobic |
|---|---|---|---|---|---|
| *M. edulis* | 1.085 g | $11.7 \times 10^{-4}$ W | $24.3 \times 10^{-4}$ W | $2.79 \times 10^{-4}$ W | 11% |
| | 1.216 | | | $1.79 \times 10^{-4}$ | 15% |
| | 0.084 | $4.34 \times 10^{-5}$ | | $7.72 \times 10^{-6}$ | 18% |
| | 0.051 | $1.12 \times 10^{-4}$ | | $7.56 \times 10^{-6}$ | 7% |
| | 0.081 | $1.13 \times 10^{-4}$ | $3.39 \times 10^{-5}$ [a,c] | | 23% |
| | | $0.54 \times 10^{-4}$ | $4.87 \times 10^{-5}$ [b,c] | $2.59 \times 10^{-5}$ [b,c] | 48% |
| *C. edule* [d] | 0.223 | $1.55 \times 10^{-4}$ | | $6.50 \times 10^{-5}$ | 42% |
| *A. islandica* | 1.458 | $7.17 \times 10^{-4}$ | | $1.68 \times 10^{-4}$ | 23% |
| | | $6.61 \times 10^{-4}$ | | | 25% |

[a] $\dfrac{\text{Heat production rate in air}}{\text{Heat production rate in water}} \times 100 = 30\text{-}63\%$

[b] Mantle cavity fluid drained then sealed.

[c] $\dfrac{\text{Anaerobic heat production rate}}{\text{Heat production rate in air}} \times 100 = 53\text{-}76\%$

[d] Two different individuals of the same size.

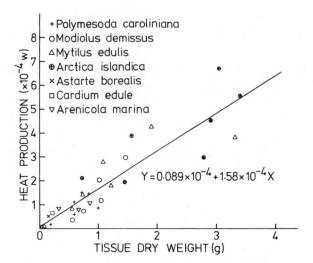

Figure 5.   Anaerobic heat production rates
versus body size at 20°C.  *Polymesoda caroliniana*
and *Modiolus demissus* were from Mobile Bay, Ala-
bama, *Mytilus edulis*, *Arctica islandica*, *Astarte
borealis* from Kiel Bight in the western Baltic,
and *Cardium edule* and *Arenicola marina* from the
North Sea.   The average weight-specific anaerobic
rate is $1.77 \times 10^{-4}$ W g$^{-1}$.

## DISCUSSION

Characteristics of Anaerobic Metabolism and Measurement
Problems

Long-term measurements of anaerobic heat production indicate
that anaerobic metabolism, like aerobic metabolism (Newell, 1972),
can be characterized by periods of active and resting or basal
metabolism; that rhythmic patterns of cyclic activity and rest
are observed in some individuals but not in others; that there are
differences in activity pattern between species as well as between
individuals of the same species; that the daily relative proportion
of active and resting metabolism varies and appears rather unpredictable; and that, consequently, even average rates for long periods of measurements are variable.

How much of the variability in anaerobic metabolic rates is
an artifact of handling and the way animals are kept anaerobic
remains to be determined.  If it is natural variability, and it
appears to be largely so, long-term measurements are required to
determine reliable average values representative of an individual's
daily metabolic rate.  Furthermore, with understanding and control
of still unknown variables, small but significant differences might
ultimately be shown among some of the seven species that now appear
to have similar weight-specific rates of anaerobic heat production.

Future refinements are essential, but the present data are sufficient to show the first order trends in benthic anaerobic metabolism.

When only resting metabolism is considered, the correlation
between anaerobic heat production rate and body size is improved
(Pamatmat, 1979). About 9% of the variability in the regression of anaerobic heat production versus size for *Polymesoda caroliniana* and *Modiolus demissus* is due to changes in physical activity. Much of the scatter in the regression of oxygen consumption
rate on body size has been shown to be the result of variability
in physical activity so the lower and upper limits in the scatter
plot are determined by uptake rates during periods of quiescence
and maximum activity, respectively (Newell, 1972). This appears
to be the case as well with anaerobic metabolism versus body
size. Newell rightly argues that metabolic measurements should
differentiate between standard and active metabolism for comparative purposes. It may also be argued, however, that the regression line for total anaerobic metabolic rates is of greater value
in energy budget calculations and is the one of interest in ecological energetics unless the relative durations of standard and
active metabolism have been established. The relationship between active and resting anaerobic metabolism needs further investigation.

To draw up a population's energy budget, it is, of course,
necessary first of all to know the relative length of time that
individuals spend aerobic and anaerobic. This appears to be quite
variable. *Arctica islandica*, for example, remains buried commonly
for one to seven days, but rarely up to 24 days (Taylor, 1976).
For this species, measurements lasting up to at least seven days
are realistic, but it is possible that an individual that would
have stayed anaerobic in nature for one day would not show the
same pattern or level of activity after one day as another individual that would have remained buried for seven days. Thus, it is
essential to know the basis for this variability and its effect
on metabolic rate. The individual *Arctica islandica* that was
kept anaerobic for ten days (Fig. 1) began to show large bursts
of heat production on the second day and repeated whatever physical activity it was doing to generate such heat about every 30 h
until the seventh day. These physical movements might be the
clam's efforts to emerge from its "buried" state beginning on the
second day. Large peaks of heat evolution have not been observed
in less than 24-h anaerobic measurements in *Arctica islandica* but
they have been observed in *Polymesoda caroliniana* (Pamatmat,
1979). One aspect of my present measurements that is subject
to criticism is that the bivalves are forced to retain their metabolic acid end products. We do not know to what degree many of
these animals get rid of metabolic byproducts while remaining
anaerobic. In the case of *Arctica islandica*, it is conceivable
that the clam could open its valves from time to time while remaining buried and eliminate some of the accumulated acids. Some
species are known to excrete lactic acid: *Nereis virens* (Scott,
1976), and *cirripedes* (Barnes *et al.*, 1963). On the other hand,
*Uca pugnax* is believed to retain practically all the lactic acid

it produces even during prolonged anaerobiosis (Teal and Carey, 1967).  Elimination of carbon dioxide is thought to be important and may be the primary reason for air-gaping (de Zwaan and Wijsman, 1976).  Upon resumption of aerobic metabolism, unexcreted metabolic end products are oxidized (= oxygen debt) and the animal experiences initial elevation of its oxygen uptake rate.  *Arctica islandica* shows a higher elevation of its initial rate of oxygen uptake the longer the period of anaerobiosis (Taylor, 1976).

One uncertainty about the present measurements concerns the heat effect of non-metabolic chemical reactions, e.g., the reaction between acidic end products and $CaCO_3$ of the shell, heat of dissociation and solution, etc.  Their combined heat effect could be relatively small or undetectable but if significant, it should be substracted from the measured values in order to obtain true metabolic rates.  In nature, cyclic periods of anaerobiosis lead to shell dissolution (Crenshaw and Neff, 1969) and the uncorrected measurements could therefore represent actual heat flux from the bivalve.

The significantly higher weight-specific metabolic rate of *Mya arenaria* may be an artifact.  This clam possesses a relatively large siphon, which normally protrudes outside the shell, even during tidal exposure of the sediment, when the animal is not ventilating and is presumably anaerobic.  For calorimetry, this species was treated like the other bivalves and was forced to fully retract its siphon for wrapping and containment in the chamber.  This state could be abnormally stressful to the clam and could account for its elevated metabolic rate.  The simultaneous myogram experiment with the one specimen of *Arctica islandica* indicates that irritation of the animal causes an elevation in its metabolic activity.  Other techniques for measuring anaerobic metabolism of *Mya arenaria* are obviously needed.

### Relationship between Anaerobic and Aerial Heat Production Rates

Anaerobic heat production rate of *Mytilus edulis* is always less than but varies from 11 to 76% of aerial heat production rate. Since anaerobic heat production rate is relatively more stable than aerial or aquatic aerobic metabolism, the variability in the percentage values must be attributed to variability in aerial heat production rate.  This probably indicates that the relative proportion of aerobic and anaerobic metabolism, and/or of active and resting metabolism, during the aerial consumption of oxygen, is variable.

### Relationship Between Anaerobic and Aquatic Aerobic Heat Production Rates

Anaerobic heat production rate is evidently also a variable percentage of aquatic heat production rate (7-48% for *Mytilus edulis*, 42% for *Cardium edule*, and 23-25% for *Arctica islandica*). The variability probably arises more from the changing relative proportion of active and resting metabolism during aerobic respiration than during anaerobic respiration.

Relationship Between Aerial and Aquatic Aerobic Metabolic Rates

Heat production rates in air are always less than but vary from 30-63% of aerobic heat production rates in water (Table 2). In terms of oxygen uptake, aerial rates by *Mytilus edulis* are 4 and 6% of the aquatic rate, but *M. galloprovincialis* showed 11 and 17%, *Mytilus californianus*, 74%, *Modiolus demissus*, 56%, *Cardium edule*, 28 and 78% (Widdows *et al.*, 1979), and *Modiolus demissus*, 63% (Kuenzler, 1961), and 66% (Booth and Mangum, 1978). The species differences are attributed to the degree of shell gape during air exposure (Widdows *et al.*, 1979). It is also likely that variability results from a changing proportion of active and resting metabolism from time to time. Furthermore, in view of an oxygen debt build-up during anaerobiosis in bivalves (Taylor, 1976; Widdows *et al.*, 1979), such percentage values would vary according to the immediate past metabolic state of the individual.

Metabolism vs. Body Size

In spite of the variability in the instantaneous and average daily rates of anaerobic heat production, and in spite of possible interspecific differences that could be hidden by intraspecific variations, the significant trend in the present data on anaerobic heat production versus body size probably represents a first order trend that can be compared with the same relationship for aerobic metabolism of well-studied species.

The relationship between oxygen consumption rate and body size has been intensively and widely studied and discussed (Zeuthen, 1947; Hemmingsen, 1960). For the majority of organisms, including metazoan poikilotherms, the relationship is described by a power function in which b, the power coefficient of body size, tends to an average value of 0.75 (Hemmingsen, 1960). The value of b for anaerobic heat production versus body size equals 1. The difference between the two functions is attributed to gas exchange limitation during aerobic metabolism with increasing body size (Pamat-mat, in press). Oxygen gas exchange is not involved in anaerobic metabolism and metabolic rate increases linearly with the mass of respiring tissue.

The aerobic metabolism of *Modiolus demissus* and *Mytilus edulis* has been carefully measured by Read (1962) who obtained b values of 0.798 ± 0.155 for *M. demissus* and 0.635 ± 0.078 for *Mytilus edulis*. Read's equations are graphed in Figure 6 together with my common linear regression equation for anaerobic metabolism and the reduction in metabolic rate in the changeover from aerobic metabolism.

The larger the individual *Mytilus edulis* that switches from aerobic to anaerobic metabolism, the smaller the decrease, e.g., 97.2% for a 50-mg mussel, decreasing to 92.9% for a 5-g mussel. The trend is the same for *Modiolus demissus*, indicating that small individuals save relatively more energy than large mussels by going anaerobic. On the average, individuals of all sizes of both species that go anaerobic save 95% of the energy that would have been lost as heat during aerobic metabolism.

The average value of 95% reduction in energy demand that has been calculated is much higher than the values indicated by the differences between actually measured rates of aerobic and anaerobic heat production (Table 2). The reason for the smaller percentage values in the measurements is probably that the experimental animals were starved. Starvation lowers oxygen uptake rates (Bayne *et al.*, 1976a) mainly because of the depression of active metabolism while standard metabolism remains relatively unaffected (Newell, 1972). The measured rates of anaerobic heat production could be reasonable approximations of anaerobic rates at 20°C in freshly collected animals, but not the measured rates of aerobic heat production.

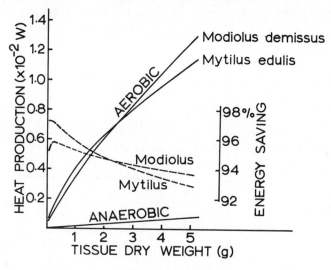

Figure 6. Comparison of aerobic and anaerobic metabolism versus body size. The curves for aerobic metabolism are from the equations of Read (1962). Oxygen uptake rates are converted to heat production rates (1 ml $O_2$ $h^{-1}$ = 5.58 x $10^{-3}$ W). The broken lines represent the difference between aerobic and anaerobic rates expressed as percent of aerobic rate. On the average there is a reduction by 95% in energy expenditure resulting from the changeover to anaerobic metabolism. In comparison, ATP yield during anaerobiosis is about 90% of the yield during aerobiosis.

Benefits from Facultative Anaerobiosis

The obvious advantage of facultative anaerobiosis is that it enables the organism to survive periods of anoxia when water ventilation is not possible (during tidal exposure, or when ventilation or pumping of water would expose the animal to deleterious substances or unpleasant conditions). For example, in the presence of noxious dissolved substances in the water during sudden changes in salinity while there is rainfall in the inter-

tidal (Kushins and Mangum, 1971), there is the risk of desiccation
at high temperature during tidal exposure (Barnes *et al.*, 1963;
Lent, 1968). Thus, by going anaerobic, the individual can avoid
stressful environmental conditions, cover up for extended times
against predator attack, or retreat below the reach of predators.
However, bivalves buried in anoxic sediment are not completely
beyond the reach of predators like *Pisaster brevispinus* (Van Vel-
huizer and Phillips, 1978) that may become anaerobic themselves.

There are isolated pieces of evidence that bivalves stored
in aerated water at low temperatures tend to remain anaerobic for
extended periods as compared to bivalves at warmer temperatures.
The shells of *Modiolus demissus* and *Crassostrea virginicus* remain closed most of the time at a storage temperature of 5°C.
Tissue analysis shows that they contain higher concentrations of
volatile fatty acids (Strength, pers. comm.), which are by-products of anaerobic metabolism (Kluytmans *et al.*, 1975). When
their mantle cavity is drained of water, making them aerobic in
air, *Modiolus demissus* shows unusually high initial rate of heat
production (unpublished) indicative of repayment of oxygen debt.
*Modiolus modiolus*, *Crassostrea virginicus*, *Mytilus edulis*, *Littorina littorea* and *L. rudis*, all facultative anaerobes, live
through freezing conditions down to -15°C (Kanwisher, 1955). Facultative anaerobiosis could somehow be related to the individual's
cold adaptation and freezing tolerance.

There may be some subtle benefits as well, arising from the
lowered energy expenditure during anaerobiosis as compared to
aerobic metabolism. De Zwaan (1977) regards facultative anaerobiosis as an energy-saving mode of respiration. During high-
temperature tidal exposure, intertidal bivalves have lower maintenance energy requirements when they switch to anaerobic metabolism. Shell-gaping (Lent, 1968) and anaerobic metabolism in bivalves might seem contradictory, but it is known that even during
periods of gaping, bivalves are only partly aerobic. Bivalves,
while consuming oxygen in air, accumulate greater oxygen debt the
longer they are exposed to air (Taylor, 1976; Widdows *et al.*,
1979).

Anaerobic metabolism, however, has been thought to be wasteful of energy stores because with the same amount of substrate
material utilized, anaerobic glycolysis yields only about 7.6% of
total energy available in glycogen (with production of lactic acid)
as compared to its total oxidation into $CO_2$ and $H_2O$ through aerobic
metabolism (Newell, 1972). ATP yield from anaerobiosis of bivalves
and polychaetes is now known to be greater than that of classical
vertebrate glycolysis (Hochachka *et al.*, 1973). A variety of metabolic end products are produced. Formation of succinic acid means
ATP yield of 3.7 moles/mole of glucose, and for propionic acid, a
possible yield of as much as 5.4 moles of ATP/mole of glucose, etc.,
in comparison with a total yield of 36 moles of ATP/mole of glucose
that is completely oxidized to $CO_2$ and $H_2O$ (Gnaiger, 1977). The
possibility of even greater anaerobic ATP yield of as much as
13 moles per mole of glucose-6-phosphate is seen by Hochachka (1975).
On the basis of succinic acid formation alone, the switch from
aerobic to anaerobic metabolism reduces the yield of ATP to the in-

dividual by $\frac{36-3.7}{36}$ x 100 = 90%. This loss is smaller than the
average reduction in respiration of about 95%. Hence, although
the individual cannot fully utilize the available energy in its
substrate during anaerobiosis, on balance it seems to do better
energetically because of its diminished maintenance cost. Furthermore, the partly oxidized substrate is not completely lost to the
individual; at least part of it is retained and oxidized later upon
resumption of aerobic metabolism. Of obvious interest is whether
there is, in fact, a saving in ATP that the anaerobic individual
puts into useful work, e.g., membrane transport and bio-synthesis.
Possible tissue growth seems contradicted by the evidence of shell
growth cessation during anaerobiosis (Lutz and Rhoads, 1977; Gordon and Carriker, 1978). Growth, however, does not have to occur
during anaerobiosis. What we need to determine is whether individuals that ingest the same amount of food grow better when continuously aerobic or when periodically anaerobic. The situation
with facultative anaerobes could be analogous to vertically migrating zooplankton that theoretically derive energetic advantage from
lowered respiration at cooler depths during the day after feeding
in warmer surface waters at night (McLaren, 1963). Of course, the
effect of anaerobiosis on the species' digestion, assimilation,
reproduction and other physiological functions also comes into question. We may ask whether anaerobic metabolism can fully meet the
physiological needs of the individual and, if not, why not?

Not all benthic organisms have such low levels of anaerobic
metabolism as the bivalves in comparison with their aerobic metabolism. *Uca pugnax* (Teal and Carey, 1967; Pamatmat, 1978), *Cirolana borealis* (de Zwaan and Skjoldal, 1979), and *Carcinus maenas*
(unpublished) show anaerobic metabolic rates of more than 50% of
their aerobic rates. These species will not derive the energetic
advantage that *Modiolus demissus* and *Mytilus edulis* do. These
species are not as adapted for extended anaerobic metabolism as the
bivalves. De Zwaan and Skjoldal (1979) distinguish at least two
groups of facultative anaerobes: 1) the group that relies primarily on glycolysis and produces lactic acid as the principal metabolic
end product; this includes *Cirolana borealis* and *Uca pugnax* and
2) the group that produces succinate, alanine, propionate, and
other substances indicating more efficient ATP yield; this includes
the bivalves. The lactate producers rely on glycolysis primarily
as a quick source of energy, while the succinate, etc. producers
are the facultative anaerobes that appear to have developed
metabolic pathways for the most efficient anaerobic utilization
of their energy stores.

Anaerobic Metabolism of Benthic Populations

Energy budgets for populations of facultative anaerobes, e.g.,
*Modiolus demissus* (Kuenzler, 1961), and *Littorina irrorata* (Odum
and Smalley, 1959), are based on measured rates of oxygen uptake,
neglect lower metabolic rates during anaerobiosis, and should tend
to overestimate total respiration of the population. Calculated
net efficiency of production, (growth + reproduction)/(growth +

reproduction + respiration), for facultative anaerobes will be
correspondingly higher than those reported in the literature if
respiration is overestimated. Unfortunately, laboratory measurements of anaerobic metabolism are not enough to give us the true
respiration values. We still need to find out the relative lengths
of time that populations are anaerobic and aerobic. It might seem
that the logical technique for determining the extent of anaerobic
metabolism in the field is through quantitative chemical analysis
of metabolic end products of field samples. The difficulty, however,
with excretory losses and the exact metabolic rate of some of these
products will have to be considered. Other problems with biochemical assessment of energy metabolism are pointed out by Gnaiger (1977).
Taylor's (1976) finding of a lack of rhythm, synchrony or uniformity of
duration in the cycle of aero-anaerobiosis among individual *Arctica islandica* suggests the probability of large individual variations in biochemical composition and the need for approaches other
than random sampling. Seasonal, tidal, short-term temporal, and
local environmental factors are probably important.

So it seems that the problem of accurately estimating population respiration of facultative anaerobes is now more difficult
than ever. When organisms are partly aerobic and partly anaerobic, as is evidently often the case at low oxygen tension or during aerial respiration, the rate of oxygen uptake may not accurately reflect total metabolism. This will depend on whether anaerobic end products are excreted or retained and subsequently oxidized.
Accordingly, oxygen uptake rates either underestimate or accurately
represent total metabolism. After aerobic shutdown, metabolic
activity obviously can no longer be measured by oxygen uptake methods, and the question remains whether the metabolic activity
during shutdown can be estimated in terms of the oxygen debt repayment upon resumption of aerobiosis.

It is possible that in spite of the widespread occurrence of
facultative anaerobiosis, in terms of the population's total energy
budget, the energy demand during anaerobiosis, or the energy yield
of anaerobic pathways amounts to only a small fraction and could
be nearly negligible. When viewed in terms of its adaptive and
survival value to the population, however, the energetics of facultative anaerobiosis could be out of proportion to its numerical
value.

ACKNOWLEDGMENTS

Much of this work was done at the University of Kiel, Federal
Republic of Germany, with support from the Sonderforschungsbereich
95 of that institution, through the invitation of Dr. H. Hinkel-
mann and Dr. B. Zeitzschel. My research there was aided by many
colleagues including W. Bengtsson, G. Graf, J. Rumohr, G. Ramm, V.
Smetacek, and others. Their assistance is gratefully acknowledged.
U.S. National Science Foundation Grant OCE77-08634 and Department
of Energy Grant EE-77-S-05-5465 provided funds for instrumentation
and the development of techniques in calorimetry.

REFERENCES

Barnes, H., D. M. Finlayson, and J. Piatigorsky. 1963. The effect
    of desiccation and anaerobic conditons on the behavior, survival
    and general metabolism of three common cirripedes. J. Anim.
    Ecol. 32: 233-252.
Bayne, B. L., C. J. Bayne, T. C. Carefoot, and R. J. Thompson.
    1976a. The physiological ecology of *Mytilus californianus*
    Conrad. 1. Metabolism and energy balance. Oecologia (Berl.)
    22: 211-228.
Bayne, B. L., C. J. Bayne, T. C. Carefoot, and R. J. Thompson.
    1976b. The physiological ecology of *Mytilus californianus* Con-
    rad. 2. Adaptations to low oxygen tension and air exposure.
    Oecologia (Berl.) 22: 229-250.
Booth, C. E., and C. P. Mangum. 1978. Oxygen uptake and transport
    in the lamellibranch mollusc *Modiolus demissus*. Physiol. Zool.
    51: 17-32.
Coleman, N. 1973. The oxygen consumption of *Mytilus edulis* in
    air. Comp. Biochem. Physiol. 45A: 393-402.
Crenshaw, M. A., and J. M. Neff. 1969. Decalcification at the
    mantle-shell interface in molluscs. Am. Zool. 9: 881-885.
Gnaiger, E. 1977. Thermodynamic considerations of invertebrate
    anoxibiosis, pp. 281-303. *In* I. Lamprecht and B. Schaarschmidt
    (eds.). Applications of calorimetry in life sciences. Walter
    de Gruyter, Berlin.
Gordon, J., and M. R. Carriker. 1978. Growth lines in a bivalve
    mollusc: subdaily patterns and dissolution of the shell.
    Science. 202: 519-521.
Hemmingsen, A. M. 1960. Energy metabolism as related to body
    size and respiratory surfaces, and its evolution. Rept. Steno
    Mem. Hosp., Copenhagen. 9: 1-110.
Hochachka, P. W. 1975. An exploration of metabolic and enzyme
    mechanisms underlying animal life without oxygen, pp. 107-137.
    *In* D. C. Malins and J. R. Sargent (eds.). Biochemical and bio-
    physical perspectives in marine biology, v. 2. Academic Press,
    London.
Hochachka, P. W., J. Fields, and T. Mustafa. 1973. Animal life
    without oxygen: basic biochemical mechanisms. Am. Zool. 13:
    543-555.
Kanwisher, J. W. 1955. Freezing in intertidal animals. Biol.
    Bull. 109: 56-63.
Kluytmans, J. H., P. R. Veenhof, and A. de Zwaan. 1975. Anaero-
    bic production of volatile fatty acids in the sea mussel *Mytilus
    edulis* L. J. Comp. Physiol. 104: 71-78.
Kuenzler, E. J. 1961. Structure and energy flow of a mussel popu-
    lation in a Georgia salt marsh. Limnol. Oceanogr. 6: 191-204.
Kushins, L. J., and C. P. Mangum. 1971. Responses to low oxygen
    conditions in two species of the mud snail *Nassarius*. Comp.
    Biochem. Physiol. 39A: 421-435.
Lent, C. M. 1968. Air-gaping by the ribbed mussel, *Modiolus de-
    missus* (Dillwynn): effects and adaptive significance. Biol.
    Bull. 134: 60-73.

Lent, C.M. 1969. Adaptations of the ribbed mussel, *Modiolus demissus* (Dillwynn), to the intertidal habitat. Am. Zool. 9: 283-292.

Lutz, R. A., and D. C. Rhoads. 1977. Anaerobiosis and a theory of growth line formation. Science 198: 1222-1227.

Mangum, C. P. 1970. Respiratory physiology in annelids. Am. Scient. 58: 641-647.

McLaren, I. A. 1963. Effects of temperature on growth of zoo-plankton, and the adaptive value of vertical migration. J. Fish. Res. Bd. Can. 20: 685-727.

Moon, T. W., and A. W. Pritchard. 1970. Metabolic adaptations in vertically separated populations of *Mytilus californianus* Conrad. J. exp. mar. Biol. Ecol. 5: 35-46.

Newell, R. C. 1972. Biology of intertidal animals. Paul Elek, London.

Odum, E. P., and A. E. Smalley. 1959. Comparison of population energy flow of a herbivorous and a deposit-feeding invertebrate in a salt-marsh ecosystem. Proc. Nat. Acad. Sci. 45: 617-622.

Pamatmat, M. M. 1978. Oxygen uptake and heat production in a metabolic conformer (*Littorina irrorata*) and a metabolic regulator (*Uca pugnax*). Marine Biology. 48: 317-325.

Pamatmat, M. M. 1979. Anaerobic heat production of bivalves (*Polymesoda caroliniana* and *Modiolus demissus*) in relation to temperature, body size, and duration of anoxia. Mar. Biol. 53: 223-229.

Read, K. R. H. 1962. Respiration of the bivalved molluscs *Mytilus edulis* L. and *Brachidontes demissus plicatulus* Lamarck as a function of size and temperature. Comp. Biochem. Physiol. 7: 89-101.

Scott, D. M. 1976. Circadian rhythm of anaerobiosis in a poly-chaete annelid. Nature (Long.). 262: 811-813.

Shick, J. M. 1976. Physiological and behavioral responses to hypoxia and hydrogen sulfide in the infaunal asteroid *Ctenodis-cus crispatus*. Marine Biology. 37: 279-289.

Taylor, A. C. 1976. Burrowing behavior and anaerobiosis in the bivalve *Arctica islandica* (L.). J. mar. biol. Ass. U.K. 56: 95-109.

Teal, J. M., and F. G. Carey. 1967. The metabolism of marsh crabs under conditions of reduced oxygen pressure. Physiol. Zool. 40: 83-91.

Van Veldhuizen, H. D., and D. W. Phillips. 1978. Prey capture by *Pisaster brevispinus* (Asteroidea: Echinodermata) on soft substrate. Mar. Biol. 48: 89-97.

Widdows, J., B. L. Bayne, D. R. Livingston, R. I. E. Newell, and P. Donkin. 1979. Physiological and biochemical responses of bivalve molluscs to exposure to air. Comp. Biochem. Physiol. 62A: 301-308.

Zeuthen, E. 1947. Body size and metabolic rate in the animal kingdom, with special regard to the marine microfauna. Compt. Rend. Lab. Carlsberg, Ser. Chim. 26: 17-161.

Zwaan, A. de. 1977. Anaerobic energy metabolism in bivalve

molluscs, pp. 103-187. *In* H. Barnes (ed.), Oceanography and
marine biology, annual review, v. 15. Aberdeen Univ. Press,
Scotland.

Zwaan, A. de, and H. R. Skjoldal. 1979. Anaerobic energy metabolism of the scavenging isopod *Cirolana borealis* (Lilljeborg).
Comp. Biochem. Physiol. In press.

Zwaan, A. de, and T. C. M. Wijsman. 1976. Anaerobic metabolism
in Bivalvia (Mollusca). Characteristics of anaerobic metabolism. Comp. Biochem. Physiol. 54B: 313-323.

POPULATION STUDIES

# An Evaluation of Experimental Analyses of Population and Community Patterns in Benthic Marine Environments

Paul K. Dayton
J. S. Oliver

ABSTRACT:  A basic objective of science is to describe and understand the mechanisms by which various natural patterns are produced and maintained.
In recent years experimental approaches have been
used with varying degrees of success in efforts to
understand the mechanistic relationships structuring ecological communities.  This essay: 1) affirms
our conviction that critical testing of specific
hypotheses is a vital component of science and that
properly controlled experiments offer the cleanest
and most powerful tests of hypotheses, and 2) discusses problems that we perceive to exist with experimentation in ecology.

    We discuss currently active paradigms and de-
lineate problems such as artifacts that occur in
popular experimental designs, alternate hypotheses
that are commonly ignored, and misinterpretations
resulting from not properly appreciating scale in
time and space.  Perhaps the most difficult problem is that preconceptions tend to flavor questions, determine research designs, and bias interpretation of the data.  The preconceptions commonly result in an emphasis on verification rather

than falsification of hypotheses, a process where-
by counter examples are ignored, alternate hypotheses brushed aside, and existing paradigms manicured.  The escape from this trap is vigorous con-
fronting of the preconceptions and paradigms.


## INTRODUCTION

Laboratory experiments have a long history in ecology.  Their
design is to hold constant all the variables except the one of interest.  Small-scale field experiments have a very different design
in that all environmental parameters are allowed to vary naturally
while the one of interest is manipulated.  Finally, "natural experiments," referring to large perturbations or, in a few cases, geographic patterns, can be used to test general hypotheses.  It is
widely appreciated that there are difficulties in interpreting laboratory research in terms of field relevance; likewise, the difficulties in finding effective controls for large-scale natural disturbances are well understood and most geographic comparisons of
some theoreticians are hopelessly difficult to control.  The success
of some field experiments as well as reviews critical of both laboratory and natural experiments (e.g., Connell, 1974; Paine, 1977)
appear to have resulted in a shift of emphasis from laboratory and
natural experimental approaches toward field experiments.
Unfortunately, small-scale field experiments are plagued by
many of the same problems common to large-scale ones, and there is
a need for more thoughtful consideration of issues revolving around
time, a real scale, and magnitude or intensity of experimentally
induced disturbance.  For example, in most communities the response
to a perturbation will depend upon the extent and/or intensity of
the perturbation, its shape, severity, season, and so forth.  The
relevant controls are similarly variable.  The potential theoretical rationale for such experimentation may include biogeography,
community stability, population regulation, behavior, or environmental physiology.  The proper control is dictated by the question
or theory under consideration, yet the experiment and its control
often address a very different scale than is being considered in
the theoretical rationale for the research. This relationship
between an explicit hypothesis and the scale of the experiment
relates to another general problem.  We concur with the spirit of
Frank's (1957) statement:  "If we make an attempt to set up a theory or hypothesis on the basis of recorded observations, we soon
notice that without any theory we do not even know what we should
observe," but a philosophic problem emerges in that the theory and
observations or tests of the theory become so intertwined and complementary that the discipline becomes a narrowing dogma rather
than a progressive science pushing ahead with new conceptual break-
throughs.  The procedural way out of this trap is to make rigorous
attempts to actually disprove the theory itself by observations and
tests.  This may sound trite, but it seems to us that much recent

ecological research is explicitly or implicitly designed to verify
rather than falsify existing preconceptions.

The objective of this review is to evaluate some of the common
methods of field experiments in marine habitats. We are especially
concerned with 1) the scale of the processes in question and the
appropriateness of the experimental scale, 2) the efforts to identify and evaluate alternate hypotheses, and 3) the problem of preconceptions. Because appropriate questions and techniques appear to
differ for hard and soft substrata, the two habitats will be considered separately.

## HARD SUBSTRATA

Because of the ease of manipulation, much relatively successful
experimentation has been done on hard substrata. The questions,
techniques, and some patterns of the results have been reviewed exhaustively (Connell, 1974; Paine, 1977). Here we discuss the problems of artifacts, alternate hypotheses, limitations imposed on the
worker by the techniques, and the influence of preconceptions.

Removal of all individuals of a particular species, such as
Paine's (1966) manual removal of the asteroid *Pisaster ochraceus*,
is probably one of the simplest manipulations possible. The null
hypotheses are usually clear, and there are few important artifacts.
When the removed species is relatively motile, individuals reinvade
and create a relatively large "edge effect." This usually dictates
that a large area be manipulated, and it can be difficult to choose
a genuinely comparable area for the control. If the two areas are
near to each other, the motile individuals in the control area may
immigrate into the experimental removal area, resulting in a "lower
than normal" density in the control area. Perhaps somewhat cleaner
are those experiments in which sessile species are removed in order
to test their roles in the community (Sutherland and Karlson, 1977;
Sutherland, 1978). The results of such experiments with fouling
associations usually suggest that, even though the removed species
may have considerable effect, the availability of larvae in the
plankton is an important parameter influencing community establishment and subsequent development. Although most hard substrata
workers recognize the importance of larval availability, the issue
is rarely addressed in the literature.

Another problem with removal experiments is that researchers
pick species which they feel do have important roles; usually
ignored are the many species with possibly minor roles. Thus the
hypothesis that most species in a community have no effect on the
other populations is rarely addressed, despite the fact that it is
central to a continuingly important current controversy in community ecology over the reality and mechanisms of successional change
(Clements, 1916; Gleason, 1927; Connell and Slatyer, 1977;
Simberloff, 1979).

Another common procedure is to regulate density of motile animals by use of barriers (Dayton, 1971; Haven, 1973) or cages
(Connell, 1961). Many artifacts attendant to such structures are
often improperly controlled. For example, the barriers, such as

Dayton's (1971) dog dishes or Haven's (1973) fences, may restrict
limpet passage, but they also interrupt the flow of sea water over
the substrata and create unnatural turbulence and small eddies.
These may function to trap and accumulate sediment, larvae and
algal spores and gametes.  They also baffle the wind and thereby
considerably reduce the effect of desiccation.  These artifacts
were evaluated at Mukay Bay, Washington (Dayton, unpublished) by
installing a series of parallel, unconnected fences perpendicular
to the surge which were the same height as the dishes, but completely open to limpets and predators.  There was a conspicuous recruitment of algae behind the fences, probably resulting from both physical deposition of reproductive products and protection from desiccation.  In this case the artifact was identified, and the comparative analysis (Dayton, 1971) was between similar dishes contrasting
different densities of limpets but having the same turbulence artifact.  Haven's (1973) analyses were similarly controlled.  But the
obvious solution to these problems is that of John Cubit (1974), who
devised a copper-based paint which molluscs will not cross.  Cubit
found spectacular limpet effects, corroborating earlier conclusions
regarding effects of limpets on at least the recruitment of algae.
     The use of cages to restrict predators has become a popular
technique.  Unfortunately cages have many effects in addition to the
exclusion of predators.  Intertidal cages obviously shade the substratum and offer considerable protection from desiccation.  In addition, they certainly function much more efficiently than dishes
or other barriers in the trapping, or, at times, exclusion of sediment, algal spores, larvae, and small predators.  Such artifacts
are difficult to control; the use of "roofs" held 1-2 cm from the
substrata approximate the shade effects, but do not duplicate desiccation because the wind can circulate underneath the roof.  Nor do
they have the same efficiency at trapping sediments or reproductive
units such as larvae, pieces of fertile drift algae, and small predators.  Sometimes cages with animals inside can be used, but often
the predator of interest does not fit in the cage, and, in any case,
a caged animal does not behave in the same fashion as it does when
not restricted.  Some of the artifacts can be partially controlled
by use of blank cages which duplicate the cage effects in areas
where the animals normally excluded by the cages are removed manually.
Here the cage and its control evaluates the cage artifact.  A very
modest attempt to remove manually all the predators was made by
Dayton (unpublished) at Portage Head and San Juan Island, Washington,
to contrast complete predator exclusion cages (Dayton, 1971), and
their possible artifacts with predator exclusion experiments without
cages.  In these non-replicated studies, no dramatic cage effect
was observed, possibly because the cages were monitored every 7-10
days and cleaned of drift and fouling material, and possibly because
this was done during a period of mild weather, so that desiccation
effects were minimal.  It is our belief that even in the intertidal
situations where cages have been very successful, these artifacts
rarely are considered but probably are not trivial.

     More difficult to control, however, is the phenomenon by which
cages and roofs offer varying degrees of protection to the small

predators including the young of the species allegedly excluded.
For example, Dayton (1971) copied Connell's (1961) model of marine
caging studies and found that from spring through fall the nemer-
teans, *Emplectonema gracile* and *Paranemertes peregrina*, and several
flatworms were often found inside cages where they were observed
eating young barnacles (see Hurley, 1975).  Also included were increased densities of polychaetes and amphipods.  The dorid opistho-
branch *Onchidoris bilamellata*, a predator specializing on barnacles,
was found inside cages during the winter, and often it stripped the
caged area of the barnacles.  Indeed, frequently only dead barnacles
and an *Onchidoris* egg mass were found inside cages.  Similarly,
small *Leptasterias hexactis* and carnivorous gastropods often were
seen in cages.  Regular monitoring allows one to make corrections
when such problems are observed.  But the probability of observing
the effects of such predators under "roofs" or in adjacent controls
is much smaller than in the cages, where the predators remain because they are protected from desiccation or where they often simply
become trapped as they grow bigger.  Clearly there is a risk in
using cages and their controls as the sole means of testing null hypotheses regarding the effects of large predators.

     Because the use of cages on hard substrata has proven to be a
powerful research tool and because they are relatively simple to in-
stall and maintain, their use is appealing.  Nevertheless, there are
many important selective pressures that cannot be defined or explored
by the use of even well-controlled cage techniques.  Workers therefore should not truncate their creative approaches toward these
problems by restricting their research to cage techniques.


SOFT SUBSTRATA

     In contrast to communities on hard substrata, which are struc-
turally dominated by large, easily-observed individuals, many soft
substrata communities are composed primarily of small, rarely-ob-
served animals which lie buried beneath the surface.  Historically,
their patterns of distribution and abundance have been correlated
with sediment type and the presence of various functional groups
which often influence the sediment and its various physicochemical
properties.  These approaches are summarized exhaustively in excellent reviews by Rhoads (1974) and Gray (1974) and are brought up to
date by recent symposia (Keegan *et al.*, 1977; Coull, 1977) and by
the other contributions to this volume.  Our objective is to consider some strengths and weaknesses of current approaches to the study
of soft-bottom infaunal communities.


Functional Groups

     Many infaunal species can be classified into broad ecological
categories such as feeding and mobility types.  These functional
groups may have important interactions and roles in the community.
A relatively distinct spatial segregation of deposit and suspension-
feeding infauna in Buzzard's Bay (Sanders, 1958, 1960) was partially

explained by the trophic group ammensalism hypothesis (Rhoads and Young, 1970). The hypothesis generally states that the activities of deposit feeders interfere with the establishment and maintenance of suspension-feeder populations. Rhoads and Young (1970) showed that deposit feeders cause a marked increase in sediment water content and a decrease in substrate stability (i.e., sediments were more easily resuspended by water currents). This high turnover and resuspension of unstable sediments may affect larval habitat selection or increase juvenile mortality of suspension feeders by clogging feeding and respiratory surfaces and by burial.

Rhoads and Young (1970) tested clogging of the trophic ammensalism hypothesis by placing the suspension-feeding clam, *Mercenaria mercenaria*, at three distances above the highly reworked Buzzard's Bay bottom. Individual growth was significantly greater (p < 0.05) at 45 and 70 cm from the bottom (1.2–1.3 mm/66 days) than on the lowest tray at 10 cm (0.6 mm/117 days). However, the sand bottom control, an intertidal group of clams, grew at rates comparable to the 10-cm treatment. This low intertidal growth was related to the difficulty of depositing calcium carbonate during low tides (Rhoads and Young, 1970). Another explanation for these results is that clam growth is enhanced at the greater distances above the reworked bottom, possibly because of an increase in optimal particle size and suspended bacteria. The similarity in growth between the intertidal control and the 10-cm treatment may involve a tradeoff between simultaneous increases in food and clogging on the 10-cm tray. Rhoads (1973) later observed an enhancement of growth in another bivalve, *Crassostrea virginica*, placed over the same reworked bottom when compared to its growth over a sandy bottom. Thus, while clogging was observed in one species, enhancement was documented for both, and there is still little evidence that suspension-feeder growth rates are lower in the reworked mud than in a comparable control. Regardless of the specific mechanism, differential mortality of suspension-feeding larvae and adults would support the trophic ammensalism hypothesis. For example, Levinton and Bambach (1970) provided field evidence for greater mortality of suspension-feeding bivalves in reworked sediments. Certainly this type of interaction is seen in the intensive work of Myers (1977) and Brenchley (1979); future work is likely to reinforce the value of the functional group concept.

Interspecific Competition

Competition is a celebrated process, but it is often difficult to demonstrate in nature. A number of studies show intraspecific spacing and aggression among soft-bottom species (e.g. Holme, 1950; Connell, 1963; Reish, 1957; Roe, 1971). Laboratory and field experiments show interspecific avoidance interactions among infaunal bivalves. Fenchel's (1975, 1977) work on interspecific character displacement is a rare example of evidence that such competitive interactions strongly affect contemporary population patterns in soft bottoms. Interference competition can broadly overlap with a number of biological disturbance processes, and without clear definitions of the resource, the demarcation between other negative species interactions and competition can be arbitrary.

Manipulative field and laboratory experiments have shown direct adult interference competition among several infaunal bivalves. Levinton (1977) observed vertical and horizontal interspecific avoidance in laboratory aquaria. Peterson (1977) hypothesized that competition was more intense among burrowing species with broadly overlapping stratification in the sediment. This hypothesis was partially supported by the disjunct horizontal distributions of species with similar vertical distributions. The thalassinid shrimp, *Callianassa californiensis* and the bivalve *Sanquinolaria nuttallii* have similar vertical patterns but are segregated from each other horizontally. Peterson (1977) removed *Callianassa* by collapsing their burrows and observed the successful recruitment of *Sanquinolaria* in the disturbed plots. These results at first appear consistent with the competition hypothesis; however, they were explained by the alternate hypothesis that *Callianassa* eat or bury the young *Sanquinolaria* (Peterson, 1977). Other potentially important alternative hypotheses were not considered. For example, clam larvae may be attracted to the experimental disturbance just as various polychaete larvae such as *Armandia brevis* apparently are attracted to modified habitats (Oliver, 1979). Although there was no *Sanquinolaria* recruitment in the controls, larval availability might also contribute to Peterson's (1977) result in a relative sense, as clam larvae tend to be more available for settlement than shrimp larvae. For example, periodic larval settlement and subsequently high mortality of bivalves are common phenomena (e.g., Muus, 1973; Oliver, 1979), while large recruitment pulses of *Callianassa* are relatively rare, at least in Bodega Harbor (Ronan, personal communication) and Elkhorn Slough, California (personal observations). The effects of larval selection and larval availability can be tested, but at this point there are several viable hypotheses to explain the avoidance of a well-developed *Callianassa* bed by *Sanquinolaria* larvae.

Although the interpretations from the *Callianassa* manipulations are equivocal, Peterson and Andre (in press) performed an elegant series of experiments that demonstrate competitive interference between two bivalve species. These species had widely overlapping vertical stratification in the sediment, but also had disjunct horizontal distributions. Peterson and Andre transplanted known numbers of the two suspension-feeding bivalves into the same sediment column and observed a significantly higher emigration of *Sanquinolaria nuttallii* in the presence of *Tresus nuttallii* than in its absence. *Sanquinolaria* growth also was reduced by the presence of *Tresus*. In a second experiment, they substituted dead shells secured to wooden poles simulating clams with extended siphons for the living *Tresus*. The results were similar. *Sanquinolaria* had a lower growth rate in the presence of these artificial clams and siphons (wood poles) than they did in their absence. The artificial clam experiment suggests that the interaction is mediated by spatial interference and not by competition for food. This is an excellent example of finding the appropriate system to elucidate a natural process. While the competition issue is blurred in one system (*Cal-*

*lianassa* and *Sanquinolaria*), Peterson and Andre provide a simple and crisp view of competition in another (*Tresus* and *Sanquinolaria*).

Ronan (1975) performed some simple transplant experiments to assess the interaction between *Callianassa californiensis* and the phoronid worm, *Phoronopsis viridis*. He carefully placed 1,000 adult phoronids into the *Callianassa* bed, leaving two plots exposed to *Callianassa* burrowing and excluding *Callianassa* from a third plot with a plastic cage. Although most of the transplanted phoronids survived, the number of broken and disturbed tubes (found at an angle of greater than 45°) was significantly greater in the plots exposed to *Callianassa* (Table 1). An average of 26% of the transplanted tubes were broken and disoriented in the uncaged plots and only 3% were disrupted in the caged plots that excluded *Callianassa*.

Table 1. Breakage and disorientation of transplanted phoronid tubes by the burrowing activities of *Callianassa*. *Callianassa* was excluded from areas containing transplanted tubes in the cage treatments. Data are from two experiments performed in the fall of 1973 and the summer of 1974 (in parenthesis) (From Ronan, 1975).

|                | Number of Tubes |          |          |
|----------------|-----------------|----------|----------|
| Tube Condition | Uncaged         | Uncaged  | Caged    |
| Normal         | 260(345)        | 292(303) | 395(310) |
| Broken         | 33( 39)         | 45( 20)  | 9( 4)    |
| Disturbed      | 89( 82)         | 64( 56)  | 4( 6)    |

Fall experiment: $x^2 = 128$, $p < 0.01$
(Summer experiment: $x^2 = 79$, $p < 0.01$)

Thus, burrowing *Callianassa* can disrupt the tubes of sedentary phoronids. Nevertheless, the two species are generally found at distinct intertidal elevations and are horizontally separated by many meters (Ronan, 1975). The upper intertidal limits of *Phoronopsis* therefore appear to be regulated very rarely by an interaction with *Callianassa*. Moreover, the phoronid transplant experiments provide no clue as to the factors that control the lower intertidal limits of *Callianassa*, which are usually well above those of *Phoronopsis*.

Ronan (1975) also examined potential interactions between the phoronid worm bed and the bivalve *Macoma secta*. Ronan found that lateral movement of *Macoma secta* was influenced more by the mass properties of the sediment (a primarily physical property) than by the presence of phoronid tubes (a biological property). Vertical burrowing was significantly depressed only at the highest phoronid densities and was deepest in the northern sandflat deposit (Table 2). On the other hand, phoronid density had no effect on *Macoma*'s horizontal movement, which was significantly less in all the treatments on southern sandflat deposit (Table 2). Apparently the physical differences between the northern and southern areas were as important to the bivalve movement as the number of phoronid tubes.

Table 2. A: Mean depth of vertical burrowing of transplanted *Macoma secta* (introduced on the sediment surface) after 60 days and B: Mean horizontal movements in 2 days (introduced to a depth of 8 cm). Means connected by horizontal lines are not significantly different (p >0.01). N = 60 m each case; clams were 1 to 6 cm (from Ronan, 1975).

A. *Macoma secta* Vertical Burrowing Depth (cm)

| Density of phoronid tubes per m² | >10,000 | <6,000 | Natural[*] 0 | Transplant[**] 0 |
|---|---|---|---|---|
| South Sandflat | 11 | 13 | 13 | 16 |
| North Sandflat | 11 | 15 | 17 | 16 |

B. *Macoma secta* Horizontal Burrowing Movements (cm)

| | | | | |
|---|---|---|---|---|
| South Sandflat | 1 | 2 | 1 | 4 |
| North Sandflat | 4 | 4 | 4 | 4 |

[*]Natural sediment from the designated sandflat with no phoronids
[**]Sediments transplanted from the opposite sandflat with no phoronids

One of the most interesting attempts to demonstrate competition among small infaunal animals is Woodin's (1974) study of a mobile polychaete worm, *Armandia brevis*, and a relatively sedentary polychaete, *Platynereis bicanaliculata*. *Platynereis* settled and constructed tubes on the mesh of her exclusion cages; this effectively reduced the number of *Platynereis* reaching the bottom beneath the cage. A striking difference in the abundances of *Platynereis* and *Armandia* between treatments and controls was interpreted as evidence for competition for space, in which the experimental reduction in *Platynereis* apparently resulted in a competitive release in *Armandia* (Woodin, 1974). An alternate hypothesis also explains these results as well as similar results from a different soft-bottom community and environment: *Armandia* larvae are attracted to the particular sedimentary habitat modified by the presence of cages. For example, when cages of identical design were exposed in a subtidal sand bottom in Elkhorn Slough, they did not exclude *Platynereis* or any other sedentary polychaete, yet the abundance of *Armandia* increased (Table 3). *Armandia* larvae settle into a wide variety of sediments (Hermans, 1978; Oliver, 1979) and are apparently attracted to disturbed bottoms (Oliver, 1979). Although possibly this is a factor of relative mobility, Brenchley's (1979) laboratory experiments

also suggest that the mobile *Armandia* will survive cage-induced deposition better than many sedentary species, such as *Lumbrineris*
*inflata*, which was not excluded by Woodin's cages but showed the
same negative correlation as *Armandia* (Woodin, 1974).  Many of the
cage artifact problems discussed for the rocky intertidal are also
present in soft-bottom environments.  Additional problems are considered in the next section.

Table 3.  The mean density of *Armandia brevis* inside and outside
of caged bottom areas after a two-month exposure period (May through
June, 1978) in the Elkhorn Slough, California (from Hulberg and Oliver,
in preparation).

|  | Number[*] Replicate Cores | ($\bar{x} \pm$ 95% CL) *Armandia* *brevis* per core | One Tailed[**] Probability (cage vs. uncaged) |
|---|---|---|---|
| Cage # 1 | 4 | 222±85 | 0.002 |
| Cage # 2 | 5 | 166±42 | 0.001 |
| Uncaged Bottom | 8 | 39±21 | – |

[*]core = 0.005 m$^2$; sieving screen size = 0.25 mm.
[**]Rank Sum Test

### Predation and Disturbance

Predation and other biological disturbances may be among the
most important processes affecting soft-bottom communities.  But to
a certain extent, this statement reflects the success of the manipulative experiments designed to evaluate these processes in other
habitats and is often taken as a matter of faith for soft substrata
work where the parallel experiments are not without problems.  It
is difficult to remove predators from soft-bottom habitats because
they are either large and mobile or infaunal and cryptic.  Predator
exclusion cages, however, have been used in soft-bottom communities,
but they do not result in an increase in a competitively dominant
prey as in other communities (Peterson, in press).  The reasons for
this are unclear, but in marked contrast to hard substrata associations, competition for space seems to have a minor role in most soft-
bottom communities.  In addition, there is a technical problem with
most of the recent caging manipulations in soft-bottom communities.
There has been a failure to control the hydrodynamic and sedimentary
habitat changes produced by cages (Virnstein, 1977b; Hulberg and
Oliver, in prep.).  As a result, although most of the existing
experimental results were interpreted as a consequence of excluding
predators, they are equally well explained by the alternate hypothesis that infaunal species respond to the habitat changes produced
by the cages.  In fact, the same density increases observed inside
cages may also occur around the outside of cages which actually

attract the fishes presumed to be the predators excluded by the cages (Fig. 1). Nevertheless, cages can be a powerful manipulative tool for studying soft bottoms (e.g., Ronan, 1975; Virnstein, 1977a, b; VanBlaricom, 1978). Virnstein's (1979a) study is an excellent example of a rigorous attempt to falsify a number of realistic alternate hypotheses. Even so, the tethering and caging of an aggressive and highly disruptive portunid crab probably mimics few natural circumstances. Furthermore, polychaete populations only increased significantly in those cages that trapped a large amount of fine sediment (up to 35% silt/clay compared to 11-19% for the other cages). Therefore, the polychaete patterns can also be explained by non-predator-related habitat changes unrelated to the abundance and activity of predators. However, the mass of evidence indicating that crab predators have a dramatic effect on the local bivalve populations (Virnstein, 1977a) is certainly convincing, although here too crab larvae might be attracted to the different sediment regime inside the cages.

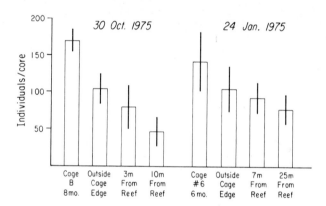

Figure 1. The mean number per 0.018 m$^2$ core of polychaete individuals inside exclusion cages, directly next to the outside-cage cage, and at various distances from the cage arrays or reef. Bars indicate 95% confidence limits (from Hulberg and Oliver, in prep.)

Summarizing, cages are valuable tools; in addition to predator exclusion experiments, they can be used to seed populations such as bivalves into new areas and hold them until they become established, they allow certain behavioral observations, they attract meiofauna, etc. But the use of cages must be tempered with an appreciation of their tendency to alter the habitat.

Some of the most informative experimental manipulations investigate the disturbance effects of benthic predators. Woodin (1978) showed that the area adjacent to a large *Diopatra cuprea* tube (polychaete) was an effective refuge for a number of smaller polychaetes from the feeding disturbance of the blue crab and the horseshoe

crab, *Limulus polyphemus*. She could mimic the refuge effect by in-
serting straws into the sediment. The physical structure of a tube
apparently stabilizes the sediment and was not disrupted by crabs.

VanBlaricom (1978) documented the infaunal colonization of the
feeding depressions made by the rays, *Urolophus halleri* and *Mylio-
batis californicus*, and also of artificial ray holes he dug which
provided identical patterns. While digging for larger prey, the
rays displace small infauna, many of which are consumed by fish.
The average life span of a ray pit is about two weeks, and approximately 25% of his study area, a wave-swept subtidal sandflat in
southern California, was in some stage of recovery from ray disturbance at any given time. A small amphipod, *Acuminodeutopus heteru-
ropus*, was among the earliest and most abundant colonists. It is
a tube builder and an active swimmer which carries a marsupial
clutch of one or two large eggs. Apparently the amphipod scoots
from one ray depression to another and utilizes the fine particulate
layer trapped by the holes. It releases a large juvenile ready to
join the pit migrations. *Acuminodeutopus* appears to be an obligate
opportunistic species, dependent on the frequent small-scale disturbances of rays. It is also the second most abundant macrofaunal
invertebrate on the highly ray-disturbed sandflat studied by
VanBlaricom (1978). On the other hand, neither *Acuminodeutopus* nor
any ecologically equivalent species inhabits similar wave-swept
sandflats in central California where ray disturbance is extremely
rare (Oliver, 1979).

## Complementary Laboratory and Field Experiments

Although the utility of laboratory observation and experimentation may be greater for small infaunal invertebrates than for larger
ones, the general apprehension of Connell (1975) about laboratory
work is unwarranted. Certainly laboratory discoveries must have
field verification, and the choice of questions and experiments that
can be approached in the laboratory depends upon the insight and experience of good natural historians. We emphasize that field experiments have the same dependence, and in neither case should this ob-
viate the need for experimentation.

Laboratory experiments overcome several problems encountered
by the study of small infaunal animals. Potentially important behavioral interactions, feeding mechanisms, mobility and searching
patterns, and other natural history observations may be difficult
or impossible to make in the field. These observations may be possible in the laboratory and allow better definition of hypotheses
which can be tested either in the laboratory or in the field. Laboratory experiments also allow critical controls that are often impossible in the field. Furthermore, well-controlled or not, many
experiments can be performed only in the laboratory.

There is a rich history of laboratory work which has demonstrated important animal-sediment relationships (reviewed by Rhoads,
1974), the ability of larvae to avoid or select certain habitat
characteristics (reviewed by Wilson, 1958; Meadows and Campbell,
1972; Gray, 1974; Scheltema, 1974), animal interactions (e.g.,

Trevallion *et al.*, 1970; Levinton, 1977; Reise, 1977; Bell and
Coull, 1978; Oliver, 1979), and many interesting trophic relationships (e.g., Newell, 1965; Hargrave, 1970; Stephens *et al.*, 1978).
Many other examples are found in this volume. Rarely, however, are
laboratory and field experiments and observations used in a complementary manner. The laboratory studies of Reise (1977) and Bell and
Coull (1978) are promising extensions of the field caging work and
have revealed interesting information.

The work summarized by Trevallion *et al.* (1970) on the ecology
and energetics of the bivalve *Tellina tenuis* and its predator, the
0-group plaice, *Pleuronectes platessa*, is an excellent example of
complementary laboratory and field approaches. They performed a
series of feeding experiments with different densities of clams and
fish in large laboratory tanks and explained the results in terms
of prey abundance and physiological condition in relation to predator abundance, growth rate, and diet in the field. They also examined the effect of increased food on clam growth in laboratory experiments. These analyses indicate that clam growth and reproduction are limited by food concentration, that siphon regeneration
after predation may reduce reproductive effort and lower stock recruitment, and that plaice have a type III or "S" shaped functional response to prey density. As a result of these studies, this
is one of the better known predator-prey systems in soft-bottom habitats.

Brenchley's (1979) study of animal-sediment relations involved
a similarly innovative series of laboratory and field experiments.
One of her most interesting experimental results is that mortality
of sedentary infauna is greater than the mortality of mobile infauna during laboratory-induced sedimentation or burial. Another very
interesting result is that the effects of sediment stabilizers such
as seagrasses, large immobile clams, etc. and the effects of sediment destabilizers such as deposit-feeders and other mobile fauna
are additive.


Larval Recruitment

The central themes in soft-sediment community research overlap
broadly: functional group interference spills into interspecific
competition, and both are related to predation and disturbance.
Physical and biological factors interact strongly, and many of the
above mechanisms may act, at least partially, through differential
larval recruitment. There are two general biological components of
the recruitment process: a) habitat selection and b) mortality
during and just after settlement. Not only are these components
rarely distinguished, but there have been few field experiments of
any aspect of larval recruitment. The natural hydrographic and sedimentary environments are especially difficult to reproduce in laboratory aquaria. To a limited extent, the problems with laboratory
experiments can be overcome with field manipulations. For example,
field experimental manipulations of sediment grade and water and
sediment motion result in consistent species-specific recruitment
patterns into artificial sedimentary habitats (plastic cups of de-

faunated sediments). These experimental recruitment patterns correspond remarkably well with the abundance patterns of established
bottom populations and with larval recruitment into large (400 m$^2$)
defaunated bottom patches (Oliver, 1979). Generally larvae settled
into the particular artificial cup habitat which was the closest
mimic to the natural sedimentary environment inhabited by the same
species (Fig. 2). Processes affecting larval recruitment are poorly understood, but may be some of the most important processes controlling the establishment and maintenance of soft-bottom communities.

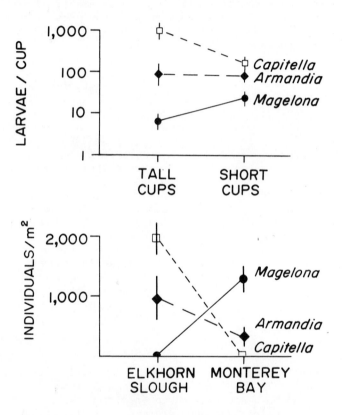

Figure 2. The mean number of polychaete
larvae settling into short (n = 10) and tall
(n = 10) plastic cups (artificial sedimentary
habitats), and the bottom population abundances of the same species in the wave-pro-
tected Elkhorn Slough (n = 80), and along the
wave-swept subtidal beach in Monterey Bay (n =
117). The tall (13 cm) and short (7 cm) cups
are filled with the same amount of defaunated
sediment and are gross mimics of these large-
scale natural sedimentary habitats (from Oliver, 1979). Bars indicate 95% confidence limits.

Commensalism

Commensalism and other symbioses are biological interactions
that have received little attention in ecologically oriented soft-
bottom studies.  The observations of Rhoads and Young (1971) are an
interesting exception.  They found a group of small tube-building
suspension feeders living on the sides of sediment mounds deposited
by the large holothuroid, *Molpadia oolitica*.  The absence of the
tube builders in the depressions between mounds may be related to
interference by surface deposit feeders or to clogging.  Irrespective of the cause, the sedentary fauna were found only in association
with the holothuroid mounds, the presence of which allows the persistence of the worms and the habitat.  A similar commensal association is found around the tubes of *Diopatra cuprea*; in addition to
the large suite of species growing on the *Diopatra* tubes themselves,
a number of small sedentary polychaetes appear dependent on the
stable substratum provided by the larger polychaete tubes (Woodin,
1978).  On the other hand, the active mounds made by squillid sto-
matopods and thalassinid shrimps in tropical lagoons are inhabited
by a large number of highly motile podocopid ostracods and the few
sedentary species are more abundant between mounds (Fig. 3).
VanBlaricom (1978) observed a commensal interaction between feeding
rays and a smaller demersal flatfish.  The flatfish consumed deep-
burrowing infauna that were exposed by the foraging rays.  There are
undoubtedly many other examples of similar commensal interactions,
including the burrow dependencies such as *Cryptomya*, pea crabs, etc.
on burrows of *Callianassa, Upogebia, Urechis, Arenicola*, etc. (e.g.,
Ricketts *et al.*, 1968; MacGinite and MacGinite, 1968), but there
has been no attempt to synthesize these numerous examples into the
current soft-bottom research thinking.  In fact, the community consequences of commensal interactions may be as important, or even
more important, as the more commonly studied negative relationships.

SCALE AND DISPERSAL

A basic goal of science is to make correct generalizations
about nature; we do experiments to elucidate a mechanism, and we
then tend to generalize as much as possible.  While the objectives
are good, the efforts to generalize often lead to problems because
events at different scales are likely to be dominated by different
processes.  For example, benthic processes observed in the laboratory, small field experiments, and large episodic events are likely
to be dominated by phenomena very different in space and time.
VanBlaricom's ray pits quickly were colonized by motile amphipods,
which moved into the small disturbed patches as adults.  But
VanBlaricom also observed a large slump of the sand bottom into a
nearby canyon which sterilized a large area of the sandflat.  The
recolonization pattern of the slump area contrasted markedly with
that of the ray pits, as the slump was invaded by polychaetes,
which disperse as larvae, and by other crustaceans, but some of the
amphipods, which disperse so quickly into the ray pits, did not become important colonists of the slump area for several years (Fig.
4).

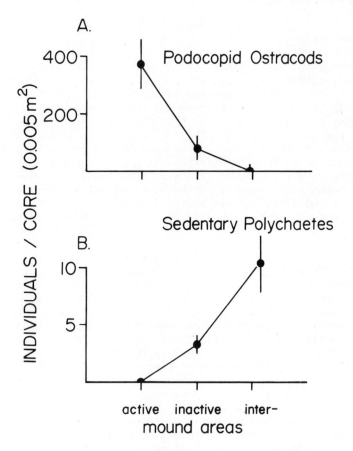

Figure 3.  The number of A. podocopid
ostracods, and, B. sedentary polychaetes found
on active and inactive crab mounds and between
mounds in the Lizard Island Lagoon, Australia.
Active mounds receive deposit from burrowing
crabs.  Data are means and 95% confidence limits (from Oliver and Slattery, in prep.).


     Temporal patterns are more difficult to understand, as they
include seasonal variation, year-to-year differences, and infrequent
episodic events.  Seasonal pulses of larval availability are reasonably well understood, but they have not received sufficient recognition in soft-bottom experimental work, which rarely controls for
larval availability.  More difficult for experimental ecologists are
big year-to-year differences in recruitment patterns (see Burkenroad, 1946).  Such patterns are known for many dominant species,
including *Mytilus californianus* (Dayton, 1971; Paine, 1974), *Balanus
cariosus* (Dayton, 1971), *Strongylocentrotus purpuratus* (Ebert, 1968;
Dayton, 1975), *Dendraster excentricus* (Merrill and Hobson, 1970;
Timko, 1975; Davis and VanBlaricom, 1978), *Renilla kollikeri* (Kastendiek, 1975; Davis and VanBlaricom, 1978), *Owenia collaris* (Fager,

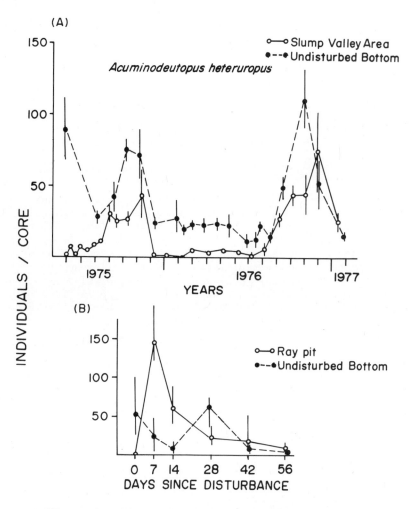

Figure 4.   A.   Recovery of the amphipod *Acumino-deutopus* in a large slump area (means per 0.018 m² core and standard error), and, B. in a small pit formed by a feeding ray (means and ranges).   The amphipods quickly colonized ray pits but only slowly entered the large slump valley (from VanBlaricom, 1978)

1964), and many species of soft-bottom bivalves (Loosanoff, 1964; Coe, 1956).   A particularly dramatic example of a rare but extreme-ly heavy settlement of the echiuroid *Listriolobus pelodes* is docu-mented by Word (1978) in which the dense (to 2 kg/m²) population completely altered the facies of the community.   Indeed, as we learn more about marine systems, these year-to-year differences appear to be general phenomena, especially for long-lived species.

Many of these patterns are linked to the dispersal biology of the relevant species.   This includes larval adaptations, physical oceanography, and migration of established fauna.   Species which

broadcast larvae commonly co-occur with species which brood their
larvae, but probably face different selective pressures, at least
for the young individuals. The potential effects of delayed metamorphosis, spawning patterns, larval behavior and settling phenomena, and consequences of current patterns on dispersal are appreciated, but essentially unknown (e.g., Vance, 1973 and especially
Strathmann, 1974, 1977). In many cases, temporal and spatial variations in recruitment may be explained by shifts in currents (Coe,
1956; Loosanoff, 1964). Adult abundance patterns of some estuarine
barnacles may be related to the vertical distribution of larvae and
the movement of fresh- and salt-water layers in large estuaries
(Bousfield, 1955).

The isolation of water masses occurs on many scales and can
have unexpected consequences. For example, simple larval collection jars (such as those of Thorson, 1946; and Reish, 1961) revealed
three consistent patterns of larval dispersal within the Elkhorn
Slough and between the slough and the adjacent Monterey Bay (Fig. 5).
The first pattern was the restriction of certain larvae from totally
different taxonomic groups (polychaetes, crabs, bivalves) to a distinct water mass in the back or upper slough. The larvae of back
slough bottom species were seldom flushed into the mouth (Fig. 5).
Smith (1973) documented the hydrographic isolation of the upper
slough and estimated a residence time of 300 days during periods of
low rainfall. On the other hand, the lower slough is affected by
a daily tidal exchange with the Monterey Bay. In the second pattern, larvae of the bottom species living at the mouth of the slough
are flushed into the bay. While these flushed species were collected in large numbers in Monterey Bay, the upper slough species never
occurred in the offshore collecting jars (Fig. 5). The third pattern involved the inability of many offshore larvae to penetrate
into the slough (Fig. 5). Similar dispersal patterns are probably
common in many coastal embayments and at least partially account
for the unpredictable settlement of many species. Although Ayers
(1956) proposed that all the larvae of the bivalve *Mya arenaria*
were flushed from a small coastal embayment in Massachusetts,
species-specific flushing and retention of larvae can be highly dependent on larval behavior, as well as circulation patterns.


EPISODIC EVENTS

Finally, ecologists tend to consider rare but large-scale episodic events as being of a geological, rather than ecological, time
scale. This arrogance is costly, however, as many populations of
benthic invertebrates are dominated by old individuals, often apparently representing a single cohort or age class and apparently
exhibiting life histories characteristic of declining populations,
allowing for massive recruitment when conditions are right. Thus,
many commonly observed patterns may result from past variations in
climate and/or large-scale oceanographic parameters. These effects
are manifested by cold kills (Dahlberg and Smith, 1970), salinity
changes from rain runoff, heavy sediment loads related to floods
(Schafter, 1972; Reineck and Singh, 1973), reductions in dissolved

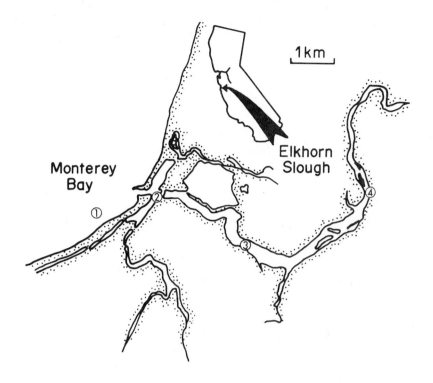

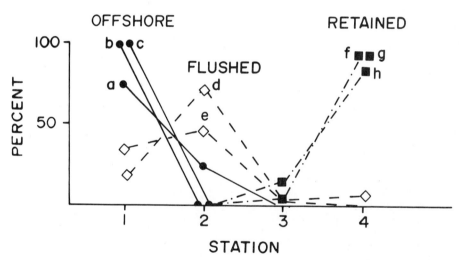

Figure 5. The percentage of larvae of representative polychaete species collected at three stations in the Elkhorn Slough
and one in the Monterey Bay. Nearly 100% of the larvae of the three
offshore species and the three back-slough species were collected
above the adult habitats. The offshore species are a. *Nothria elegans*, b. *Magelona sacculata*, and c. *Amaeana occidentalis*; the species
living at the slough mouth (flushed) are d. *Armandia brevis* and e.
*Capitella capitata*; and the back-slough species (retained) are f.
*Ctenodrilus serratus*, g. *Ophryotrocha* sp. and h. *Streblospia benedicti*. Data are from jars exposed at 14-day intervals during the
late spring and early summer (from Jong and Oliver, unpublished
data).

oxygen (Leppäkoski, 1971), severe red tides (Simon and Daver, 1972), tropical storms (Boesch *et al.*, 1976), unusual cold (Eggleston and Hickman, 1972), and unusually low tides (Glynn, 1968; Yamaguchi, 1975; Loya, 1976). Conspicuous shifts in relative abundance can be seen in cores from anaerobic basins (Soutar and Isaacs, 1974). Slight shifts of large currents such as the California Current of the eastern Pacific may have resulted in the "warm-water years" in the late 1950's, which brought tropical fishes to southern California, and may have contributed to the disappearance of large kelp forests (North, 1971). Similarly, intertidal and shallow coral reefs may be structured largely by periodic typhoons (Connell, 1978). Certainly the assumption of constancy of an Antarctic sponge community (Dayton *et al.*, 1974) has turned out to be incorrect (Dayton, 1979), and it now appears that the habitat may have been subjected to grounded ice within the last few hundred years (unpublished ages of bivalve shells).

## PRECONCEPTIONS

The last but probably most important problem we discuss is the sensitive and difficult issue of preconceptions. To a considerable extent preconceptions permeate science as they determine the questions and influence the design of experiments, their controls, and their interpretations. The preconception issue is a persistent component of ecology and can be seen in most field experimentation in which the investigators implicitly or explicitly tend to support or verify their hypotheses rather than making a rigorous effort to negate them. This is a difficult issue because in the early stage an hypothesis may be new and stimulating and basically true; thus an investigator who experimentally negates reasonable alternate hypotheses is following progressive scientific methods. Eventually, however, a subtle change can occur and the alternate hypotheses become weak or trivial and the investigator would seem to be buttressing the hypothesis. While this research contributes interesting and important details, it is at this point that the hypothesis becomes a paradigm (*in sensu* Kuhn, 1970), at least to the investigator or a subset of colleagues. This may be true of much rocky intertidal work (Dayton, 1971 through Lubchenco and Menge, 1978, and Peterson, 1979) which refines basic premises of key biological interactions. Certainly much of the research discussed in this essay seems designed to verify rather than negate hypotheses, and it is clear that many reasonable alternative hypotheses often are not considered.

The interpretation and analysis of observations and experiments also can be biased by preconceptions. Even such an excellent and influential paper as Connell (1970) might not be free of this problem. This paper, for example, presents the hypothesis that barnacles living in the high intertidal zone, in which they enjoy a refuge from effective predation, have a sufficient reproductive output to sustain the barnacle recruitment in the lower zones where barnacles usually are consumed by predators. This hypothesis could be negated by calculating a net reproductive rate significantly less than unity (Leslie, 1966). Connell did calculate a life table

(Table 14; in Connell, 1970) with this objective; unfortunately the $m_x$ column appears to have been influenced by efforts to verify rather than negate the hypothesis. For example, the actual measurement of the size-dependent fecundity was determined from a regression uncritically calculated from four barnacles presumably collected at Ladysmith Harbor, Vancouver Island, Canada (Barnes and Barnes, 1956). Ladysmith Harbor is an extremely rich *Balanus glandula* habitat very dissimilar to the study areas on San Juan Island. It seems likely that Barnes and Barnes (1956) collected the four barnacles from a lower intertidal zone where, in addition to the richer habitat, they would have more feeding opportunity and thus be all the more likely to be more fecund than the barnacles in the upper refuge zone addressed by Connell. Thus this component of the $m_x$ column is probably an overestimate of the fecundity; nonetheless, Connell then doubled this value in the actual calculation of his $m_x$ column on the assumption that all *Balanus glandula* release two broods of larvae per year. He offers limited evidence that some barnacles can release a second brood in the early fall which then settles in the late fall or early winter. With good feeding conditions for the barnacles, it is possible for barnacles to release multiple broods, but Barnes and Barnes (1956) argue that it does not happen often. Yet this assumption, so critical to the life table, easily can be tested by comparing the spring and late fall settlements, which are not similar in the San Juan Islands. It also can be negated with a hammer, as was done in a limited manner by Dayton (1969-1970; unpublished) who found that many *B. glandula* in the high intertidal refuge appear sterile, probably because 1) they are immersed so rarely that they barely feed enough to sustain basal metabolism, much less produce reproductive products and 2) many are too far from their neighbors to allow cross fertilization. For these reasons the $m_x$ column in Table 14 (Connell, 1970) seems to have systematic errors which may double the real values. Thus the failure to negate the hypothesis that these upper refuge zone barnacles are capable of maintaining the entire population obscures the possibility that the San Juan Island *Balanus glandula* population is maintained by larvae advected from elsewhere. If this is true, it would be especially interesting because these barnacles represent an example of Elton's "key industry species" (see Birkland, 1974) in that they are the most characteristic rocky intertidal species in the region and support many species of predators.

The objective of this discussion has been to emphasize that such preconceptions (or paradigms of Kuhn, 1970) are common and can influence even the most successful scientists, yet they have great cost: the loss of reality. Similar preconceptions have influenced many other hard- and soft-bottom projects and will continue to do so, as they dominate science from the conceptualization of the question, design of the research program, and interpretation of the results (see Gould, 1978, for a well-documented example). The most effective progressive solution to this is to consciously and rigorously attempt to disprove and destroy the working hypothesis rather than attempt to "support" or "verify" it as we see done overtly or covertly in many papers in our field.

DISCUSSION

We affirm our conviction that thoughtfully conceived and carefully controlled field experiments offer the most satisfactory tests
of hypotheses posed to clarify the evolutionary mechanisms resulting
in observed patterns of distribution and abundance. We fully concur
with the arguments of Connell (1974) and Paine (1977) extolling the
importance of field experimentation as the most powerful use of scientific methods in ecology. The purpose of this essay is to offer
suggestions by which these methods can be improved.

It seems to us that much recent soft substrata research is pat-
terned after paradigms developed in the rocky intertidal. We believe that many of the parallels are incorrect. For example, rocky
intertidal workers focus on the large algae, barnacles and mussels
and their predators, all of which are easily observed and manipulated; ignored are the groups of small animals such as flatworms, nematodes and other pseudocoelomates, insects, polychaetes, small molluscs, crustacea, etc. It is assumed (but never tested) that the
patterns of these smaller species conform to those of the larger
species, the dynamics of which are becoming well known. The soft
substrata workers, on the other hand, usually are not able to deal
with the dynamics of the parallel large animals such as *Tresus*,
*Saxidomus*, *Callianassa*, *Molpadia*, *Dendraster*, *Diopatra*, etc.; thus
most soft substrata research focuses on the dynamics of the populations of small infaunal species. Obviously all species have both
potentially limiting resources and predators, and there are thus
grossly parallel processes working in both hard- and soft-bottom
communities. But the systems are very different, and the techniques
necessary to study these processes must be developed independently,
as it is clear that hard-bottom approaches such as caging are often
inappropriate to soft substrata. And, indeed, many important rocky
intertidal processes such as space competition appear irrelevant to
small infaunal populations (Peterson, in press).

Biological communities are structured by processes acting within a broad range of spatial and temporal scales. Spatial patterns
vary through small to very large scales and tend to be mediated by
dispersal phenomena. In marine systems, the dispersal phenomena
result from integration of specific life history adaptations and
complex oceanographic processes. While absolutely crucial to the
maintenance of the community, this integration is extremely difficult to understand or to investigate and has never been successfully unraveled. The temporal patterns are similarly mediated through
life history adaptations and oceanographic and climatic processes,
the integration of which determines the availability of larval prop-
agules. While larvae of some species are predictably, if seasonally, available for recruitment, most species have patterns of temporal periodicity much too long for ecological study. Because these
scale problems are so difficult, there is a tendency to ignore them
and focus on patterns of larval recruitment and survival. Unfortunately the integration of the spatial and temporal scales and their
often physical oceanographic and climatological driving forces is
usually critically important to the understanding of the entire
pattern. While these problems of scale remain difficult, we urge

that they be integrated into all levels of benthic research to the maximum extent possible.

Finally, we argue that both correlative pattern matching and seemingly rigorous experimentation can lead into a trap of perpetuating preconceived ideas. This trap can be seen in almost all scientific research where the emphasis is on the verification rather than falsification of hypotheses. The verification of ideas may be the most treacherous trap in science, as counter-examples are overlooked, alternate hypotheses brushed aside, and existing paradigms manicured. The successful advance of science and the proper use of experimentation depend upon rigorous attempts to falsify hypotheses.

ACKNOWLEDGMENTS

This essay is dedicated to the memory of Tom Ronan, a good friend and promising young scientist who died before his time. We thank Chris Jong of the Moss Landing Marine Laboratories for larval settlement data from the Elkhorn Slough. We are especially grateful to Charles Peterson, Jim Porter, Lisa Levin and Glenn VanBlaricom for constructive criticism of the manuscript, and to Ken Tenore and Bruce Coull for their patience.

REFERENCES

Ayers, J. C. 1956. Population dynamics of the marine clam, *Mya arenaria*. Limnol. Oceanogr. 1: 26-34.

Barnes, H., and M. Barnes. 1956. The general biology of *Balanus glandula*. Darwin. Pac. Sci. 10: 415-422.

Bell, S. S., and B. C. Coull. 1978. Field evidence that shrimp predation regulates meiofauna. Oecologia. 35: 141-148.

Birkland, C. 1974. Interaction between a sea pen and seven of its predators. Ecol. Monogr. 44: 211-232.

Boesch, D. F., R. J. Diaz, and R. W. Virnstein. 1976. Effects of tropical storm Agnes on soft-bottom macrobenthic communities of the Jones and York estuaries and the lower Chesapeake Bay. Chesapeake Science. 17: 240-259.

Bousfield, E. L. 1955. Ecological control of the occurrence of barnacles in the Miramichi estuary. Nat. Mus. Canada, Bull. No. 137: 69 pp.

Brenchley, G. A. 1979. On the regulation of marine infaunal assemblages at the morphological level: a study of the interactions between sediment stabilizers, destabilizers and their sedimentary environment. Ph.D. Thesis, The John Hopkins University, 265 pp.

Burkenroad, D. D. 1946. Fluctuations in abundance of marine animals. Science. 103: 684-686.

Clements, F. E. 1916. Plant succession. Carnegie Inst. Wash. Publ. 242: 512 pp.

Coe, W. R. 1956. Fluctuations in populations of littoral marine invertebrates. J. Mar. Res. 15: 212-232.

Connell, J. H.  1961.  Effects of competition, predation by *Thais lapillus* and other factors on natural populations of the barnacle *Balanus balanoides*.  Ecol. Monogr. 31:  61-104.

Connell, J. H.  1963.  Territorial behavior and dispersion in some marine invertebrates.  Res. Popul. Ecol. 5:  87-101.

Connell, J. H.  1970.  A predator-prey system in the marine intertidal region.  I.  *Balanus glandula* and several predator species of *Thais*.  Ecol. Monogr. 40:  49-78.

Connell, J. H.  1974.  Ecology:  field experiments in marine ecology.  Experimental Marine Biology, pp. 21-54, Academic Press.

Connell, J. H.  1978.  Diversity in tropical rain forests and coral reefs.  Science. 199:  1302-1310.

Connell, J. H., and R. O. Slatyer.  1977.  Mechanisms of succession in natural communities and their role in community stability and organization.  Am. Nat. 111:  1119-1144.

Coull, B. C.  1977.  Ecology of marine benthos.  The Belle W. Baruch Library in Marine Science, No. 6, Univ. So. Carolina Press, Columbia, South Carolina, 467 pp.

Cubit, J.  1974.  Interactions of seasonally changing physical factors and grazing affecting high intertidal communities.  Ph.D. Thesis, Univ. Oregon, Eugene, Oregon.

Dahlberg, M. D., and Smith, F. G.  1970.  Mortality of estuarine animals due to cold on the Georgia coast.  Ecology. 51:  931-933.

Davis, N., and G. R. VanBlaricom.  1978.  Spatial and temporal heterogeneity in a sand bottom epifaunal community of invertebrates in shallow water.  Limnol. Oceanogr. 23:  417-427.

Dayton, P. K.  1971.  Competition, disturbance, and community organization:  the provision and subsequent utilization of space in a rocky intertidal community.  Ecol. Monogr. 41:  351-389.

Dayton, P. K.  1975.  Experimental evaluation of ecological dominance in a rocky intertidal algal community.  Ecol. Monogr. 45:  137-159.

Dayton, P. K.  1979.  Observations of growth, dispersal and population dynamics of some sponges in McMurdo Sound, Antarctica.  Proceeding International Colloquium on Sponge Biology, Paris, 1978.  In press.

Dayton, P. K., G. A. Robilliard, R. T. Paine, and L. B. Dayton.  1974.  Biological accommodation in the benthic community at McMurdo Sound, Antarctica.  Ecol. Monogr. 44:  105-128.

Ebert, T. A.  1968.  Growth rates of the sea urchin *Strongylocentrotus purpuratus* related to food availability and spine abrasion.  Ecol. 49:  1075-1091.

Eggleston, D., and R. W. Hickman.  1972.  Mass stranding of mulloscs at Te Waewae Bay, Southland, New Zealand.  New Zealand Journal of Marine and Freshwater Research. 6:  379-382.

Fager, E. W.  1964.  Marine sediments:  effects of a tube-building polychaete.  Science. 143:  356-359.

Fenchel, T.  1975.  Character displacement and coexistence in mud snails (Hydrobiidae).  Oecologia. 20:  19-32.

Fenchel, T.  1977.  Competition, coexistence and character displacement in mud snails (Hydrobiidae).  *In* Ecology of Marine Benthos, B. C. Coull (ed.), Univ. So. Carolina Press, Columbia, South Carolina, pp. 229-243.

Frank, P. 1957. Philosophy of Science, The link between science and philosophy. Prentice-Hall, Inc. Englewood Cliffs, N.J. 394 pp.

Gleason, H. A. 1927. Further views on the succession concept. Ecol. 8: 299-326.

Glynn, P. W. 1968. Mass mortalities of echinoids and other reef flat organisms coincident with midday, low water exposures in Puerto Rico. Mar. Biol. 1: 226-243.

Gould, S. J. 1978. Morton's ranking of races by cranial capacity. Science. 200: 503-509.

Gray, J. S. 1974. Animal-sediment relationships. *In* Oceanogr. Mar. Biol. Ann. Rev., Barnes (ed.). 12: 223-261.

Hargrave, B. T. 1970. The effect of deposit-feeding amphipods on the metabolism of benthic microflora. Limnol Oceanogr. 15: 21-30.

Haven, S. B. 1973. Competition for food between the intertidal gastropods *Acmaea scabra* and *Acmaea digitalis*. Ecol. 54. 143-151.

Hermans, C. O. 1978. Metamorphosis in the opheliid polychaete *Armandia brevis*. *In* Settlement and Metamorphosis of Marine Invertebrate larvae, Chia and Rice (eds.), Elsevier North-Holland Biomedical Press, 113-126 pp.

Holme, N. A. 1950. Population-dispersion in *Tellina tenuis* Da Costa. J. Mar. Biol. Assn. U.K. 29: 267-280.

Hurley, A. C. 1975. The establishment of populations of *Balanus pacificus* Pilsbry (Cirripedia) and their elimination by predatory turbellaria. Journal of Animal Ecology. 44: 521-532.

Kastendiek, J. E. 1975. The role of behavior and interspecific interaction in determining the distribution and abundance of *Renilla koelikeri*, a member of a subtidal sand bottom community, Ph.D. Thesis, Univ. Calif., Los Angeles, 194 pp.

Keegan, B. F., P. O. Ceidigh, and P. J. S. Boaden. 1977. Biology of benthic organisms. Pergamon Press, New York, 630 pp.

Kuhn, T. S. 1970. The structure of scientific revolutions. 2nd rev. ed., Chicago: Univ. Chicago Press.

Leslie, P. H. 1966. The intrinsic rate of increase and the overlap of successive generations in a population of guillemots (*Uria aalge* Pont.) J. Anim. Ecol. 35: 291-301.

Leppäkoski, E. 1971. Benthic recolonization of the Börnholm Basin (southern Baltic) in 1969-1971. Thallassia Jugoslavica. 1: 171-179.

Levinton, J. S. 1977. Ecology of shallow water deposit-feeding communities Quisset Harbor, Massachusetts. *In* Ecology of Marine Benthos, B. C. Coull (ed.), Univ. So. Carolina Press, Columbia, So. Carolina, 191-227 pp.

Levinton, J. S., and R. K. Bambach. 1970. Some ecological aspects of bivalve mortality patterns. Amer. J. Sci. 268: 97-112.

Loosanoff, V. I. 1964. Variations in time and intensity of settling of the starfish, *Asterias forbesi*, in Long Island Sound during a twenty-five-year period. Biol. Bull. 126: 423-439.

Loya, Y. 1976. Recolonization of Red Sea Corals affected by natural catastrophies and man-made perturbations. Ecology. 57: 278-289.

Lubchenco, J., and B. A. Menge. 1978. Community development and persistence in a low rocky intertidal zone. Ecol. Monogr. 59: 67-94.

MacGinitie, G. E., and N. MacGinitie. 1968. Natural history of marine animals. McGraw-Hill, New York, 523 pp.

Meadows, P. S. M., and J. I. Campbell. 1972. Habitat selection by aquatic invertebrates. Adv. Mar. Biol. 10: 271-382.

Merrill, R. J., and E. D. Hobson. 1970. Field observations of *Dendraster excentricus*, a sand dollar of western North America. Am. Midl. Nat. 83: 595-624.

Muus, K. 1973. Settling, growth and mortality of young bivalves in the Øresund. Ophelia. 12: 79-116.

Myers, A. C. 1977. Sediment processing in a marine subtidal sandy bottom community. I. Physical aspects. Journal of Marine Research. 35: 609-632.

Newell, R. 1965. The role of detritus in the nutrition of two marine deposit feeders, the prosobranch *Hydrobia ulvae* and the bivalve *Macoma balthica*. Zoo. Soc. Lond. Proceed. 144: 25-45.

North, W. J. 1971 The biology of giant kelp beds (*Macrocystis*) in California. Nova Hedwigia Z. Kryptogamented Suppl. 32: 600 pp.

Oliver, J. S. 1979. Processes affecting the organization of soft-bottom communities in Monterey Bay, California and McMurdo Sound, Antarctica. Ph.D. Thesis, University of California, San Diego.

Paine, R. T. 1966. Food web complexity and species diversity. Amer. Natur. 100: 65-75.

Paine, R. T. 1974. Intertidal community structure: experimental studies on the relationship between a dominant competitor and its principal predator. Oecologia. 15: 93-120.

Paine, R. T. 1977. Controlled manipulations in the marine intertidal zone, and their contributions to ecological theory. *In* The Changing Scenes in Natural Sciences, Academy of Natural Sciences, Special Publ. 12: pp. 245-270.

Peterson, C. H. 1977. Competitive organization of the soft-bottom macrobenthic communities of southern California lagoons. Mar. Biol. 43: 343-359.

Peterson, C. H. 1979. The importance of predation and competition in organizing the intertidal epifaunal communities of Barnegat Inlet, New Jersey. Oecologia. 39: 1-24.

Peterson, C. H. in press. Predation, competitive exclusion, and diversity in soft-sediment benthic communities of estuaries and lagoons. *In* Ecological Processes in Coastal and Marine Systems, R. J. Livingston (ed.), Plemun Press, New York.

Peterson, C. H., and S. V. Andre in press. An experimental analysis of interspecific competition among marine filter feeders in a soft-sediment environment. Ecology. 61.

Reineck, H. E., and I. E. Singh. 1973. Depositional sedimentary environments: with reference to terrigenous clastics. Springer-Verlag, New York, Heidelberg, Berlin, 439 pp.

Reise, K. 1977. Predator exclusion experiments in an intertidal mudflat. Helgolander wiss. Meeresunters. 30: 263-271.

Reish, D. J. 1957. The life history of the polychaetous annelid *Neanthes caudata* (delle Chiaje), including a summary of development in the family Nereidae. Pac. Sci. 11: 216-228.

Reish, D. J. 1961. The use of the sediment bottle collector for monitoring polluted marine waters. Cal. Fish Game. 47: 261-272.

Rhoads, D. K. 1973. The influence of deposit-feeding benthos on water turbidity and nutrient recycling. Am. J. Sci. 273: 1-27.

Rhoads, D. K. 1974. Organisms-sediment relations on the muddy seafloor. *In* Oceanogr. Mar. Biol. Ann. Rev., H. Barnes (ed.). 12: 263-300.

Rhoads, D. C., and D. K. Young. 1970. The influence of deposit-feeding organisms on sediment stability and community trophic structure. J. Mar. Res. 28: 150-178.

Rhoads, D. C., and D. K. Young. 1971. Animal sediment relations in Cape Cod Bay, Massachusetts. II. Reworking by *Molpadia oolitica* (Holothuroidea). Mar. Biol. 11: 255-261.

Ricketts, E. F., J. Calvin, and J. Hedgpeth. 1968. Between Pacific Tides. Stanford University Press, Stanford, California.

Roe, P. 1971. Life history and predator-prey interactions of the nemertean *Paranemertes peregrina* Coe. Ph.D. Thesis, Univ. Washington, 129 pp.

Ronan, T. E. 1975. Structure and paleoecological aspects of a modern marine soft-sediment community: an experimental field study. Ph.D. Thesis, Univ. Calif., Davis, 200 pp.

Sanders, H. L. 1958. Benthic studies in Buzzard's Bay. I. Animal-sediment relationships. Limnol. Oceanogr. 3: 245-258.

Sanders, H. L. 1960. Benthic studies in Buzzard's Bay. III. The structure of the soft-bottom community. Limnol. Oceanogr. 5: 138-153.

Schafer, W. 1972. Ecology and paleoecology of marine environments. Oliver and Boyd, Edinburgh, 568 pp.

Scheltema, R. S. 1974. Biological interactions determining settlement of marine invertebrates. Thalassia Jugosl. 10: 263-296.

Simberloff, D. 1979. A succession of paradigms in ecology: idealism to materialism and probabilism. Synthese, in press.

Simon, J. M., and D. M. Daver. 1972. A quantitative evaluation of red tide-induced mass mortalities of benthic invertebrates in Tampa Bay, Florida. Environmental Letters. 3: 229-234.

Smith, R. E. 1973. The hydrography of Elkhorn Slough, a shallow California coastal embayment. Moss Landing Marine Lab. Tech. Pub. 73-2.

Soutar, A., and J. D. Isaacs. 1974. Abundance of pelagic fish during the 19th and 20th centuries as recorded in anaerobic sediments off the Californias. Fish. Bull. 72: 257-273.

Stephens, G. C., M. J. Volk, and S. H. Wright. 1978. Transepidermal accumulation of naturally occurring amino acids in the sand dollar, *Dendraster excentricus*. Biol. Bull. 154: 335-347.

Strathmann, R. R. 1974. The spread of sibling larvae of sedentary marine invertebrates. Am. Nat. 103: 29-44.

Strathmann, R. R. 1977. Toward understanding complex life cycles of benthic invertebrates. *In* The Ecology of Fouling Communities, J. D. Castlow (ed.), Proceedings U.S. - U.S.S.R. workshop within program "Biological productivity and biochemistry of the world oceans," 1-20 pp.

Sutherland, J. P. 1978. Functional roles of *Schizoparella* and *Styela* in the fouling community in Beaufort, North Carolina. Ecol. 59: 257-264.

Sutherland, J. P., and R. H. Karlson.  1977.  Development and sta-
    bility of the fouling community at Beaufort, North Carolina.
    Ecol. Monogr. 47:  425-446.
Thorson, G.  1946.  Reproduction and larval development of Danish
    marine bottom invertebrates with special reference to planktonic
    larvae in the sound (Øresund).  Medd. Komm. Danm. Fiskeriog
    Havunders., Ser. Plankton. 4:  1-523.
Timko, P.  1975.  High density aggregation in *Dendraster excentricus*
    (Esehscholtz):  Analysis of strategies and benefits concerning
    growth, age structure, feeding, hydrodynamics, and reproduction.
    Ph.D. Thesis, Univ. Calif., Los Angeles, 323 pp.
Trevallion, A., J. H. Steele, and R. Edwards.  1970.  The dynamics
    of a benthic bivalve.  *In* Marine Food Chains, J. H. Steele (ed.),
    Oliver and Boyd, Edinburgh, 285-295 pp.
VanBlaricom, G. R.  1978.  Disturbance, predation, and resource al-
    location in a high-energy sublittoral sand-bottom ecosystem:
    experimental analysis of critical structuring processes for the
    infaunal community.  Ph.D. Dissertation, University of California,
    San Diego, 328 pp.
Vance, R.  1973.  On reproductive strategies in marine benthic in-
    vertebrates.  American Naturalist. 107:  339-352.
Virnstein, R. W.  1977a.  The importance of predation by crabs and
    fishes on benthic infauna in Chesapeake Bay.  Ecol. 58:  1199-
    1217.
Virnstein, R. W.  1977b.  Predator caging experiments in soft sedi-
    ments:  caution advised.  *In* M. L. Wiley (ed.) Estuarine Inter-
    actions, Academic Press, New York.
Wilson, D. P.  1958.  Some problems in larval ecology related to
    the localized distribution of bottom animals.  *In* Buzzati-
    Traverso, A.A. Perspectives in Marine Biology, 87-90.  Univ.
    Calif. Press.
Woodin, S. A.  1974.  Polychaete abundance patterns in a marine soft
    sediment environment:  the importance of biological interactions.
    Ecol. Monogr. 44:  171-187.
Woodin, S. A.  1978.  Refuges, disturbance, and community structure:
    a marine soft-bottom example.  Ecol. 59:  274-284.
Word, J.  1978.  The infaunal index.  *In* W. Bascom, (ed.), Coastal
    Water Research Project Annual Report, 1978.  Southern California
    Coastal Water Research Project, El Segundo, California, 253 pp.
Yamaguchi, M.  1975.  Sea level fluctuations and mass mortalities
    of reef animals in Guam, Mariana Islands.  Micronesia. 11:  227-
    243.

# Predation by Demersal Fish and its Impact on the Dynamics of Macrobenthos

Wolf E. Arntz

ABSTRACT: Since 1968, investigations have been carried out in the western Baltic on inter-relationships of the dynamics of macrobenthos and demersal fish. These studies have involved: 1) investigations of over 5,000 stomach and gut analyses to quantify the food (including seasonal changes) of cod, whiting, dab, plaice, flounder and some less important fish species; 2) survey of infaunal macrobenthos over eight years (1968-1971 and 1975-1978); and 3) a three-year experimental study on dynamics and production of macrobenthos at the "Benthosgarten" station. This paper also includes fish data published annually by the International Council for the Exploration of the Sea and from other studies carried out in Kiel Bay.

The interaction of macrobenthos and demersal fish is discussed, particularly regarding the effects of selective predation. Differences in predation intensity from year to year, resulting in reduced population levels of macrobenthos, were observed, but the long-term dynamics of the more important benthic food species in the west-

ern Baltic were seemingly not influenced by the
year-class strength of demersal fish.  Likewise,
the year-class strength of benthos in different
years did not affect the size and production of
the demersal fish stocks in the area.  A number
of possible reasons for this apparent lack of
correlation are discussed.


INTRODUCTION

    The benthic subsystem in the western Baltic is comparatively
simple (Arntz, 1978a).  Some 150 macrobenthic species have been
described from the area, but many of them have to be considered  as
temporal immigrants (Kühlmorgen-Hille, 1963).  Less than 20 species
are important as food for demersal fish.  Of some 40 species of demersal fish collected so far, only five are commercially important
(cod, *Gadus morhua*; dab, *Limanda limanda*; whiting, *Merlangius mer-
langus*; flounder, *Platichthys flesus*; and plaice, *Pleuronectes
platessa*).  Cod dominates the demersal fish, comprising 85% of the
total standing biomass.
    The structurally uncomplicated ecosystem of the western Baltic
Sea lends itself to evaluating benthos-demersal fish interactions.
To what extent are the dynamics of macrobenthos, at least the more
important food species, biologically controlled by fish predation?
Are demersal fish food-limited?  An answer to these questions might
shed some light on the relative importance of biological interactions in temperate benthic communities.


MATERIALS AND METHODS

    This paper is based on data derived from three different
sources:
    1) An eight-year (1968-71 and 1975-78) benthic survey was car-
ried out in two neighboring trawling areas, Süderfahrt and
Millionenviertel, in the northern part of Kiel Bay (cf. Arntz,
1978a).  This area is very uniform in depth (22-24 m) and sediment
type (muddy sand and sandy mud).
    Three samples were taken bimonthly at each of seven stations
by means of a $0.1m^2$ modified Van Veen grab (Dybern *et al.*, 1976).
The samples were washed on a 1 mm screen and preserved in 4% buffered formalin.  In the laboratory, they were sorted within a few
weeks using Pauly's (1973) design, and the molluscs then preserved
in 70% alcohol.  All weights presented in this paper are based on
preserved wet weight measurements.  For some considerations, the
biomass figures were converted to ash-free dry weight using
Brunswig's (1973) and my own unpublished conversion factors.
    The benthos data of all stations lumped together for each
sampling date (i.e., each point represents 1 $m^2$ [to which the 2.1 $m^2$
$\triangleq$ 21 grabs were reduced]).  The variability of the more common species
is low compared to the inner part of the Bay (cf. Arntz and Brunswig,

in press). For example, the standard deviation from the mean is less than 70% for the seven most common species, and probably reflects true changes in numbers and biomass.

2) The data on fish abundance are derived from 10 years of quantitative fishing at Süderfahrt and Millionenviertel and additional data from trawling areas in Kiel Bay (cf. Arntz, 1971). I or Prof. Thurow, from the Bundesforschungsanstalt für Fischerei, and his collaborators collected the fish samples using the research vessels "Hermann Wattenberg", "Alkor", "Littorina", "Solea" and, in a few cases, different fish cutters. Every two months two 1-h hauls were collected at each station; sometimes samples were taken throughout a 24-h period to measure diel changes. More than 1,000 hauls were carried out in the investigation period. An ICES 50 ft. standard trawl (mesh size in codend 12 mm) was used but in many cases, mainly after 1971, an 80 ft trawl (same mesh size) was used instead. For these latter catches, the fish numbers were reduced by a factor obtained from comparative hauls: cod 0.83, whiting, 0.67, plaice 0.59, flounder 0.37, dab unchanged (Arntz, unpublished.)

Day and night catches differed substantially in the area of investigation. Total number of fishes per 1-h haul with the 50 ft trawl averaged 206 ± 123 (std. dev.) throughout a 24-h sampling period at Süderfahrt, February 9-10, 1976. Numbers of total fish were highest (243-481 per trawl) between 1400 and 2200 h compared to 64-172 per trawl between 2300 and 1400 h. Cod was the major contributor to this diel change. The relative abundance of cod averaged 143 ± 94 (std. dev.). Abundance was highest (151-349) in the afternoon and early evening (1400-2200 h) compared to abundances of 42-105 from 2300-1400 h. Whiting also showed higher relative abundance (37 to 74) between 1400-1730 h compared to 2-18 during the rest of the sampling period. The changes of plaice, flounder and dab were less consistent. Most of the hauls were taken during daytime. Only in one area (Hohwachter Bucht) were most of the hauls taken at night.

Annual fluctuations of age classes of cod were determined from data of Thurow and ICES data of "virtual population analysis" from Subdivision 22 (whole Belt Sea).

Stomach (gut) content data were collected through one year for each of the more important species. With the exception of four-bearded rockling (*Onos cimbrius*) and long rough dab (*Hippoglossoides platessoides*), which were mainly sampled in another trawling area (Vejsnas Rinne), only data from Süderfahrt and Millionenviertel, where we had benthos data, were used to determine interactions in the benthic subsystem. Sample sizes were: juvenile cod, 730; adult cod, 1153; dab 1077; (all from 1968); flounder 327; plaice 390; whiting 1185; (all from 1971); long rough dab, 151; rockling, 65 (both "rare" species from years 1971-78). We were also able to compare these data to previously published information (Arntz, 1971b, 1974, 1978b, c); the data on plaice, whiting and some rarer species will be published soon (Arntz and Finger, in prep.).

In the case of gadoid fish, only stomach contents were studied; for flatfish, both stomachs and midguts were used. Each stomach (gut) was cut out on board ship, immediately after being caught, and preserved individually in 10% buffered formalin. The sorting into species and taxonomic groups was carried out later in the laboratory.

Ivlev's (1961) methods were used to study food selection and food coincidence (see Arntz, 1978a). Studying the correlation between benthos and fish data, the data series that correlated positively according to the Spearman rank correlation (Elliott, 1971) were tested for auto-correlation by means of a time series analysis (Pfanzagl, 1963, modified by Probst, 1973). Because the new series generated during the randomization process by permutation produced too low a number of new combinations, one of the requirements of this test was not quite fulfilled. For this reason, the results are only considered as a confirmation of the rank correlations.

3) Additional data were derived from a three-year experimental field study on dynamics and production of macrobenthos, carried out in the "Hausgarten" area of Kiel Bay at 20 m depths. Large (1.5 $m^2$ surface) "Eternit" containers were placed on the bottom in December 1976 and one container was removed every two months during the following three years. The total results of this experiment will be published shortly (Arntz and Rumohr, in prep.).


RESULTS


Predation, Food Selection, Food Coincidence, and Seasonality of Food Intake

Macrobenthos contribute a very large share of the food of demersal fish in the western Baltic (Fig. 1). Only for large cod and for whiting did fish provide a share comparable to that of macrobenthos. For the other demersal fish, the macrobenthos share (by weight) between 69% for long rough dab and 99% for flounder. The contribution of molluscs and fish increased while that of crustaceans decreased and, excluding whiting, the percentage of polychaetes remained fairly stable with increasing size class of the demersal fish (Figure 2 presents the data for whiting and plaice; for cod and flounder, c.f. Arntz 1974, 1978c). Zooplankton was not a significant food source for demersal fish > 10 cm in length.

There were differences in food preference by the demersal fish, but even so predation was heaviest on a very limited number of macrobenthic species (Table 1). The food species most heavily preyed upon was the cumacean *Diastylis rathkei*, followed by the bivalve *Cyprina islandica*, the polychaetes *Nephthys* spp., *Terebellides stroemi* and *Harmothoe sarsi*, and another bivalve *Abra alba*. If the weight percentage in the stomachs of whiting and the guts of plaice are compared to the respective weight percentage in the benthos (using Ivlev's (1961) "electivity index" $E = \frac{s - b}{s + b}$ , where s = percentage in the stomachs and b = percentage in the benthos), some food items, mainly small crustaceans, were highly selected by both fish species while others, mainly thick-shelled molluscs of the genus *Astarte* and brittle stars, were strongly rejected (Fig. 3). Polychaetes were selected by plaice and rejected by whiting; these two demersal fish have extremely different feeding habits. A much

more similar picture occurred between cod and dab. Data on food
selection by cod and dab were given by Arntz (1978a) who also discussed some shortcomings of Ivlev's method.

The strong selection of only a limited number of food species
leads to a high degree of food overlap among some species of co-
occurring demersal fish and also among different size groups within
species (Fig. 4). A similarity index (Shorygin's good coincidence
index) was used to illustrate intra- and interspecific food overlap
among size groups and species of co-occurring demersal fish (see
Arntz, 1978a). Index values range from zero (no overlap) to 100
percent (total food overlap). Generally, the intraspecific values
of food coincidence are higher than the interspecific ones, and the
values between neighboring size groups are higher than those between
size groups farther apart (Fig. 4a, b).

Feeding (both quantity and taxonomic composition) by demersal
fish is subject to a diel rhyrhm (see Arntz and Brunswig, 1975).
Fish and molluscs are fed upon mainly in winter and spring, while
crustaceans and polychaetes are fed upon mainly in summer and autumn
(Fig. 5; for dab, c.f. Arntz, 1971b).

These data indicate tight interrelations between macrobenthos
and demersal fish in the western Baltic; they also show that predation pressure is high both on a small number of food species, and at
certain times of the year.


Dynamics of Macrobenthic Food Species


Macrobenthos data from the Süderfahrt/Millionenviertel area
are available from 1968-1972 and 1975-1978, i.e., for eight complete
years. Tables 2 and 3 present annual means of abundance and biomass, and the eight-year average for total macrobenthos and some
of the more important species. Mean total abundance/m$^2$, as retained by a 1 mm screen, varied between 955 and 1801 and mean total
biomass between 30 g and 45 g ash-free dry weight (AFDW). The
large bivalves *Cyprina islandica* and *Astarte* (3 spp.) comprised 80-
90% (AFDW) of the total macrobenthic biomass. If these bivalves
are excluded, the biomass of the remaining macrobenthos varied between 3.99 g and 6.11 g AFDW. However, the relative share of some
of the more important fish food species and their dominance (by
weight) varied considerably (Fig. 6; for the absolute values cf.
Table 3). Because production (excluding *Cyprina* and *Astarte*) is
in most cases several times the biomass (Arntz, 1971a), the variations in production must have been even greater.

Three different types of fluctuation were observed: 1) some
"conservative" species and groups (e.g. *Astarte* and *Nephthys* spp.)
fluctuated very little; 2) some species (e.g. *Abra alba* and *Terebellides stroemi*), called "irregular seasonal species" (Arntz, in
press), exhibit very wide fluctuations; 3) "regular seasonal species" like *Diastylis rathkei* fluctuated less dramatically although
there were differences in the amplitude of the yearly peaks. The
most common pattern observed, most clearly shown by the brittle
star *Ophiura albida*, was lower densities between 1968 and 1972 and

higher densities from 1975 to 1978. Figure 7 presents in more detail changes in density and biomass of three dominant food species
that typify the three groups referred to above.

Thus there is no common pattern to temporal fluctuations of
the dominant fish food species in the area. In the northern part
of Kiel Bay abiotic factors do not play a destructive role as in
the inner parts of the Bay, where anoxia leads to more or less regular depletion of the fauna and subsequent recolonization (Arntz,
in press). In this area, therefore, the variability of the benthic
fauna might be caused, at least to a certain extent, by fluctuations in the numbers of fish predators. Even if benthic fluctuations should be caused by factors other than predation by fish, the
varying amount and quality of fish food in the different years
should have an impact on the size or production of the demersal
fish stocks in the area.


Variations in the Demersal Fish Stocks

Annual mean abundances of demersal fish for eight years (1968-
1971 and 1975-1978) at the Süderfahrt/Millionenviertel area for the
whole time of investigation show that cod was the overall dominant
species, followed by dab and whiting (Table 4). The cod's dominance by weight was much higher because cod individuals are heavier
than those dab and whiting of comparable length. Demersal fish
fluctuations from the overall mean were less pronounced than those
exhibited by benthos, and do not show the cycling that characterized the benthos. Except for whiting, changes in the abundances of
the demersal fish followed that of the benthos, with low densities
up to 1972 and higher densities during 1975 to 1978.

Additional data of Thurow (unpubl.) on cod age read from otoliths (Fig. 8), even though from the whole of Kiel Bay, show that
there were large differences in the annual mean numbers of 0 to III
group cod during this period. The means do not contain hauls from
the spawning period (January-April) when a larger percentage of cod
does not feed and when mass concentrations in some trawling areas
occur. This could have biased the mean values. Also, in the case
of age group 0, the annual average only contains specimens from
July onwards because the trawl only occasionally catches small specimens of the new year-class during the first seven months of the year.
Anyway, the data on age group 0 are less reliable than those of the
other age groups, as can be seen in several cases from the numbers
of age group I in the subsequent year, e.g., in 1978. This is a
problem for much of the available fisheries statistics. Only from
its third year do the abundance data of a year-class become reliable
because they are based on commercial catches. Total landings by
weight in subdivision 22 remained comparatively stable during 1968
to 1977. The catches were around 30,000 metric tons (28,212 to 33,
289 with the exception of a high of 38,233 in 1973). Unfortunately,
we have no macrobenthic data for 1973.

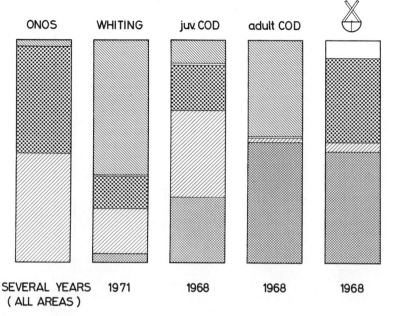

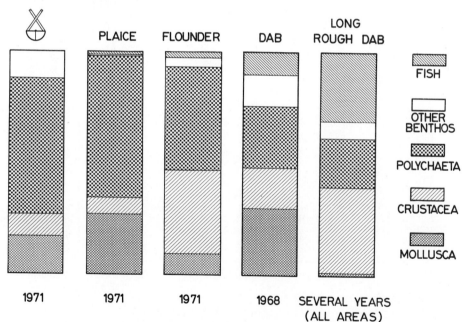

Figure 1.   Food composition in the macrobenthos and in the stomachs (guts) of demersal fish from Süderfahrt/Millionen-viertel trawling area with the exception of data for four-bearded rockling and long rough dab from Whole Kiel Bay.  Note: *Cyprina islandica* is included in the stomach contents but not contained in the composition of benthos.

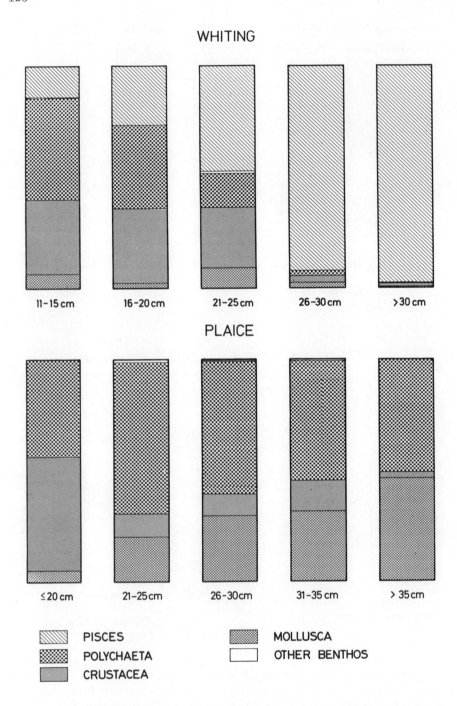

Figure 2. Changes in the diet with increasing size
of whiting and plaice from Süderfahrt/Millionenviertel
trawling area in 1971.

Table 1. Percent share by weight (annual mean) of the most important food species in cod and dab (1968), in whiting, plaice and flounder (1971), and in long rough dab and four-bearded rockling (data from 1971-1978) from Süderfahrt/Millionenviertel trawling area (except long rough dab and rockling from whole of Kiel Bay). + = < 1%.

| Food species | Juv. cod | Adult cod | Dab | Flounder | Whiting | Plaice | Rockling | Long rough dab |
|---|---|---|---|---|---|---|---|---|
| Abra alba | + | 1.1 | 14.8 | 7.4 | + | 19.6 | - | + |
| Cyprina islandica | 27.3 | 52.9 | 14.1 | - | 3.5 | 3.6 | - | - |
| Diastylis rathkei | 31.0 | 1.5 | 17.2 | 37.4 | 16.1 | 8.1 | 40.2 | 32.3 |
| Gastrosaccus spinifer | 3.3 | + | 1.6 | + | 1.0 | + | 2.7 | + |
| Harmothoe sarsi | 8.2 | + | 3.8 | 1.0 | + | + | 1.5 | 2.4 |
| Nephthys spp. | 7.4 | 1.0 | 18.9 | 14.0 | 11.8 | 46.5 | 43.0 | 12.3 |
| Terebellides stroemi | + | - | + | 30.5 | + | 11.0 | + | + |
| Pherusa plumosa | 2.3 | + | + | + | + | 1.7 | - | - |
| Pectinaria koreni | + | + | 1.5 | 1.6 | + | 2.4 | + | + |
| Ophiura albida | + | - | 10.9 | + | + | - | - | 9.5 |
| Pomatoschistus minutus | 10.5 | 1.1 | 1.4 | 1.0 | + | 1.2 | + | 26.9 |
| Clupeidae | + | 17.6 | + | - | 58.9 | - | - | - |
| Gadidae | - | 24.6 | - | - | - | - | - | - |

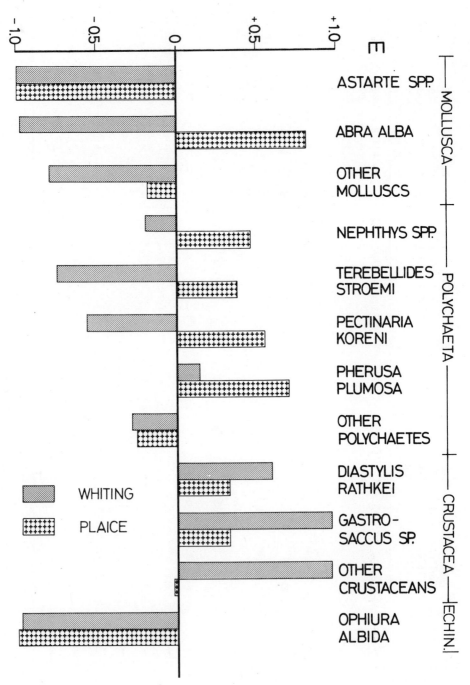

Figure 3.   Food selection (Ivlev's electivity index E) by whiting and plaice from Süderfahrt/Millionenviertel trawling area, 1971.

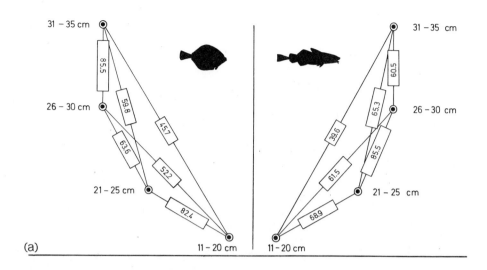

(a)

Figure 4.   Intraspecific (a) and interspecific (b) food overlap (Shorygin's food coincidence index) in cod and dab from Süder-
fahrt/Millionenviertel trawling area (annual mean 1968; redrawn
from Arntz, 1978a).   See text for explanation.

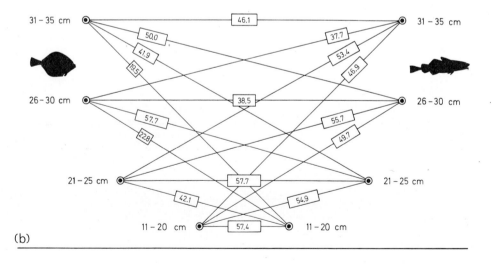

(b)

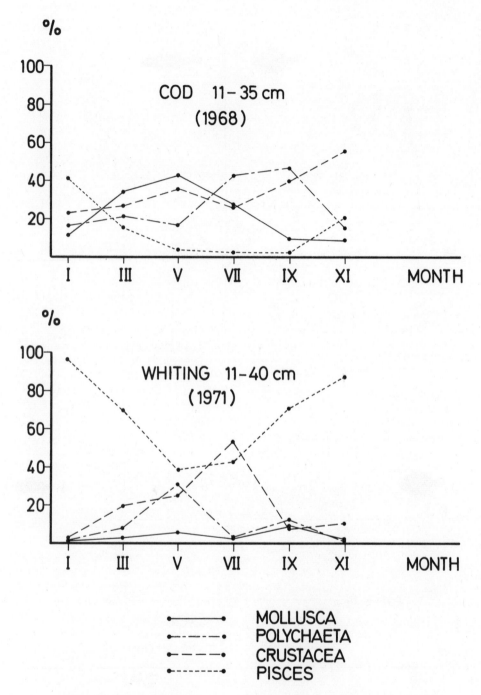

Figure 5.  Seasonal food intake by cod (1968, above) and whiting (1971, below) from whole Kiel Bay.

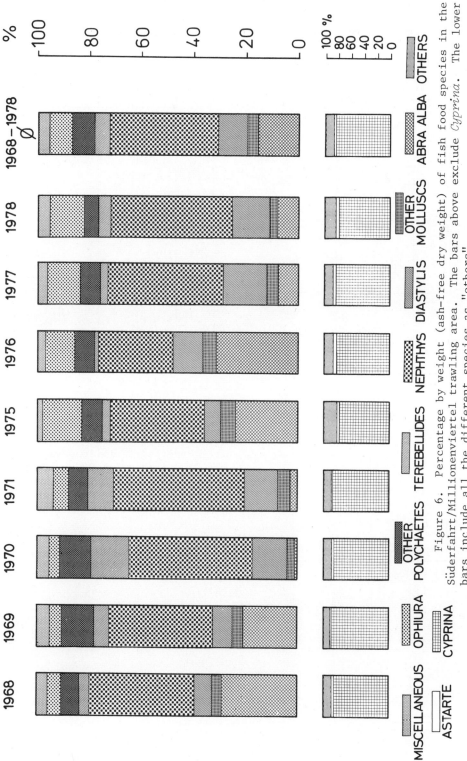

Figure 6. Percentage by weight (ash-free dry weight) of fish food species in the Süderfahrt/Millionenviertel trawling area. The bars above exclude *Cyprina*. The lower bars include all the different species as "others".

**TEREBELLIDES STROEMI**

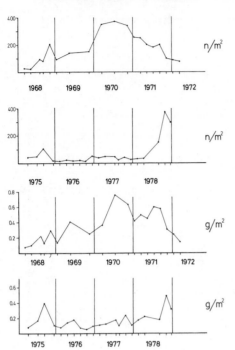

**DIASTYLIS RATHKEI**

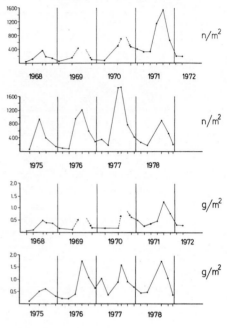

**NEPHTHYS** SPP.

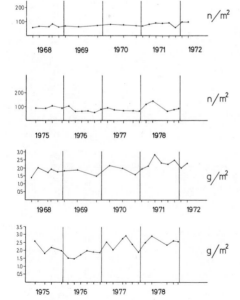

Figure 7. Changes in abundance (density per $m^2$) of three macrobenthic fish food species from Süderfahrt/Millionenviertel trawling area. In 1969 and 1970, sampling failed to catch the summer maxima of *Diastylis rathkei*.

Table 2. Abundance of macrobenthos in the Süderfahrt/Millionenviertel trawling area. Annual means for the years 1968 to 1971 and 1975 to 1978 and mean for the investigation period. All values per m².

| | 1968 | 1969 | 1970 | 1971 | 1975 | 1976 | 1977 | 1978 | Ø 1968 – 1978 and s |
|---|---|---|---|---|---|---|---|---|---|
| *Abra alba* | 229 | 124 | 14 | 40 | 324 | 472 | 129 | 86 | 177 ± 156 |
| *Cyprina islandica* juv. | 32 | 44 | 28 | 27 | 29 | 44 | 44 | 37 | 36 ± 8 |
| *Cyprina islandica* > 15 mm | 10 | 19 | 8 | 5 | 10 | 18 | 18 | 15 | 13 ± 5 |
| *Astarte* spp. | 131 | 97 | 98 | 87 | 124 | 121 | 99 | 125 | 110 ± 17 |
| Other molluscs | 13 | 32 | 28 | 47 | 69 | 98 | 63 | 41 | 49 ± 27 |
| *Diastylis rathkei* | 169 | 228 | 493 | 707 | 383 | 543 | 728 | 466 | 465 ± 202 |
| *Gastrosaccus spinifer* | 3 | 5 | 1 | 3 | 4 | 3 | 2 | 2 | 3 ± 1 |
| Other crustaceans | 5 | 31 | 50 | 47 | 34 | 21 | 23 | 51 | 33 ± 16 |
| *Nephthys* spp. | 67 | 70 | 78 | 86 | 93 | 74 | 77 | 96 | 80 ± 11 |
| *Terebellides stroemi* | 86 | 141 | 349 | 197 | 55 | 25 | 39 | 180 | 134 ± 108 |
| *Pherusa plumosa* | 3 | 4 | 2 | 4 | 3 | 2 | 4 | 3 | 3 ± 1 |
| *Pectinaria koreni* | 13 | 37 | 31 | 12 | 48 | 66 | 40 | 20 | 28 ± 23 |
| Other polychaetes | 120 | 350 | 480 | 372 | 282 | 177 | 125 | 116 | 253 ± 139 |
| *Ophiura albida* | 68 | 47 | 29 | 59 | 180 | 151 | 182 | 169 | 98 ± 62 |
| Other fauna+ | 16 | 13 | 14 | 27 | 12 | 4 | 3 | 3 | 10 ± 9 |
| Σ | 955 | 1223 | 1695 | 1715 | 1640 | 1801 | 1558 | 1395 | 1498 ± 288 |

+excl. ascidians

Table 3.  Biomass of macrobenthos in the Süderfahrt/Millionen-
viertel trawling area.  Values are annual means for the years 1968 to
1971 and 1975 to 1978 given as wet weight including molluscan shells
and conversion to ash-free dry weight (AFDW) per m$^2$.

|  | 1968 | 1969 | 1970 | 1971 | 1975 | 1976 |
|---|---|---|---|---|---|---|
| *Abra alba* | 20.7 | 14.8 | 0.7 | 2.1 | 23.9 | 31.9 |
| *Cyprina islandica* | 792. | 984. | 711. | 978. | 580. | 798. |
| *Astarte* spp. | 73. | 41. | 67. | 60. | 55. | 70. |
| Other molluscs | 4.1 | 4.4 | 2.7 | 5.4 | 8.2 | 8.1 |
| *Diastylis rathkei* | 2.1 | 2.3 | 3.8 | 4.3 | 2.7 | 5.1 |
| *Gastrosaccus spinifer* | 0.03 | 0.04 | 0.01 | 0.02 | 0.02 | 0.02 |
| Other crustaceans | 0.00 | 0.07 | 0.11 | 0.07 | 0.08 | 0.11 |
| *Nephthys* spp. | 13.5 | 13.3 | 14.6 | 18.0 | 16.6 | 13.5 |
| *Terebellides stroemi* | 1.8 | 2.9 | 6.5 | 5.2 | 2.1 | 1.1 |
| *Pherusa plumosa* | 0.4 | 0.3 | 0.2 | 0.3 | 0.2 | 0.5 |
| *Pectinaria koreni* | 0.4 | 2.0 | 1.3 | 0.7 | 1.5 | 3.0 |
| Other polychaetes | 2.8 | 4.3 | 4.3 | 3.1 | 4.2 | 2.7 |
| *Ophiura albida* | 2.2 | 2.2 | 1.8 | 3.0 | 9.9 | 7.5 |
| Other fauna (excl. ascidians) | 1.8 | 1.9 | 1.5 | 2.5 | 0.9 | 1.4 |
| Σ wet weight | 914.8 | 1073.5 | 816.5 | 1082.7 | 705.3 | 942.9 |
| Σ AFDW | 37.44 | 44.50 | 33.77 | 44.96 | 30.27 | 39.40 |
| Σ excl. *Cyprina* wet weight | 122.8 | 89.5 | 105.5 | 104.7 | 125.3 | 144.9 |
| Σ excl. *Cyprina* AFDW | 5.76 | 5.14 | 5.33 | 5.84 | 7.07 | 7.51 |
| Σ excl. *Cyprina* and *Astarte* wet weight | 49.8 | 48.5 | 37.5 | 44.7 | 70.3 | 74.9 |
| Σ excl. *Cyprina* and *Astarte*, AFDW | 4.30 | 4.32 | 3.99 | 4.64 | 5.97 | 6.11 |

Source of conversion factors:  [+]Arntz, unpubl., [++]Brunswig, 1973,
[+++]Estimate

Table 3 (continued)

| 1977 | 1978 | Ø 1968–1978 ± 1 std dv (wet weight) | Conversion factor to AFDW | Converted long-term means AFDW |
|---|---|---|---|---|
| 7.0 | 7.8 | 13.6 ± 11.1 | 0.06[+] | 0.82 |
| 810. | 635. | 786. ± 145. | 0.04[+] | 31.44 |
| 80. | 84. | 66. ± 14. | 0.02[+] | 1.32 |
| 6.1 | 4.3 | 5.4 ± 2.0 | 0.04[+++] | 0.22 |
| 6.4 | 5.8 | 4.1 ± 1.63 | 0.14[++] | 0.57 |
| 0.02 | 0.02 | 0.02 ± 0.01 | 0.18[++] | <0.01 |
| 0.16 | 0.27 | 0.11 ± 0.08 | 0.16[+++] | 0.02 |
| 18.7 | 20.6 | 16.1 ± 2.8 | 0.13[++] | 2.09 |
| 1.6 | 3.0 | 3.0 ± 1.9 | 0.09[++] | 0.27 |
| 0.8 | 0.6 | 0.4 ± 0.2 | 0.07[++] | 0.03 |
| 2.0 | 0.6 | 1.4 ± 0.9 | 0.06[++] | 0.08 |
| 2.5 | 2.6 | 3.3 ± 0.8 | 0.09[+++] | 0.30 |
| 7.6 | 8.4 | 5.3 ± 3.3 | 0.09[++] | 0.48 |
| 1.4 | 1.8 | 1.7 ± 0.5 | 0.10[+++] | 0.17 |
| | | | | |
| 944.3 | 744.8 | 906.4 ± 135.1 | | |
| 39.40 | 32.78 | | | 37.82 |
| 134.3 | 139.8 | 120.4 ± 19.4 | | |
| 7.00 | 7.38 | | | 6.38 |
| 54.3 | 55.8 | 54.4 ± 12.6 | | |
| 5.40 | 5.70 | | | 5.05 |

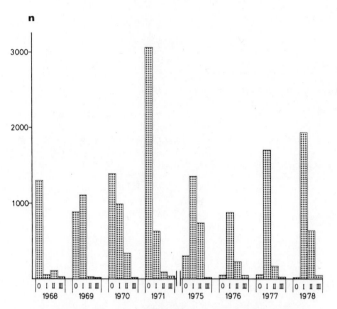

Figure 8.   Abundance of age groups 0–111
of cod given as number per 10 h trawl (50 ft.
standard trawl), from whole Kiel Bay.  Compiled
from Thurow's (unpublished) data; ages read from
otoliths.

Table 4.   Annual mean values (nos/h trawl with a 50 ft standard trawl)
of the most common demersal fish at the Süderfahrt/Millionenviertel area.

| Year | No. of hours trawled | Cod | Whiting | Dab | Flounder + Plaice | |
|---|---|---|---|---|---|---|
| 1968 | 38 | 140 | 35 | 22 | 8 | 205 |
| 1969 | 26 | 151 | 43 | 34 | 7 | 235 |
| 1970 | 34 | 183 | 28 | 90 | 9 | 310 |
| 1971 | 50 | 179 | 46 | 72 | 6 | 303 |
| 1975 | 34 | 210 | 27 | 71 | 14 | 322 |
| 1976 | 36 | 184 | 29 | 70 | 13 | 296 |
| 1977 | 30 | 273 | 43 | 87 | 13 | 416 |
| 1978 | 24 | 250 | 48 | 191 | 8 | 497 |
| Mean all years | | 196 ± 46 | 37 ± 7 | 80 ± 51 | 10 ± 3 | 323 ± 94 |

Comparison of Benthos and Fish Data

Dynamic interactions between benthos and demersal fish could
occur in various ways.  First, for a given year, a high recruitment
of fish could produce an abundant year class and heavier predation.
This could reduce the standing density of the prey populations.
Second, relating fish of age group I to the benthos of the preceding
year, the survival and growth of a certain year class of fish might
be better if there is more benthic food available.  Third, relating
the fish to the benthos of the subsequent year, increased year
classes of demersal fish might impact the benthos in the following
years, e.g., by decimating the population size to a level that limits recruitment and recovery.  A first comparison of the fluctuations in abundance and biomass of macrobenthos with the fluctuations in fish abundance (Tables 2-4) gives the impression that
there might be a common pattern for a number of fish predators and
benthic prey:  lower standing crops up to 1971, and higher standing
crops between 1975 and 1978.  There are exceptions, like *Terebellides stroemi*, and some of the smaller crustaceans and polychaetes.
For whiting, the picture is that of a continuous change between
high and low years.

However, there was an almost complete lack of statistical correlation between the annual mean values of benthos density and biomass (including separate comparisons for *Diastylis*) and the abundance of demersal fish (Table 5).  Of all possible comparisons (62)
six of the seven significant correlations were interrelations between cod and dab and their most important food, *Diastylis rathkei*.
These were positive correlations that suggest a common response to
other factors (e.g., favorable hydrographic conditions and a rich
plankton production might lead to more *Diastylis* and more cod in the
same year) or attraction (if there are more *Diastylis*, more cod
gather in the area to feed on them).  There is definitely no tendency towards the kind of counterbalance (good fish years ≙ weak
benthos years and vice versa) we might expect if food were limited.

Quite unexpectedly, there was hardly any significant correlation at all between the numbers of individual age groups of cod in
different years and the density and biomass of *Diastylis rathkei* in
the same, previous, or subsequent years.  It is mainly age groups
0 to II of cod that feed on this species.  The only significant
Spearman correlation (cod AG I: *Diastylis* biomass) turned out not
to be parallel.

To determine if cod production varied among years, we plotted
Thurow's unpublished growth data of cod using Pauly and Gaschütz's
(in press) seasonal growth computation formula (Fig. 9).  We also
plotted the original monthly values of each age group for the different years.  A single sample test for randomness (Siegel, 1956) indicated that in 1973 and 1974 cod mean growth deviated from randomness.
Unfortunately, these are just the years where we have no benthos data.
However, the slight deviations in 1973 and 1974 hardly provide an
argument for better growth of cod in years when their density is low

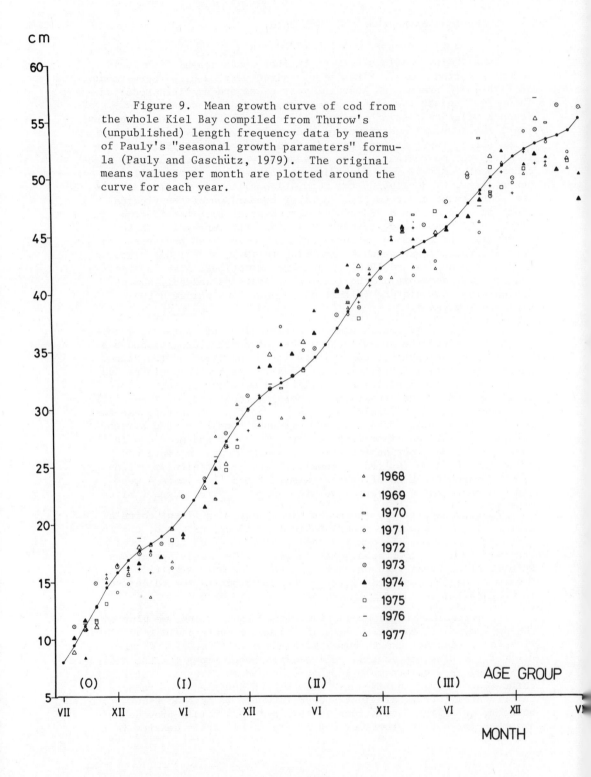

Figure 9. Mean growth curve of cod from the whole Kiel Bay compiled from Thurow's (unpublished) length frequency data by means of Pauly's "seasonal growth parameters" formula (Pauly and Gaschütz, 1979). The original means values per month are plotted around the curve for each year.

Table 5. Correlation (Spearman's rank correlation coefficient, $r_s$) between the annual means of abundance of demersal fish in the trawl and the abundance and biomass of macrobenthos. Significant correlations are marked with a (+): those correlations which proved to be parallel according to Probst's (1973) time series analysis are underlined.

| | Macrobenthic Abundance | Macrobenthic Biomass | Diastylis A | Diastylis B |
|---|---|---|---|---|
| Σ Fish KB | 0.167 | -0.048 | | |
| Σ Fish S/M | 0.595 | 0.381 | | |
| Σ Fish KB Previous year | -0.214 | 0.179 | | |
| Σ Fish KB Subsequent year | 0.771 | 0.086 | | |
| Σ Fish S/M Previous year | 0.571 | 0.679 | | |
| Σ Fish S/M Subsequent year | 0.371 | 0.829(+) | | |
| Cod KB | | | 0.738(+) | 0.405 |
| Cod S/M | | | 0.381 | 0.929(+) |
| Dab KB | | | 0.810(+) | 0.833(+) |
| Dab S/M | | | 0.833(+) | 0.619 |
| Dab S/M Previous year | | | 0.750(+) | 0.464 |
| Dab S/M Subsequent year | | | -0.086 | 0.771 |
| Cod KB Previous year | | | 0.607 | 0.179 |
| Cod KB Subsequent year | | | 0.771 | 0.200 |
| Cod S/M Previous year | | | 0.536 | 0.679 |
| Cod S/M Subsequent year | | | 0.429 | 0.829 |
| Cod AG 0-III KB ++ | | | 0.000 | 0.107 |
| Cod AG 0-III KB ++ Previous year | | | 0.214 | -0.143 |
| Cod AG 0-III KB ++ Subsequent year | | | -0.714 | 0.371 |
| Cod AG I KB ++ | | | 0.500 | 0.857(+) |
| Cod AG I KB ++ Previous year | | | 0.071 | 0.429 |
| Cod AG I KB ++ Subsequent year | | | 0.771 | 0.486 |
| Cod AG 0 KB ++ | | | -0.357 | -0.619 |
| Cod AG 0 KB ++ Subsequent year | | | 0.086 | -0.886 |
| Cod AG I +++ | | | -0.300 | -0.900 |
| Cod AG I +++ Previous year | | | -0.300 | 0.100 |
| Cod AG I Subsequent year | | | -0.800 | 0.400 |
| Cod AG II +++ | | | -0.700 | -0.500 |
| Cod AG II +++ Previous year | | | -0.500 | -0.700 |
| Cod AG II +++ 2 years before | | | -0.800 | -0.800 |

Note: All correlations given in this Table refer to the data given above (Tables 2 to 4). Thurow's data are marked with ++, ICES (Subdivision 22) data with +++. KB = Kiel Bay, all trawling areas; S/M = Süderfahrt/Millionenviertel. For the significance limits of Spearman's correlation cf. Elliott (1971), p. 121.

(in fact, only the density of the age class I was low in 1973; see Thurow, 1974). Also, there is no hint that growth of cod was better in years when there was a high benthic abundance.

Changes in the abundance of demersal fish and benthos in the western Baltic, despite very close predator-prey interactions, are not related. Periodic changes in recruitment and growth and subsequent production of demersal fish seem to be independent of fluctuations in their main food resource. Long-term fluctuations in the benthos do not seem to be related to changes in predation pressure. Although this outcome is quite consistent with the theory that abiotic parameters are most important in physically controlled environments, it is somewhat problematic in light of our knowledge edge about the energy flow in temperate marine ecosystems from caging experiments.

## DISCUSSION

There is no doubt that the much higher mortality of benthic species like *Diastylis rathkei* that are preyed upon by demersal fish was mainly caused by fish predation. Conversely, populations of non-prey species remained virtually unchanged (e.g., the deep-living clam *Mya truncata*; Arntz and Rumohr, in prep.). Predation is size-class dependent. For example, heavy predation on *Diastylis* started at an early point of population development when the numbers increased rapidly and continued to do so in spite of largely reduced numbers, until late autumn. The predation probably continued because the individual weight of the prey still increased and thus biomass values remained high. An 80% mortality of *Diastylis* between June and December was not unusual. This was shown by our field experiments at 20 m depth; for this species, mortality due to senescence occurs only after one year (Zimmer, 1933) (Fig. 10).

The best way to document the effect of predation on benthos is the use of predator exclusion experiments. Blegvad (1928) carried out the first experiment of this type in shallow water of the Limfjord. He reported a 60-fold increase in standing crop under the cages after one summer. In a similar study in the German Wadden Sea, Reise (1977) found an infaunal biomass four times higher under cage shelters (1 mm mesh) from March-June, and a 23-fold increase from July-October. In Reise's study, large-meshed cages (20 mm mesh) had no effect similar to cages (30 mm mesh) placed on the bottom of Kiel Bay by Arntz (1977) because small predators entered the exclosures. However, in another experiment, 20 mm mesh led to a significantly higher survival of large infauna (Reise, 1978). The importance of small predators, mainly decapods, is also stressed by Virnstein (1978). In the Chesapeake Bay, the densities of all infaunal species increased under exclosures (12 mm mesh) while predators (the decapod *Callinectes* and the sciaenid fish *Leiostomus*) in enclosure cages reduced infaunal density (Virnstein, 1977). Andersen and Ursin (1977), in their model of the North Sea, confirmed that small fish and invertebrate predators can be just as important as large fish.

There is no doubt that the shape of seasonal population curves of many macrobenthic species is influenced by predation and that predation pressure keeps many population levels below the carrying capacity of the environment (Virnstein, 1977). On the other hand, our comparison showed that, unexpectedly, benthos in our area were not so limited as to allow for a measurable impact on the demersal fish populations in Rice Bay. Predation by these fish might not have been heavy enough to account for the observed fluctuations of benthos. Perhaps in our area predation pressure does not primarily determine benthic dynamics nor is fish production mainly determined by standing prey (i.e., food availability).

Energy budgets published so far on marine and brackish water ecosystems of temperate latitudes do not indicate that the fish feeding on the benthos are inefficient (Steele, 1974; Arntz and Brunswig, 1975; Ankar, 1977; Wolff and de Wolf, 1977; Rosenberg *et al.*, 1977; Arntz, 1978a). Mills (1975), referring to Steele's (1974) model of the North Sea, points out that invertebrate carnivores (shrimp, crabs, starfish) must also utilize some of the benthic production; with this consideration, there might not be enough food resources left to cover the needs of the demersal fish.

We can roughly check the magnitude of the fish-benthos relation in the northern Kiel Bay, based on the values presented in Tables 1-3. Assuming a 50% trawl catch efficiency and a much lower efficiency for juvenile cod and gobies, we estimate density at one fish per 30 $m^2$. If an average fish weighs 150 g and has a food intake of 3% of its body weight throughout the year (see Arntz, 1974), then the demersal fish would consume 55 $g/m^2$ of which ~ 30 g (2.8 g AFDW) are "normal benthos" (i.e., excl. *Cyprina, Astarte* and fish), 12.5 $g/m^2$ (2.4 g AFDW) are *Cyprina* soft parts, and 12.5 g (2.2 g AFDW) are fish.

This figure has yet to be compared with the production of macrobenthos. We are only gradually getting more reliable production data for the most important species from our field experiments. In 1977, the annual production (measured by Crisp's 1971 method) of *Diastylis rathkei* in the large "Benthosgarten" containers amounted to 7.5 $g/m^2$ (1.05 g AFDW); in 1978, it was 48.9 $g/m^2$ (6.85 g AFDW). The annual mean biomass values were 0.31 and 1.54 g AFDW respectively. Thus, the P/B ratio of *Diastylis* was 3.4 and 4.5 in the two investigation years.

The annual mean biomass of the "normal benthos" (again without *Cyprina* and *Astarte*) referred to above is 54.4 g (5.05 g AFDW). Assuming production of this group (composed mainly of short-lived polychaetes, crustaceans, and *Abra alba*) is three times the biomass, there would be an annual production of 160 $g/m^2$ (~ 15 g AFDW), of which 20% would be consumed by the commercial fish. So, although demersal fish biomass in the western Baltic is low due to fishing pressure, the fish seem to consume a considerable part of benthic production. *Cyprina islandica*, which has a low $P/\bar{B}$, is not included in these calculations because we do not yet have any reliable production estimates for this species. Preliminary data indicate that for *Cyprina* total mortality $\underline{\Lambda}$ production because it maintains a stable standing stock over long periods of time. With a $P/\bar{B} = 3$, which might be overestimated if our estimates of macro-

benthic production are not too low, we would expect a tighter in-
terrelation between macrobenthos and demersal fish than indicated
by the results of this study.  However, small fish predators which
we might not have effectively sampled, and invertebrate predators
(mainly the starfish *Asterias rubens*, in shallow water, the shorecrab
*Carcinus maenas*, and the shrimp *Crangon crangon*) as a whole might be
much more important than, and thus mask, any effect of the commercial
fish.  This would be in contrast to what has been assumed so far for
the *Abra alba* community where *Carcinus* is uncommon (Arntz, 1977) and
where *Asterias* did not seem to play a great part (Arntz, 1971a and
data in Tables 1 and 2).  This has, however, recently been doubted
by Anger *et al.* (1977) and by Nauen (1978) whose findings indicate
that we may have underestimated the role of *Asterias*.  *Asterias*
might exploit the benthic food resource when fish are scarce because
they can remain in a "waiting stage" for a long time under suboptimal
food conditions (Nauen, 1978).  However, Nauen might overestimate
consumption by *Asterias* in the *Abra alba* community.  Nauen estimated
*Asterias* annual consumption between 9 and 52 g wet wt/m$^2$.  If this
is added to the consumption by fish, and if predation by small carnivores is also included (*Harmothoe sarsi*, *Halicryptus spinulosus*),
there seems to be maximum exploitation of macrobenthic secondary
production of the western Baltic.

However, we cannot exclude the possibility that even abiotic
factors considerably influence the dynamics of macrobenthos.  Abiotic
factors are important in the western Baltic (Arntz, *et al.*, 1976).  As
far as the northern part of Kiel Bay is concerned, the main factor
is an occasionally very strong inflow of more saline water from the
northern Baltic Sea.  In fact, there were quite a number of very strong
inflows in the fall of 1975 and the spring of 1976 that might explain
the high mean biomass, especially of *Abra Alba* and *Ophiura albida*,
observed for these years.  As far as the influence of fish predation
on benthos is concerned, despite considerable predation and the fact
that production in our benthic system does not appear wasteful, there
is not noticeable impact by fish beyond the annual cropping effect
that keeps the biomass values relatively low.

The causes for fluctuations in year classes in temperate demersal fish populations have been widely discussed because they are of
great importance for commercial fisheries.  In spite of apparent
fluctuations in some of the more important fisheries (e.g., in the
North Sea , cf. Hempel, 1978), the reason is obscure.  With pelagic
fish, it is generally accepted that the key to strong year classes
must be in the survival of the larvae (Daan, 1978; Cushing, 1975;
Hempel, 1977), and this may hold true for demersal fish as well
(Jones, 1978).  But it is unclear whether the critical factor(s)
is larval "match or mismatch" with their appropriate food (Cushing,
1975) or critical food density as proposed by Jones (1973) for gadoid larvae, or (additional?) predation on the eggs and larvae by
pelagic fish as suspected by Andersen and Ursin (1977) and Ursin
(1979).  Cushing's match or mismatch theory stresses the role of
the abiotic impact.  The larval food consists of phyto- and zooplankton; in temperate latitudes the growth of these food organisms
are triggered by abiotic environmental conditions.  A nice example
of Cushing's theory has been given by Thurow (1974) who showed that

despite a normal spawning population and regular spawning in March,
the 1972 year-class of cod failed nearly completely, apparently due
to shortage of food in the critical period (Fig. 11; the decline of
food is indicated by an arrow). Possibly the decisive events that
define the year-class strength of a fish population happen during
the larval stage, when these fish do not yet feed on benthos. It
might be that differences in food availability no longer play an
important part by the time the fish have switched over to benthic
food. This would explain why differences in benthos had apparently
no impact on the number of fish in our area. Still, benthos could
have had an influence on fish growth, but only if there was competition for food. Apparently, despite a high degree of food overlap
between the demersal fish, there is no severe competition in the
Western Baltic.

The results of this study do not solve the problem raised,
e.g., by Gray (1974, 1977), to what extent biological interactions
are an important factor in temperate benthic communities. Perhaps
this first approach, using annual mean values, was too simple. At
least this paper shows that the interrelations are anything but simple, even in a limited subsystem with a comparatively poor fauna.
It also gives an idea of how important our experiments in the "Benthosgarten" area of Kiel Bay are for good estimates of trophic
transfer. As Reise (1978) has pointed out, experiments manipulating
the natural environment are the only promising way to solve these
problems. On the other hand, combined fish and benthos data from
eight years sampling are rare and seemed worth a comparison.

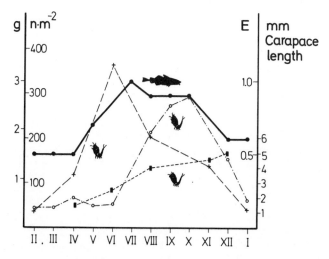

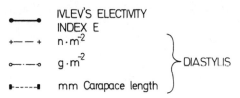

Figure 10. Changes
in density, biomass, and
size of *Diastylis rathkei*
from the Süderfahrt/Milli-
onenviertel area in 1968
and their size selection
by small (21 to 25 cm) cod.

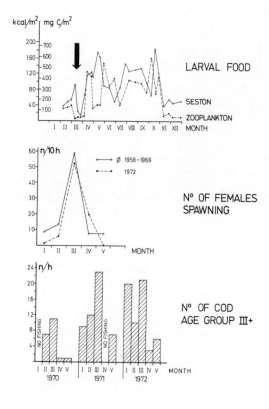

Figure 11. "Mismatch of cod larvae and larval food measured as seston and zooplankton, in spring 1972. Top shows food available (water column below 1 $m^2$); middle shows number of spawning females per 10 h trawls with data of 1972 compared to mean of 1958-1969; bottom number is that of mature cod per 1 h trawl between January and May in the years 1970-72. All data are from Thurow (1974).

ACKNOWLEDGEMENTS

I would like to give my sincere thanks to a number of colleagues who have contributed a great deal to this publication: to Prof. F. Thurow for letting me use his unpublished cod data; to Dr. D. Pauly and Dr. B. Probst for their help in statistics; to I. Finger and P. Prutz for working up a very large number of samples and their untiring assistance in completing the manuscript; and to the crews of the research vessels who always did their best to meet my hunger for samples. My thanks are also due to a number of reviewers (Drs. D. Pauly, H. Sanders, F. Thurow, E. Ursin and R. Virnstein) who eliminated the inconsistencies in the first draft of the manuscript.

REFERENCES

Andersen, K. P. and E. Ursin. 1977. A multispecies extension to the Beverton and Holt theory of fishing with accounts of phosphorus circulation and primary production. Medd. Danm. Fiskeri-og Havundersøgelser. 7: 319-435.

Anger, K., U. Rogal, G. Schriever, and C. Valentin. 1977. *In situ* investigations on the echinoderm *Asterias rubens* as a predator of soft-bottom communities in the western Baltic Sea. Helgol. wiss. Meeresunters. 29: 439-459.

Ankar, S. 1977. The soft bottom ecosystem of the northern Baltic proper with special reference to the macrofauna. Contrib. Askö Lab. Univ. Stockholm. 19: 1-62.

Arntz, W. 1971a. Biomasse und Produktion des Makrobenthos in den tieferen Teilen der Kieler Bucht im Jahr 1968. Kieler Meeresforsch. 27: 36-72.

Arntz, W. 1971b. Die Nahrung der Kliesche (*Limanda limanda* (L.)) in der Kieler Bucht. Ber. dt. wiss. KommnMeeresforsch. 22: 129-183.

Arntz, W. 1974. Die Nahrung juveniler Dorsche (*Gadus morhua* L.) in der Kieler Bucht. Ber. dt. wiss. KommnMeeresforsch. 23: 97-120.

Arntz, W. 1977. Results and problems of an "unsuccessful" benthos cage predation experiment (western Baltic), pp. 31-44. *In* B. F. Keegan, P. O. Céidigh, and P. J. S. Boaden (eds.), Biology of Benthic Organisms. Pergamon Press, New York.

Arntz, W. 1978a. The "upper part" of the benthic food web: The role of macrobenthos in the western Baltic. Rapp. P.-v.Réun. Cons. int. Explor. Mer. 173: 85-100.

Arntz, W. 1978b. The food of adult cod (*Gadus morhua* L.) in the western Baltic. Meeresforsch. 26: 60-69.

Arntz, W. 1978c. Predation on benthos by flounders, *Platichthys flesus* (L.) in the deeper parts of Kiel Bay. Meeresforsch. 26: 70-78.

Arntz, W. In press. Zonation and dynamics of macrobenthos biomass in an area stressed by oxygen deficiency. Stress Ecology (ed. G. Barrett and R. Rosenberg), Environm. Cons.

Arntz, W. and D. Brunswig. 1975. Studies on structure and dynamics of macrobenthos in the western Baltic carried out by the joint research programme "Interaction sea-sea bottom" (SFB 95 - Kiel). 10th European Symp. on Marine Biology, Ostend, Belgium. 2: 17-42.

Arntz, W. and D. Brunswig. In press. Zonation of macrobenthos in the Kiel Bay channel system and its implications for demersal fish. 4th Symp. Baltic Marine Biologists, Gdańsk, Oct. 1975.

Arntz, W., D. Brunswig, and M. Sarnthein. 1976. Zonierung von Mollusken und Schill im Rinnensystem der Kieler Bucht (westliche Ostsee). Senckenbergiana marit. 8: 189-269.

Blegvad, H. 1928. Quantitative investigations of bottom invertebrates in the Limfjord 1910-1927 with special reference to the plaice food. Rep. Danish biol. Sta. 34: 33-52.

Brunswig, D. 1973. Der Nahrungswert makrobenthischer Fischnährtiere in der Kieler Bucht. Dipl. thesis Univ. Kiel: 1-65.

Crisp, D. J. 1971. Energy flow measurements. Pages 197-279. *In* Methods for the Study of Marine Benthos (eds. N. A. Holme and A. D. McIntyre). Blackwell, Oxford.

Cushing, D. H. 1975. Marine ecology and fisheries. Cambridge Univ. Press, Cambridge: 1-278.

Daan, N. 1978. Changes in the cod stocks and cod fisheries in the North Sea. Rapp. P. -v. Réun. Cons. int. Explor. Mer. 172: 39-57.

Dybern, B. I., H. Ackefors, and R. Elmgren. 1976. Recommendations

on methods for marine biological studies in the Baltic Sea.   The
Baltic Marine Biologists. 1:   1-98.

Elliott, J. M.   1971.   Some methods for the statistical analysis of
samples of benthic invertebrates.   Freshwater biological association, Ambleside (UK).   Scient. Publ. No. 25:   1-148.

Gray, J. S.   1974.   Animal-sediment relationships.   Oceanogr. Mar.
Biol. Ann. Rev. 12:   223-261.

Gray, J. S.   1977.   The stability of benthic ecosystems.   Helgol.
wiss. Meeresunters. 30:   427-444.

Hempel, G.   1977.   Biologische Probleme der Befishchung mariner
Ökosysteme.   Naturwiss. 64:   200-206.

Hempel, G.   1978.   North Sea fisheries and fish stocks – a review
of recent changes.   Rapp. P. -v. Réun. Cons. int. Explor. Mer.
173:   145-167.

Ices, J. E.   1977-78.   Report of the working group on assessment
of demersal stocks in the Baltic.   ICES, Copenhagen.

Ivlev, V. S.   1961.   Experimental ecology of the feeding of fish.
Yale Univ. Press, New Haven.   1-302.

Jones, R.   1973.   Density dependent mortality of the numbers of
cod and haddock.   Rapp. P. -v. Réun. Cons. int. Explor. Mer.
164:   156-173.

Jones, R.   1978.   Competition and co-existence with particular reference to gadoid fish species.   Rapp. P. -v. Réun. Cons. int.
Explor. Mer. 172:   292-300.

Kühlmorgen-Hille, G.   1963.   Quantitative Untersuchungen der
Bodenfauna in der Kieler Bucht und ihre jahreszeitlichen
Veränderungen.   Kieler Meeresforsch. 19:   42-66.

Mills, E. L.   1975.   Benthic organisms and the structure of marine
ecosystems.   J. Fish. Res. Bd. Can. 32:   1657-1663.

Nauen, C.   1978.   Population dynamics of the seastar, *Asterias
rubens* L., an important competitor to demersal fish stocks in
Kiel Bay.   ICES, C. M.   1978/E:12.   1-11.

Pauly, D.   1973.   Über ein Gerät zur Vorsortierung von Benthospro-
ben.   Ber. dt. wiss. KommnMeeresforsch. 22:   1458-1460.

Pauly, D. and G. Gaschütz.   In press.   A simple method for fitting
oscillating length growth data, with a programme for pocket cal-
culators.   Ices, C. M.   1979.   1-26.

Pfanzagl, J.   1963.   Über die Parallelität von Zeitreihen.
Metrika. 6:   100-113.

Probst, B.   1973.   Auswertung ökologischer Parameter, die 1966 auf
Feuerschiff P 8 beobachtet wurden.   Dipl. thesis. Univ. Kiel.

Reise, K.   1977.   Predator exclusion experiments in an intertidal
mud flat.   Helgol. wiss. Meeresunters. 30:   263-271.

Reise, K.   1978.   Experiments on epibenthic predation in the Wadden
Sea.   Helgol. wiss. Meeresunters. 31:   55-101.

Rosenberg, R., I. Olsson, and E. Ölundh.   1977.   Energy flow model
of an oxygen-deficient estuary on the Swedish west coast.   Mar.
Biol. 42:   99-107.

Siegel, S.   1956.   Nonparametric statistics.   McGraw-Hill, Tokyo.
1-312.

Steele, J. H.   1974.   The structure of marine ecosystems.   Harvard
Univ. Press, Cambridge, Mass.   1-128.

Thorson, G.   1972.   Enforschung des Meeres.   Kindler, München:   1-
253.

Thurow, F. 1974. Zur Stärke des Dorschjahrganges 1972 in der westlichen Ostsee. Ber. dt. wiss. KommnMeeresforsch. 23: 129–136.

Ursin, E. 1979. Nordseemodell. Eine neue Grundlage für die Regulierung der Fischerei. Inf. Fischw. 26: 3–9.

Virnstein, R. W. 1977. The importance of predation by crabs and fishes on benthic infauna in Chesapeake Bay. Ecology. 58: 1199–1217.

Virnstein, R. W. 1978. Predator caging experiments in soft sediments: caution advised. Pages 261–273. *In* Estuarine Interactions. M. L. Wiley (ed.). Academic Press, New York.

Wolff, W. J. and L. de Wolf. 1977. Biomass and production of zoobenthos in the Grevelingen estuary, the Netherlands. Estuar. Coast. Mar. Sci. 5: 1–24.

Zimmer, C. 1933. Cumacea. *In* Grimpe, G. and E. Wagler, Tierwelt der Nord- und Ostsee. 10 g: 1–150.

# Population Growth and Trophic Interactions Among Free-Living Marine Nematodes

Daniel M. Alongi
John H. Tietjen

ABSTRACT: Gnotobiotic experiments were conducted
with three species of marine nematodes (the selective deposit-feeders *Monhystera disjuncta* and *Diplolaimella* sp., and the epistrate feeder *Chromadorina germanica*) to examine species interactions
on bacterial and algal diets. On diets of bacteria or the chlorophyte, *Dunaliella* sp., growth
rates of *Monhystera disjuncta* and *Diplolaimella*
sp. were significantly lower when the worms were
grown together than when grown separately. On
bacteria, neither nematode species was totally
eliminated by the presence of the other; on *Dunaliella* sp., however, *M. disjuncta* was totally
incapable of growth in the presence of *Diplolaimella* sp. Neither species was capable of sus-
tained growth on the diatoms *Cylindrotheca closterium* and *Nitzschia* sp.

The growth of *Chromadorina germanica* was unaffected by the presence of either *Diplolaimella*
sp. or *Monhystera disjuncta*. *C. germanica* grew
best on the diatom *Cylindrotheca closterium* and
did not grow on *Dunaliella* sp. In contrast with
a prior study (Deutsch, 1978), *C. germanica* was

151

maintained on bacteria.  It was necessary for the
bacteria to be present on a solid substrate, however, and not in suspension.  *C. germanica* was
able to co-exist with *M. disjuncta* and *Diplolaimella* sp. on bacteria by rasping the bacteria
(and small pieces of a cereal substratum to which
the bacteria were attached) with its teeth, while
the latter nematode species ingested cells that
were either unattached or attached to extremely
small (< 3 µm) particles.

These experiments demonstrated significant
competitive interactions among nematodes with
near-identical buccal morphologies (*Monhystera
disjuncta* and *Diplolaimella* sp.), interactions
that may partially explain the low species diversity among nematodes inhabiting muddy sediments
where competition for low-variety food resources
may be intense.

## INTRODUCTION

To further understand factors which regulate the population
distribution and structure of marine meiobenthos, it is important
to obtain quantitative and qualitative information on the trophic
interactions among these organisms.  At present, little is known
about either the realized niche or the operational habitat of individual species of marine meiofauna.  Past studies on the trophic
ecology of marine nematodes have focused mainly on their feeding
habits, nutrition, and life histories (Perkins, 1958; Chitwood and
Murphy, 1964; Hopper and Meyers, 1966a, 1966b, 1967; V. Thun, 1968;
Hopper *et al.*, 1973; Gerlach and Schrage, 1971, 1972; Ott and
Schiemer, 1973; Tietjen, 1967; Tietjen and Lee, 1972, 1973, 1977a,
1977b; Wieser and Schiemer, 1977; Deutsch, 1978).  Although resource partitioning is one mechanism by which closely related species may coexist (Schoener, 1974; Rapport and Turner, 1977; Ivester
and Coull, 1977; Tietjen and Lee, 1977a), there have been few studies conducted on meiofauna under gnotobiotic conditions where these
principles have been tested (Muller, 1975; Muller and Lee, 1977;
Tietjen and Lee, 1977a).

The aim of the present research was to examine the growth
rates of marine nematode populations, both in single species culture and in mixed culture, on different diets to observe the extent of interaction among these species, in particular, competitive
interaction.  Interactions among populations of different species
of marine meiofauna have not been studied under controlled conditions, either *in situ* or in the laboratory.  The species selected
were *Chromadorina germanica* (Bütschli), *Monhystera disjuncta* (Bastian) and *Diplolaimella* sp., which were all isolated from a salt
marsh.

MATERIALS AND METHODS

The nematodes were taken from stock cultures originally isolated from a salt marsh at Towd Point, North Sea Harbor in Southampton, New York. *Monhystera disjuncta* and *Diplolaimella* sp. both lack buccal armature and feed by simple suction via a muscular esophagus. The buccal cavity of both species has a diameter of 1 to 2 μm and the species are classified as selective deposit-feeders according to the feeding scheme of Wieser (1953). *Chromadorina germanica's* buccal cavity is slightly larger (4-6 μm), armed with three small teeth, and is classed as an epistrate feeder. The life histories of *M. disjuncta* and *C. germanica* are known (Gerlach and Schrage, 1971; Tietjen and Lee, 1977b).

All experiments were gnotobiotic and conducted in 25 cm$^2$ plastic tissue culture flasks (FALCON). The flasks were incubated at constant temperature (23$^{\circ}$C) and salinity (26 $^{\circ}$/oo) under a light regime of 18 h light/6 h dark in a Sherer environmental chamber (Model CEL 4-4).

Cultures of nematodes that utilized algae as a food source were maintained in 0.45 μm HA Millipore-filtered, autoclaved Erdscheiber medium (Lee *et al*, 1970). To reduce the level of bacteria present in the culture vessels, an antimycotic/antibiotic mixture (penicillin 10,000 Ü·ml$^{-1}$, Fungizone 25 μg·ml$^{-1}$, Streptomycin 10,000 μg) (Grand Island Biological Company, Grand Island, N. Y. 14072) was added to the medium. The algal cells were obtained from stock cultures grown on "S" medium (Lee *et al.*, 1970).

The media for cultures that utilized bacteria as potential food consisted of 0.45 μm HA Millipore-filtered, autoclaved sea water and cereal (Gerber Products Co. Mfr., Fremont, Michigan 49412). (See Table 1 for weights of cereal used in experiments). Flasks were covered with aluminum foil to retard algal growth. Bacteria were harvested from stock cultures grown axenically on enriched artificial sea water medium (Lee *et al.*, 1970).

The three nematode species were grown separately on a variety of foods. The foods chosen included three algal and two bacterial species originally isolated from the aufwuchs at Towd Point, New York. (See Table 2 for food species descriptions). Each feeding experiment was conducted using a series of six replicate flasks.

Three nematode interaction experiments were performed: 1) *Monhystera disjuncta* and *Diplolaimella* sp., 2) *M. disjuncta* and *Chromadorina germanica*, and 3) *Diplolaimella* sp. and *C. germanica*. Nine replicate flasks per food source were used in all three experiments.

The following inoculation procedures were employed throughout the course of this study: nematodes from stock cultures were gently pipetted and aseptically washed in a 9-hole spot plate containing sterile sea water. An inoculum of 40-60 nematodes was introduced into each culture vessel containing fresh medium with potential food organisms. (The same sample size for each nematode species was also utilized in the competition flasks). The initial bacterial and algal inoculum ranged in density from 1.1 x 10$^5$ to 1.2 x 10$^6$ cells·ml$^{-1}$. The concentration of each food species was measured by cell counts with a hemocytometer.

Table 1. Weights of cereal used as bacterial substrate in nematode feeding and species interaction experiments.

CEREAL WEIGHTS (gram/flask)

A) Single Species Controls:

| | Trial 1[*] | Trial 2[*] | Mean Weight |
|---|---|---|---|
| 1. *Diplolaimella* sp. | 0.05395 | 0.08874 | 0.07135 |
| 2. *Monhystera disjuncta* | 0.06965 | 0.06437 | 0.06701 |
| 3. *Chromadorina germanica* | 0.07113 | 0.06839 | 0.06976 |

B) Interaction Experiments:

| | Trial 1[*] | Trial 2[*] | Trial 3[*] | Mean Weight |
|---|---|---|---|---|
| 1. *Diplolaimella* sp.– *Monhystera disjuncta* | 0.13474 | 0.13369 | 0.13226 | 0.13356 |
| *Diplolaimella* sp.– *Chromadorina germanica* | 0.18465 | 0.17279 | 0.18673 | 0.18136 |
| *Monhystera disjuncta Chromadorina germanica* | 0.11912 | 0.12365 | 0.11284 | 0.11854 |

[*]represents the mean weight of cereal in 3 replicate experiments

Table 2.  Code numbers, species names, and sizes of food particles used in feeding and interaction experiments.

| Code Number | Species | Approximate Dimensions | |
|---|---|---|---|
| | | Length, μm | Width, μm |
| **Bacteria** | | | |
| A5-7 | *Pseudomonas* sp. 1 | 2.1 | 0.8 |
| D2-5 | *Pseudomonas* sp. 2 | 2.2 | 0.6 |
| **Diatoms** | | | |
| W510 | *Nitzschia* sp. | 4.5 | 2.0 |
| 709 | *Cylindrotheca closterium* | 10.5 | 2.2 |
| **Chlorophyte** | | | |
| 55 | *Dunaliella* sp. | 3.8 | 2.0 |

All experiments were terminated after 28 days.  The nematodes were enumerated within a 20 cm$^2$ area per culture vessel every 2-3 days using a dissecting microscope.  In the competition experiments *Diplolaimella* sp. and *M. disjuncta* were distinguished on the basis of tail morphology; *Diplolaimella* sp. has a whip-like tail, whereas *M. disjuncta's* tail is characteristically shorter and cylindrical. The morphology of *Chromadorina germanica* was readily distinguishable from that of the former two species.

The reproductive potential of the nematodes was examined and the daily intrinsic rate of natural increase, r, calculated according to:

$$r = \frac{1}{t} \ln \frac{N_t}{N_o} >$$

where r= intrinsic rate of natural increase $N_o$= population size at time zero $N_t$= population size at time t.  The time period over which r was calculated was 28 days for all experiments.

RESULTS

The population growth rates of all three nematode species in single and mixed culture on all foods are given in Figures 1-4 and r values for the organisms are given in Table 3.

Table 3. Intrinsic rate of natural increase (r) of nematodes in single and mixed species culture. Each value represents the mean of six (single species culture) or nine (mixed culture) replicate experiments.

| Nematode Species | Food | | | | |
| --- | --- | --- | --- | --- | --- |
| | Bacteria | | *Dunaliella* sp. | Algae | |
| | A5–7 | D2–5 | | *Nitzschia* sp. | *Cylindrotheca closterium* |
| **Single Species Culture:** | | | | | |
| *Diplolaimella* sp. | 0.096 | 0.086 | 0.117 | 0 | 0 |
| *Monhystera disjuncta* | 0.099 | 0.093 | 0.098 | 0 | 0 |
| *Chromadorina germanica* | 0.064 | 0.058 | 0 | 0.007 | 0.118 |
| **Mixed Species Culture:** | | | | | |
| *Diplolaimella* sp. | 0.063 | 0.065 | 0.077 | 0 | 0 |
| vs. | | | | | |
| *Monhystera disjuncta* | 0.061 | 0.070 | 0 | 0 | 0 |
| *Diplolaimella* sp. | 0.101 | 0.096 | 0.136 | 0.080 | 0 |
| vs. | | | | | |
| *Chromadorina germanica* | 0.099 | 0.097 | 0 | 0.088 | 0.121 |
| *Monhystera disjuncta* | 0.110 | 0.091 | 0.133 | 0 | 0 |
| vs. | | | | | |
| *Chromadorina germanica* | 0.096 | 0.081 | 0 | 0.009 | 0.143 |

### Population Growth in Single Species Culture

*Diplolaimella* sp. grew on the bacteria (r = 0.096 on A5-7; r = 0.086 on D2-5) and the chlorophyte, *Dunaliella* sp. (r = 0.117). Maximum population density was attained on a diet of bacterium A5-7 in 22 days; on D2-5 and *Dunaliella* sp., peak densities occurred after 28 days. The nematode exhibited poor growth on the diatoms *Nitzschia* sp. and *Cylindrotheca closterium*, maintaining growth for only 6 and 12 days, respectively (Fig. 1).

Population growth of *Monhystera disjuncta* was similar to *Diplolaimella* sp. on each of the five foods (Table 3). On diets of A5-7, D2-5 and *Dunaliella* sp., maximum population growth was reached in 26 days, and *M. disjuncta* did not appear to utilize either *Nitzschia* sp. or *C. closterium* (Fig. 1).

*Chromadorina germanica* grew on both bacteria species (r = 0.064 on A5-7; r = 0.058 on D2-5), achieving maximum density at 21 days on D2-5 and 26 days on A5-7. The nematode grew poorly on *Dunaliella* sp. (maintaining growth for only 14 days) and *Nitzschia* sp. (r = 0.007 after 28 days). Growth was fastest on a diet of the diatom *C. closterium* (r = 0.118 at 28 days).

### Population Interaction Experiments: *Diplolaimella* sp. - *Monhystera disjuncta* Experiment

On both bacterial strains, the growth of each nematode species was significantly poorer (Mann-Whitney U-Test, p < .05) when the worms were grown together than when grown separately (Fig. 2 and Table 3). On D2-5 and A5-7, the respective r values were 0.065 and 0.063 for *Diplolaimella* sp. and 0.061 and 0.070 for *Monhystera disjuncta*.

Because the diatoms *Nitzschia* sp. and *C. closterium* were not utilized by either nematode, the only alga on which interaction experiments were conducted was the chlorophyte, *Dunaliella* sp. On *Dunaliella* sp., *M. disjuncta* maintained growth for only 16 days in the presence of *Diplolaimella* sp. In addition, the growth of *Diplolaimella* sp. was significantly lower in the presence of *M. disjuncta* than when it was grown separately (Fig. 2, Table 3). Maximum density of *Diplolaimella* sp. on *Dunaliella* sp. was $530 \cdot 20$ cm$^{-2}$ in the presence of *M. disjuncta*; when grown alone, it was $1700 \cdot 20$ cm$^2$.

### *Diplolaimella* sp. - *Chromadorina germanica* Experiment

On the bacteria and chlorophyte diets, the growth of *Diplolaimella* sp. in the presence of *Chromadorina germanica* was similar to that observed in solitary culture (Fig. 3). For this organism, establishment of maximum density on the above (3) foods occurred in 29 days. *C. germanica* grew better on each bacterial species in the presence of *Diplolaimella* sp. (r = 0.097 on D2-5 and r = 0.099 on A5-7) than when cultured singly (Table 3). An r value of 0.121 (at 29 days) was calculated for *C. germanica* on the *C. closterium* diet, which was similar to that calculated for *C. germanica* in single species culture.

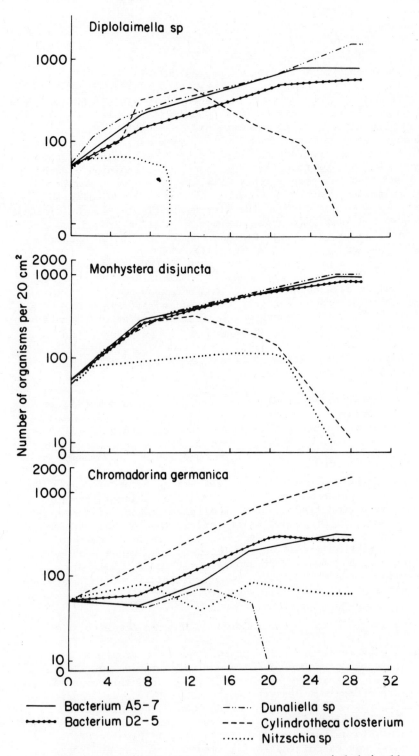

Figure 1. Population growth curves for *Diplolaimella* sp., *Monhystera disjuncta* and *Chromadorina germanica* in single species culture on selected bacterial and algal diets.

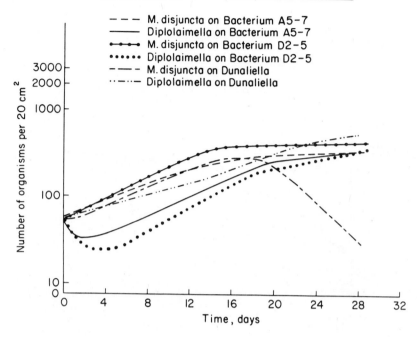

Figure 2.  Population growth curves for *Diplolaimella*
sp., and *Monhystera disjuncta* in mixed species culture on
selected bacterial and algal diets.

Both species of nematodes grew on a diet of *Nitzschia* sp., although no growth was observed for either worm on that diet when
cultured alone.  This was due to carryover of bacteria from the
stock cultures of *C. germanica* and *Diplolaimella* sp. and subsequent
serial washings.  Although an antibiotic/antimycotic was added to
the culture medium, nematode sensitivity to antibiotics (Lee *et al*;
1970; Tietjen *et al.*, 1970; Tietjen and Lee, 1973) limits the concentration usable.  The concentration of the added mixture was insufficient to totally eliminate the elevated bacteria populations that grew
as a result of the additional amount of substrate present in this experiment (Table 1).  *C. germanica* reached a maximum population ($\bar{x}$ =
520) in 21 days; *Diplolaimella* sp. attained peak density ($\bar{x}$ = 570) in
29 days.

*Monhystera disjuncta-Chromadorina germanica Experiment*

*Monhystera disjuncta* grew equally well on bacteria (r = 0.110
on A5-7; r = 0.091 on D2-5) and the green alga (r = 0.133 at 28
days) in the presence of *Chromadorina germanica* as alone (Fig. 4,
Table 3).  *C. closterium* and *Nitzschia* sp., however, were evidently
not ingested or assimilated by *M. disjuncta*, which maintained growth
on the diatoms for only one and eight days, respectively.

In culture with *M. disjuncta*, *C. germanica* grew better on bacteria (r = 0.096 on A5-7; r = 0.081 on D2-5) than when grown separately (Fig. 4 and Table 3).  On the bacteria D2-5 and A5-7, peak
densities of the chromadorid were attained at 29 and 20 days, re-

spectively.  As in single species culture, *C. germanica* grew poorly
on diets of *Dunaliella* sp. and *Nitzschia* sp. and best on the diatom,
*Cylindrotheca closterium* (Fig. 4, Table 3).

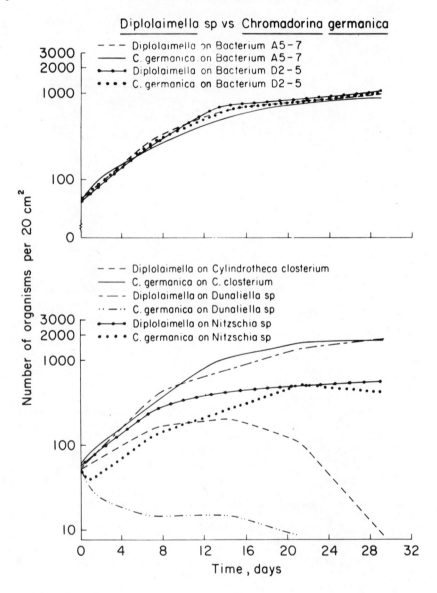

Figure 3.  Population growth curves for *Diplolaimella*
sp., and *Chromadorina germanica* in mixed species culture on
selected bacterial and algal diets.

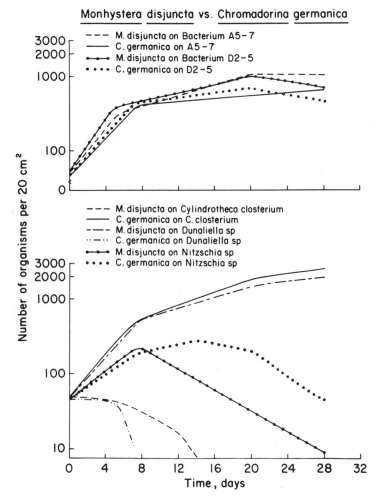

Figure 4. Population growth curves for *Monhystera disjuncta* and *Chromadorina germanica* in mixed
species culture on selected bacterial and algal diets.

DISCUSSION

Significant population interactions occurred only among nematodes of the same trophic type. Interspecific competition on all
three foods (D2-5, A5-7 and 55) was observed between *Diplolaimella*
sp. and *Monhystera disjuncta* in the mixed species cultures. On the
basis of their feeding behavior, both species probably occupy the
same niche in the natural environment. Because both nematodes possess a similar buccal morphology (*sensu* Wieser, 1953), and ingest
food particles of the same size, competition resulted. On bacteria,
neither species completely excludes the other, suggesting that both
monhysterids feed equally well on these microbes.

On the *Dunaliella* sp. diet, the decline of *M. disjuncta* densities in the presence of *Diplolaimella* sp. appears to indicate competitive exclusion. However, while *Diplolaimella* sp. appeared to be the more efficient chlorophyte consumer, this nematode was influenced by the presence of *M. disjuncta*. The severe decline of *M. disjuncta* may also have been caused by a toxic combination or total buildup of algal and nematode waste products (*e.g.*, $NH_3$), and not just interspecific competition. In solitary and mixed species culture, carrying capacity for the nematodes may well be a function of their tolerance to large numbers of algae and their respective secondary metabolites.

The growth of *Chromadorina germanica* was unaffected by the presence of either *Diplolaimella* sp. or *Monhystera disjuncta*. On algal foods, *C. germanica* apparently did not compete with *Diplolaimella* sp. and *M. disjuncta* because they ingested cells that are too large (i.e., the diatom *Cylindrotheca closterium* is approximately 10.5 µm by 2.2 µm) for the other nematodes to physically handle. *Chromadorina germanica* also coexisted on bacteria with the other two species, and microscopic examination revealed one of the mechanisms by which this coexistence may occur: spatial partitioning of the resources within the culture flasks. *C. germanica* was observed almost always adhering to the cereal flakes, whereas *Diplolaimella* sp. and *M. disjuncta* were usually found dispersed throughout the flask. *C. germanica* rasps only the bacteria cells present on the flakes; the monhysterids either suck in the bacteria from the overlying bacterial suspension or from the bottom of the flask.

In an earlier study, Deutsch (1978) observed that *Diplolaimella* sp. failed to survive on diets of diatoms but did survive on diets of bacteria and chlorophytes. She ascribed this to the fact that *Diplolaimella* sp., which feeds by the sucking action of the esophagus (as does *Monhystera disjuncta*), was unable to ingest cells larger than 3.9 µm. In both the present study and her study, the nematodes lacking buccal armature consistently failed to grow on food cells larger than 4 µm, thereby lending credence to the hypothesis that such nematodes are limited to bacteria, small chlorophytes and other small food particles.

Deutsch also observed, however, that *Chromadorina germanica* failed to survive on diets of bacteria, which was not the case in the present study. She cited the buccal armature of *C. germanica* as the cause, indicating that small cells ( < 2.9 µm in diameter) are not able to be properly held by the teeth during the ingestion process, and, in fact, probably bounce off the teeth during ingestion. However, the apparent conflict in results is resolvable - Deutsch's feeding experiments with bacteria did not involve the use of cereal as a substrate.

In her feeding experiments, most of the bacteria were in suspension and were, therefore, incapable of being ingested by *Chromadorina germanica*. Detrital-bacterial particles larger than 3-5 µm can be physically ingested by *C. germanica*, whereas single bacterial cells or clumps of cells in suspension cannot. The presence of the teeth in the stoma of *C. germanica* for the rasping and/or penetration of food cells would require a particle of at least 3 µm.

The population size of *C. germanica* was positively correlated with the amount of cereal flakes present in the culture flasks. With a mean weight of 0.06976 g/flask in the control flasks, the nematode exhibited a maximum population of $\bar{x}$ = 300 on D2-5 and $\bar{x}$ = 350 on A5-7. In the mixed species cultures, with flasks containing an average weight of 0.14448 g/flask (~2X), the population density nearly doubled for the worm on both bacterial strains.

With more substrate in the flask, more bacteria were present to create an increased food supply for both nematode trophic types. Although *Diplolaimella* sp. and *M. disjuncta* utilized a greater area of the flask than *C. germanica*, their numbers did not significantly increase in the presence of additional cereal. This might be explained in the following manner: With an increase in the bacterial population, *C. germanica* probably cropped a large amount of the additional bacteria attached to the flakes. This allowed for only a small increase in the bacterial concentration within the overlying culture medium, thereby enabling *M. disjuncta* and *Diplolaimella* sp. to increase their densities only slightly.

In field studies, numerous investigators have observed variability in nematode distribution, diversity and dominance within different sediment types (Wieser, 1960; Warwick and Buchanan, 1970; Ward, 1973; Lorenzen, 1974; Juario, 1975; Heip and Decraemer, 1974; Tietjen, 1969, 1976, 1977, in press). Our experimental data fortify present explanations of nematode population distribution and structure in mud and sandy deposits. For example, Tietjen (1977) ascribed high species dominance in shallow muds of Long Island Sound to intense competition among deposit-feeders. We have experimentally shown that competition and limited niche separation does indeed occur among nematode deposit-feeders, which in turn, would lead to competitive exclusion, high species dominance and low species diversity. The latter are characteristic of nematode populations in fine sediments (Wieser, 1960; Warwick and Buchanan, 1970; Warwick, 1971; Boucher, 1973; Ward, 1973; Lorenzen, 1974; Juario, 1975; Heip and Decraemer, 1974; Tietjen, 1976, 1977, in press).

In coarse deposits, where a more diverse food supply exists, epigrowth feeders scrape the food cells (i.e., algae, bacteria) from the sand grains, while the deposit-feeders crawl throughout the interstices in search of prey. Although the nematodes may be feeding on the same resources, both trophic types may spatially partition their environment to avoid excessive competitive interactions. Neither feeding type has a competitive advantage, and thus, low dominance results. Lessened competition, which results not only in low dominance but also in increased species richness, allows for a more diverse fauna to occur in sands. This fractionation also explains the observed high species diversity of nematodes in sandy deposits. Our evidence that the deposit-feeders (*Diplolaimella* sp. and *M. disjuncta*) and the chromadorid are able to utilize different portions of a micro-environment (culture flask) supports field observations.

Undoubtedly, there are other subtle mechanisms besides resource partitioning and particle size selection (i.e., selective ingestion) that allow species to coexist in a community. For example, nematodes may be biochemically selective with regard to the food particles they ingest (Jennings and Deutsch, 1975; Deutsch, 1978). This study of nematode trophic dynamics suggests that more intensive study is needed in this neglected aspect of marine benthic ecology before our knowledge of such diverse assemblages becomes axiomatic.

## ACKNOWLEDGMENTS

Contribution No. 109 from the Institute of Marine and Atmospheric Sciences, City University of New York. This research was supported by National Science Foundation Grant No. OCE78-09962.

## REFERENCES

Boucher, G. 1973. Premieres données ecologiques sur les nematodes libres marine d'une station de vas cotiere de Banyuls. Vie Milieu. 23: 69-100.

Chitwood, B. G. and D. G. Murphy. 1964. Observations on two marine monhysterids - their classification, cultivation and behavior. Trans. Am. Soc. 83: 311-329.

Deutsch, A. 1978. Gut structure and digestive physiology of the free-living marine nematodes, *Chromadorina germanica* (Bütschli, 1874) and *Diplolaimella* sp. Biol. Bull. 155: 317-335.

Gerlach, S. A. and M. Schrage. 1971. Life cycles in marine meiobenthos. Experiments at various temperatures with *Monhystera disjuncta* and *Theristus pertenuis* (Nematoda) Mar. Biol. 9: 272-280.

Gerlach, S. A. and M. Schrage. 1972. Life cycles at low temperatures in some free-living marine nematodes. Veröff. Inst. Meeresforsch Bremerh. 14: 5-11.

Heip, C. and W. Decraemer. 1974. The diversity of nematode communities in the southern North Sea. J. mar. biol. Ass. U.K. 54: 251-255.

Hopper, B. E. and S. P. Meyers. 1966a. Aspects of the life cycles of marine nematodes. Helgoländer wiss. Meeresunters. 13: 444-449.

Hopper, B. E. and S. P. Meyers. 1966b. Observations on the bionomics of the marine nematode, *Metoncholaimus* sp. Nature. 209: 899-900.

Hopper, B. E. and S. P. Meyers. 1967. Population studies on benthic nematodes within a sub-tropical sea grass community. Mar. Biol. 1: 85-96.

Hopper, B. E., J. W. Fell, and R. C. Cefalu. 1973. Effect of temperature on life cycles of nematodes associated with the mangrove (*Rhizophora mangle*) detrital system. Mar. Biol. 23: 293-296.

Ivester, M. S. and B. C. Coull. 1977. Niche fraction studies of two sympatric species of *Enhydrosoma* (Copepoda, Harpacticoida) Mikrofauna Meeres. 61: 137-151.

Jennings, J. B. and A. Deutsch. 1975. Occurrence and possible adaptive significance of beta-glucuronidase and arylamidase (leucine aminopeptidase) activity in two species of marine nematodes. Comp. Biochem. Physiol. 52A: 611-614.

Juario, J. V. 1975. Nematode species composition and seasonal fluctuation of a sublittoral meiofauna community in the German Bight. Veröff. Inst. Meeresforsch. Bremerh. 15: 283-337.

Lee, J. J., J. H. Tietjen, R. J. Stone, W. A. Muller, J. Rullman, and M. McEnery. 1970. The cultivation and physiological ecology of members of salt marsh epiphytic communities. Helgoländer wiss. Meeresforsch. 20: 136-156.

Lorenzen, S. 1974. Die Nematodenfauna der sublitoralen Region der Deutschen Bucht, inbesondere im Titan-Abwassergebiet bei Helgoland. Veröff. Inst. Meeresforsch Bremerh. 14: 305-327.

Muller, W. A. 1975. Competition for food and other niche-related studies of three species of salt marsh foraminifera. Mar. Biol. 31: 339-351.

Muller, W. A. and J. J. Lee. 1977. Biological interactions and the realized niche of *Euplotes vannus* from the salt marsh aufwuchs. J. Protozool. 24(4): 523-527.

Ott, J. and F. Schiemer. 1973. Respiration and anaerobiosis of free-living nematodes from marine and limnic sediments. Neth. J. Sea Res. 7: 233-243.

Perkins, E. J. 1958. The food relationships of the microbenthos, with particular reference to that found at Whitstable, Kent. Ann. Mag. Nat. Hist. Ser. 13: 64-77.

Rapport, D. J. and J. E. Turner. 1977. Economic models in ecology. Science. 195: 367-373.

Rullman, J., A. Greengart, J. Trompetee. 1970. Gnotobiotic culture and physiological ecology of the marine nematode, *Rhabdetes Marina* Bastian. Limnol. Oceanogr. 15: 535-543.

Schoener, T. W. 1974. Resource partitioning in ecological communities. Science. 185: 27-38.

Tietjen, J. H. 1967. Observations on the ecology of the marine nematode *Monhystera filicaudata* Allgen 1929. Trans. Amer. Micros. Soc. 86: 304-306.

Tietjen, J. H. 1969. The ecology of shallow water meiofauna in two New England estuaries. Oecologia. 2: 251-291.

Tietjen, J. H. 1973. Life history and feeding habits of the marine nematode *Chromadora macrolaimoides* Steiner. Oecologica. 12: 303-314.

Tietjen, J. H. 1976. Distribution and species diversity of deep sea nematodes off North Carolina. Deep Sea Res. 23: 755-768.

Tietjen, J. H. 1977a. Feeding behavior of marine nematodes. *In* Ecology of Marine Benthos, B. C. Coull, (ed.). Univ. South Carolina Press, Columbia, S.C. pp. 22-36.

Tietjen, J. H. 1977. Population distribution and structure of the free-living nematodes of Long Island Sound. Mar. Biol. 43: 123-136.

Tietjen, J. H. 1979. Population structure and species composition of the free-living nematodes inhabiting sands of the New York Bight Apex. Est. Coast. Mar. Sci. (In press).

Tietjen, J. H. and J. J. Lee. 1972. Life cycles of marine nematodes. Influence of temperature and salinity on the development of *Monhystera denticulata*. Oecologica. 10: 167-176.

Tietjen, J. H. and J. J. Lee. 1977b. Life cycles of marine nematodes. Influence of temperature and salinity on the reproductive potential of *Chromadorina germanica* Bütschli. Mikro. Meeres. 61: 257-264.

von Thun, W. 1968. Autokolögische Untersuchungen an freilenden nematoden des brackwassers. Ph.D. Thesis, Kiel University.

Ward, A. R. 1973. Studies on the sublittoral free-living nematodes of Liverpool Bay. I. The structure and distribution of the nematode populations. J. mar. biol. Ass. U.K. 50: 129-146.

Warwick, R. M. 1971. Nematode associations in the Exe estuary. J. mar. biol. Ass. U.K. 51: 439-454.

Warwick, R. M. and J. B. Buchanan. 1970. The meiofauna off the coast of Northumberland. I. The structure of the nematode populations. J. mar. biol. Ass. U.K. 50: 129-146.

Wieser, W. 1953. Die Beziehung zwischen Mundhohlengestalt, Ernahrungswiese und Vorkommen bei freibenden marinen Nematoden. Ark. Zool. (Ser 2) 4: 439-484.

Wieser, W. 1960. Benthic studies in Buzzard's Bay. II. The meiofauna. Limnol. Oceanogr. 5: 121-137.

Wieser, W. and F. Schiemer. 1977. The ecophysiology of some marine nematodes from Bermuda: seasonal aspects. J. Exp. Mar. Biol. Ecol. 26: 97-106.

# The Influence of Competition and Predation on Production of Meiobenthic Copepods

**Carlo Heip**

ABSTRACT:  An assemblage of meiobenthic copepods
was monitored twice monthly between 1968 and 1976
in a shallow brackish water habitat.  Four species
dominated and accounted for nearly all the production.  All species have only one or two generations per year and each peak has similar dynamic
characteristics.  Production is determined primarily by the number of generations, and it is suggested that competition and predation determine
the life-history characteristics.

## INTRODUCTION

Increasing species diversity over time is a common property of
communities where disturbances do not occur frequently enough to
cause important spatial or temporal heterogeneity.  Theoretical
ecologists consider competition and predation to have the most profound effect on the community diversity in temperate latitudes
(McArthur, 1972), but there is still much discussion on the relative importance of each in the process.  For example, Menge and
Sutherland (1976) suggest that competition is more important in
higher trophic levels, while predation regulates the number of species at lower trophic levels.

There has also been much discussion on the role of energy and matter flow through ecosystems as a structuring force. Recent developments in thermodynamic theory of open systems has provided firm theoretical basis for these ideas, i.e., when systems are brought far from equilibrium, they can only be maintained by a flow of energy and matter which is itself a source of order. This new order has a stability regime characterized by periodicities and cycles and the cycle is therefore the fundamental unit in systems organization (Nicolis and Prigogine, 1977).

Although a possible link between trophic structure and species interactions has been implicit in much theoretical work, the different approaches converge only slowly, and in an important recent book the approach to community ecology has remained largely empirical (May, 1976).

Much effort in aquatic biology has been devoted to measurement of production of populations and communities, rather than to species interactions. The parameter of interest is usually biomass of the population rather than its density, and interactions are ignored or expressed as a mortality term without further explanation. Although there is, of course, a relationship between biomass and density, they are not equivalent in age-structured populations in which the age structure is not stable.

The purpose of this paper is to illustrate how life-history characteristics can be explained by competitive or predator-prey interactions and how these characteristics influence production of populations. It is a purely empirical study and I will nowhere try to prove causal relationships or to model them. The interest of this study is the fact that the populations have been studied over so many years that some patterns are discerned which can be explained in some ways. I acknowledge that some of the evidence presented is rather speculative. This will partly be remedied in forthcoming papers in which many points made here will be elaborated in more detail.

MATERIALS AND METHODS

The animals from this study are meiobenthic copepods from a brackish water pond in northern Belgium (Heip, 1976). Their density was estimated from fortnightly samples taken from 12 August 1968 to 29 December 1976 with a core covering a surface area of 6 $cm^2$. According to season one (in winter) or two (in summer) samples were counted. The data were smoothed using the running median of three successive estimates and the trendline was calculated using a weighted running mean of three samples (Velleman, 1977). In so doing, variance introduced by spatial patchiness is considered as noise which is filtered by the low-pass filter applied here. However, as the procedure is unusual in ecology, all calculations were performed on the raw data as well, but there was never a large difference (i.e., more than 10–20%) between them. Rates of increase ($r = 1/t \ln N_t/N_o$) will be consistently too low when calculated from trendlines, but the bias is small: the average overall peak for *Tachidius discipes* was r = 0.042 per day for the raw data as well as for trendline data with and without smoothing; for *Parony-*

*chocamptus nanus*, with less well-defined peaks, the values are r = 0.042 (raw data), r = 0.038 (trendline values without smoothing) and r = 0.036 (trendline values with smoothing).

As this paper is based on yearly averages, rates and calendar data, which are not influenced by the procedure, the additional precision in the single estimates obtained by counting two or more samples each time was judged to be insufficient to justify the large amount of work needed to sort out and count additional samples. However, estimation of the confidence limits of single estimates is possible. In all the populations that were studied (including ostracods and copepods), the coefficient of variation $s/\bar{x}$ becomes constant and equal to 0.27 when $\bar{x}$ gets large. The relationship between $s/\bar{x}$ and $\bar{x}$ is given by (Heip, 1975b):

$$s/\bar{x} = \frac{3.27 + 0.27\,\bar{x}}{1 + \bar{x}}$$

From this *a priori* knowledge of the standard deviation confidence limits can be easily calculated. When density is extremely low ($\bar{x} \leq 10$), the 95% confidence limits are of the same magnitude as the mean (e.g., $10 \pm 10.6$ for one sample, $10 \pm 7.5$ for two samples and $10 \pm 6.1$ for three samples); for higher densities confidence limits are a fraction of the mean, even for only one sample (e.g., $100 \pm 58.7$ for one sample, $100 \pm 41.5$ for two samples and $100 \pm 33.9$ for three samples). Doubling the effort (counting two samples instead of one) increases precision by 29%; increasing the effort with 50% (counting three samples instead of two) increases precision by only 18%.

Biomass of individual species was determined by weighing a sample of adults with a Mettler ME22 microbalance. Copepodites were assumed to have an eighth of this weight, as there are five copepodite stages each approximately doubling in weight as they moult (Vinc and Heip, in press). All weights are dry weights, i.e., after drying for two hours in an oven at $110^{\circ}$C. Nauplii were neglected as their density was difficult, and their weight impossible, to estimate accurately. Also, they could not be attributed to certain species without great effort. Their production is nearly negligible anyway, because their biomass is, on average, two orders of magnitude lower than that of adults, and they are short-lived; also, much of their production adds up in the production of copepodites and adults.

Production estimates were obtained from P/B-ratios using Banse and Mosher's equation for macrobenthic animals (communicated at the Sixth International Helgoland Symposium, 1976): log P/B = -0.16 - 0.35 log B (in which P and B are in kcal). They remarked that copepods fall below this line; when accepting this and taking the slope calculated by Banse and Mosher and the estimate P/B = 3 calculated for *Canuella perplexa* (Heip, Thielemans and Huysseune, in prep.), I obtained the following equation: log P/$\bar{B}$ = -1.13 - 0.35 log ($6B.10^{-6}$) or log P/$\bar{B}$ = 0.70 - 0.35 log B, in which B is in µg.

RESULTS

Four species account for 90–100% of the total copepod density
and biomass and their densities (smoothed) are illustrated in Fig.
1. Biomass is very similar but has been omitted. Two species
(*Paronychocamptus nanus* and *Canuella perplexa*) are mainly detritivores as adults while the two others (*Tachidius discipes* and *Halicyclops magniceps*) are primarily herbivores. The first species
group (*P. nanus; C. perplexa*) shows rather complicated cycles with
several peaks during the year, while the second group (*T. discipes;
H. magniceps*) shows simple cycles with but one peak in most years.
Other species may be rare but regular (*Nitocra typica* and *Mesochra
lilljeborgi*) or irregular (*Amphiascoides debilis* and *Schizopera
compacta*). These cyclicities were recently discussed (Heip, 1979).

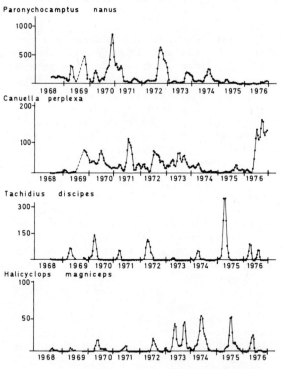

Figure 1. Density of four species
of copepods (No./10 cm$^2$).

The dynamics of all species are characterized by peaks in
which at a certain moment density starts to increase exponentially
and then decreases, again exponentially. When these characteristics are used for the description of the peaks, three important conclusions emerge (Heip, 1978 and in prep.) (for 1968-74):

1) the start of the increase is very similar from year to year
for each species (with some interesting exceptions) but differs

widely among species belonging to the same trophic category.

2) the average realized rate of increase does not differ greatly among species; it is between 0.040–0.052 per day as calculated
from the raw data, which corresponds to doubling times of 13–17
days.  Only the large *Canuella perplexa* has a smaller rate (r = 0.021)
per day.

3) the average duration of the increase is similar for the
different species and varies between 51 days (*Canuella perplexa*)
and 77 days (*Amphiascoides debilis*).

As both the rate and the duration of the increase are of the
same magnitude in all species, it follows that abundance differences are caused by either different mortality or number of peaks
(i.e., probably the number of generations).  Both explanations seem
true.  Mortality in all species is much higher in summer than in
winter (when it could be compared).  However, within a season it is
of the same magnitude in the different species.  This indicates
that it is the number of generations per year which determines the
relative abundance of these species.

The potential number of generations per year is large since
generation time is of the order of several weeks for the species
which have been investigated (Heip and Smol, 1976a).  However, the
actual number of generations appears to be much smaller as evidenced by the percentage of females carrying eggs (Fig. 2).  For
both *Paronychocamptus nanus* and *Canuella perplexa* this is a bimodal
curve in most years.  For *Tachidius discipes* and *Halicyclops magniceps* it is unimodal, with a tendency to become bimodal in later
years for *H. magniceps*.  This would indicate only one or two generations annually.  Another clue can be found in the percentage of
males.  Since mortality is larger in males than in females (Heip,
1978 and in preparation) overwintering populations of *Paronychocamptus nanus* consist mainly of females when they start reproduction in spring.  The bimodal curve of the percentage of females
carrying eggs is followed (with a lag) by an equivalent bimodal
curve in percentage of males.  This lag has been calculated by
cross-correlation between the two series which is at a maximum for
lags of six weeks and 26 weeks, corresponding to two generations.
Thus the number of generations is much less than expected from laboratory cultures.

There is a remarkable constancy in the date at which species
start to increase in density:  species start to increase in the
same month or even week year after year, a phenomenon which has also been detected in the meiobenthic ostracod *Cyprideis torosa*
(Heip, 1976).  Total numbers start to increase earliest in *Tachidius discipes* (beginning of February); followed by *Paronychocamptus
nanus* (end of March onwards), *Halicyclops magniceps* (beginning of
May onwards) and *Canuella perplexa* (mid-June onwards).  In *Tachidius discipes* and *Halicyclops magniceps* the increase is from zero
in most years, whereas in the other species it is from a minimum
value.  However, in the case of both *T. discipes* and *H. magniceps*
this increase in total numbers is followed, but not preceded by, an
increase in the percentage of females carrying eggs.  In *P. nanus*
and *C. perplexa* it is followed and preceded by such an increase.
Indeed, the percentage of females carrying eggs starts to increase

in mid-February in *P. nanus*, in the beginning of March in *T. disci-pes*, mid-April in *C. perplexa* and the beginning of July in *H. mag-niceps*. There is a second increase for *P. nanus* and *C. perplexa* in August.

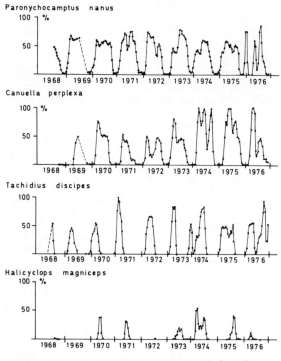

Figure 2. Percentage of females carrying eggs of four species of cope-pods (of total females).

Although the details differ somewhat from year to year and will be explored fully in forthcoming papers, the scheme is re-markably constant throughout the years. The increase in density of *P. nanus* and *C. perplexa* follows an increase of egg-carrying fe-males six to eight weeks prior and can be explained as the result of this reproductive activity. The increase of *T. discipes* and *H. magniceps* is more difficult to explain. In the case of *H. magni-ceps* it may be due to reproduction in the previous year, with hi-bernation of the adults in the burrows of the polychaete *Nereis di-versicolor* deep in the sediment (Heip, 1975a). This wintering gen-eration produces new adults about 1½-2 months after it emerges in spring, exactly what happens in both other species, wintering as free-living adult females.

*Tachidius discipes* is an enigma. Either it winters in ex-tremely low densities (and sometimes single adults were found in winter); or, it produces a wintering stage that is not easily de-tected, such as resting eggs or small nauplii with delayed develop-ment (Coull and Dudley, 1976). Such life-stages were not found,

but, if they occur, the strategy of *T. discipes* would then be quite
similar to that of *H. magniceps*, i.e., the production of life
stages wintering in protected circumstances. This problem requires
further investigation.

In most years *Halicyclops magniceps* disappears from the habitat in October, when the environmental conditions become adverse,
but this is not true for *Tachidius discipes*. *T. discipes* disappears in June or July, when temperature is still high and when it
can reproduce at a high rate (Heip and Smol, 1976a), in fact at a
higher rate than either *Paronychocamptus nanus* and *Canuella perplexa* which at that time produce their summer generation. The disappearance of *Tachidius discipes* in early summer cannot be explained
by abiotic conditions but can be correlated with predation from the
polyp *Protohydra leuckarti* (Heip and Smol, 1976b). Further analysis of this predator-prey relationship (Herman and Heip, in prep.)
showed that, contrary to expectation, predation is not a regulatory mechanism, even though the numbers of *Tachidius discipes* and
*Protohydra leuckarti* are dependent upon each other.

The polyp appears later in the habitat than its prey, since it
does not reproduce below 10°C. Apparently *T. discipes* density has
to be high enough before *P. leuckarti* develops; thus evolution towards early and rapid development is expected.

That predation does not regulate density is most clearly seen
in 1976, when for some reason *P. leuckarti* is absent but *T. discipes* nevertheless disappears in exactly the same way as in other
years. Its later reappearance in that year is due to a different
phenomenon, i.e., a catastrophic event in the form of the drying
up of the pool in the hottest summer of the century, creating an
ecological vacuum.

Summarizing, an important life-history characteristic of *Tachidius discipes* in this environment is the creation of a temporal
refuge against predation, probably by producing a stage which is
invulnerable; by doing so, maximum population sizes are not attained.

Within both species groups there are clear temporal separa-
tions between species which probably prevent competition. This is
important, as it is clear that nauplii will have quite different
niches than adults and that competitive relationships may differ
according to age. Whether this separation is caused by competition
in the past, or prevents actual competition but evolved differently, is immaterial. In both species groups the smaller species pre-
cedes the larger one in succession.

That competitive interactions occur can also be inferred after
*Amphiascoides debilis* successfully entered the system in 1972.
This species occurred in low densities in 1969 but disappeared
until its first peak in July 1972, followed by a second one in September and a third one in December. All of these peaks in this
rapid succession are relatively small. A fourth burst follows in
March 1973, and this is highly successful (44,000 ind./m$^2$ in June).
Density remains low through the winter but a new spring peak follows in March 1974 (51,000 ind/m$^2$ in May), exactly one year after
the first one. The appearance of this species is accompanied by a
shift to earlier reproduction in *Paronychocamptus nanus*: from
1969-1973 the percentage of females carrying eggs increased in ear-

ly February but in 1974 the increase started in January.  In 1973
and 1974 density of *P. nanus* is much lower than in the preceding
years (Table 1) and in 1975, when reproduction is earlier, *P. nanus*
and *A. debilis* have extremely low densities and biomass (Table 1).
Perhaps their competition early in the year prevents maximum population size.  It is in 1975 that *Tachidius discipes* reaches its
highest density and biomass of all the years.  Since it is an early
species as well, some life stage might profit from the extremely low abundance of the two other species.

The two different strategies followed by the two species
groups have important consequences for their production.  As *Paronychocamptus nanus* and *Canuella perplexa* occur over the whole year
and maintain active populations in winter, their production, and,
therefore, their part in community metabolism, will be much higher
than that of the other species (Table 2).  The four species form a series of decreasing biomass in which each term is about twice as
large as its follower in the series.  Similar patterns in the relative abundance of species are usually explained in terms of competition (May 1976).


Table 1.  Mean annual biomass and start of increasing percentage
of females carrying eggs in three successive years

|  | 1973 | 1974 | 1975 |
|---|---|---|---|
| *Paronychocamptus nanus* | 46 mg/m$^2$<br>13 Feb | 47 mg/m$^2$<br>10 Jan | 14 mg/m$^2$<br>2 Jan |
| *Amphiascoides debilis* | 7 mg/m$^2$<br>1 Feb | 16 mg/m$^2$<br>24 Jan | 4 mg/m$^2$<br>16 Jan |
| *Tachidius discipes* | 4 mg/m$^2$<br>11 Apr. | 18 mg/m$^2$<br>21 Mar | 67 mg/m$^2$<br>30 Jan |


Table 2.  Mean biomass and production per year of the four most
abundant copepod species during 1970–1976 (n = 185)

|  | $\bar{B}$ (mg/m$^2$) | $\bar{P}$ (mg/m$^2$/yr) | s/$\bar{x}$ |
|---|---|---|---|
| *Canuella perplexa* (4.2 µg) | 93 | 281 | 0.80 |
| *Paronychocamptus nanus* (0.6 µg) | 47 | 279 | 1.23 |
| *Tachidius discipes* (2.0 µg) | 27 | 105 | 2.23 |
| *Halicyclops magniceps* (3.0 µg) | 15 | 52 | 1.45 |
| Total Copepods | 195 | 769 | 0.53 |

DISCUSSION

Species of similar size have similar life histories: they increase at a similar rate and during about the same time. They differ in the number of peaks, which correspond to generations, during a year. *Tachidius discipes* and *Halicyclops magniceps* have only one generation annually because of environmental conditions: avoidance of predation for the former, low temperature for the latter. Production of these species can only be half that of species having two generations, all else being equal. In fact, as *T. discipes* is much more dependent on how quickly it can occupy its habitat, and on how much competition it will encounter in doing so, biomass and production of this species are variable from year to year. The mean annual biomass of *Paronychocamptus nanus* decreases slowly from 79 mg/m$^2$ in 1969 to 47 mg/m$^2$ in 1974, and then dramatically to only 14 mg/m$^2$ in 1975 and 8 mg/m$^2$ in 1976. This is accompanied by the appearance in 1973 of *Amphiascoides debilis*, a shift towards earlier egg-laying in *P. nanus* in 1974 and poor reproduction of both species in 1975. The vacuum created is filled up by the opportunistic species *T. discipes*, which has an overall average biomass of 105 mg/m$^2$ but attains 261 mg/m$^2$ in 1976. As its life cycle, adjusted to predation, permits the development of only one generation, the ecological vacuum reappears after the disappearance of that generation. *T. discipes* is then succeeded by *Halicyclops magniceps*, with much higher biomass from 1973 to 1976 than in previous years.

Competition cannot be proven in these cases, but is a reasonable hypothesis. Other explanations are possible; however, it is clear that production of these populations is determined mostly by characteristics of the life cycles of the species, far more than by details of development or by abiotic environmental parameters such as temperature, and these life cycles can be interpreted as resulting from species interactions. Whether life cycles are evolutionary-adaptive is, of course, one of the central problems in contemporary ecology. It is clear that restriction of reproduction to only one generation annually when the potential number of generations is at least 10-20 implies that production is only a by-product of reproductive processes. Copepods then are not the typical r-strategists they have been presumed to be (Heip, 1974), a view that has been stressed recently by several authors (Coull and Dudley, 1976; Hoppenheit, 1978; Warwick, 1980).

Although species interactions are important in determining the production of the populations involved, they influence community production far less. Mean biomass and production of the total copepod taxocene are very stable, with only slight differences in mean values over the years and small coefficients of variation (Heip, in prep.). This stability can be related to trophic relationships within the system and the amount of energy and matter flowing through the system rather than to species. When biomass is stabilized, a trend towards larger species in succession will result in lower production of the taxocene, as production is weight-dependent. There is some indication that this is indeed the case, as the biomass of an average copepod increases throughout the years (Heip, in prep.).

ACKNOWLEDGMENTS

    I gratefully acknowledge the help of Wies Gijselinck and Peter
Herman in sorting and counting part of the samples. I am also
grateful to the reviewers who made this paper less obscure than it
originally was. The author acknowledges a grant, "Bevoegdverklaard
Navorser" of the Belgian National Science Foundation (NFWO) and of
the Belgian Fund of Collective Fundamental Research (FKFO), Grant
No. 2.0010.78.

REFERENCES

Coull, B. C. and B. Dudley. 1976. Delayed naupliar development of
    meiobenthic copepods. Biol. Bull. 150: 38–46.
Heip, C. 1974. A note on the life span of r-strategists. Biol.
    Jb. Dodonaea. 42: 117–120.
Heip, C. 1975a. Hibernation in the copepod *Halicyclops magniceps*
    (Lilljeborg, 1853). Crustaceana. 28: 311–313.
Heip, C. 1975b. On the significance of aggregation in some ben-
    thic marine invertebrates. Proc. 9th. Europ. mar. biol. Symp. H.
    Barnes (ed.). 527–538.
Heip, C. 1976. The life cycle of *Cyrpideis torosa* (Crustacea,
    Ostracoda). Oecologia (Berl.). 24: 229–245.
Heip, C. 1978. Diversiteit, stabiliteit en de niche van Copepoda
    van een brakwaterhabitat. Aggregaatsthesis State University of
    Gent. 471 pp. (In Dutch, unpublished).
Heip, C. 1979. Density and diversity of meiobenthic copepods:
    the oscillatory behavior of population and community parameters.
    *In* E. Naylor and R. Hartnoll (eds.): Proc. 13th. E.M.B.S. Symposi-
    um (in press).
Heip, C. and N. Smol. 1976a. Influence of temperature on the re-
    productive potential of two brackish water harpacticoids. Mar.
    Biol. 35: 327–334.
Heip, C. and N. Smol. 1976b. On the importance of *Protohydra
    leuckarti* as a predator of meiobenthic populations. In G.
    Persoone and E. Jaspers (eds): Proc. 10th E.M.B.S. Symposium
    Vol. 2: 285–296.
Hoppenheit, M. 1978. On the dynamics of exploited populations of
    *Tisbe holothuriae* (Copepoda, Harpacticoida). Helgoländer wiss.
    Meeresunters. 31: 285–297.
McArthur, R. M. 1972. Geographical Ecology. Harper and Row, New
    York.
May, R. M. 1976. Theoretical Ecology. Principles and Applica-
    tions. Blackwell Scientific Publications, Oxford.
Menge, B. A. and J. P. Sutherland. 1976. Species diversity gra-
    dients: synthesis of the roles of predation, competition and
    spatial heterogeneity. Amer. Natur. 110: 351–369.
Nicolis, G. and I. Prigogine. 1977. Self-organization in non-
    equilibrium systems. From dissipative structures to order
    through fluctuations. Wiley and Sons, New York.
Velleman, P. F. 1977. Robust nonlinear data smoothers: defini-
    tions and recommendations. Proc. Natl. Acad. Sci. U.S. 74:
    434–436.

Vincx, M. and C. Heip.  Larval development and biology of *Canuella perplexa* T. and A. Scott, 1893 (Copepoda, Harpacticoida).  Cah. Biol. Mar. (in press).

Warwick, R. M.  1980.  Population dynamics and secondary production of benthos. pp.    to      *In* K. R. Tenore and B. C. Coull (eds.).  Marine Benthic Dynamics.  Univ. of South Carolina Press, Columbia.

# Experimental Evidence for a Model of Juvenile Macrofauna-Meiofauna Interactions

Susan S. Bell
Bruce C. Coull

ABSTRACT: Juvenile macrofauna may comprise a significant portion of the total meiofauna biomass,
but no information exists on possible biological
interactions between the permanent meiofauna and
juvenile macrofauna (temporary meiofauna). Using
manipulative caging techniques, we tested the hypothesis that juvenile macrofauna-meiofauna interactions influence abundance patterns of both these
groups. Increases in permanent meiofauna densities inside cages were associated with density
decreases of juveniles of the polychaete, *Streblospio benedicti*. In control (non-caged) areas
*S. benedicti* juveniles increased while permanent
meiofauna decreased.

Our data indicate that permanent meiofauna
have a negative effect on juvenile *Streblospio
benedicti*. The mechanism of this interaction is
unknown but may be due to permanent meiofauna
changing sediment properties, being more efficient at food utilization and/or aggressive interactions between the meiobenthic polychaete
*Manayunkia aestuarina* and *Streblospio benedicti*.
In any case, there is decreased recruitment/sur-

vival of newly-settled *S. benedicti* when permanent meiofauna densities are high.

   We present a model that explores tempo-
rary/permanent meiofauna relationships and suggest
that such interactions are important in structuring both meiofaunal and macrofaunal assemblages.

INTRODUCTION

   Meiobenthologists commonly record seasonal influxes of juvenile
macrofauna (temporary meiofauna *sensu* McIntyre, 1964) such as bivalves,
polychaetes and gastropods into the meiofauna community. However,
these "temporary meiofauna" are often neglected in any discussion of
meiofauna community structure (e.g. Coull and Fleeger, 1977) or energy
flow probably due to their ephemeral nature and/or low numerical abundance compared to other taxa (permanent meiofauna *sensu* McIntyre, 1964).
On the other hand, juvenile macrofauna may comprise a significant proportion of meiofaunal biomass (Thorson, 1966) or meiofaunal ATP (Yingst,
1978) (see Table 1). These observations suggest that an organism need
not be numerically dominant to exert an important effect on community
structure or on the transfer and packaging of energy. Presently, however, there exists no synthesis or discussion of the role played by
juvenile macrofauna while in the meiofaunal size range or the effect
temporary meiofauna may have on the survivorship of permanent meiofauna.
   From a macrofaunal perspective, the importance of the "meiofaunal"
stage of juvenile macrofauna has been discussed by Muus (1966) and
Thorson (1966). Juvenile macrofauna are generally not encountered
in life history or survey studies by macrobenthologists since most
young stages pass through the processing sieves (usually ca. 0.5 mm).
Therefore differential recruitment or survival of juvenile macrofauna
has been only indirectly studied by measuring density and/or species
changes in adult populations (e.g., McCall, 1977; Brousseau, 1978;
Virnstein, 1978). Few others have attempted to directly measure
either mortality of juvenile macrofauna or juvenile macrofauna–meiofauna associations while these temporary forms remain within the
meiofaunal size range (but see Muus, 1966, 1973; Shaffer, 1978).
   Juvenile macrofauna–meiofauna interactions may be important processes which influence establishment of macrofauna or meiofauna given
that: 1) juvenile macrofauna appear in the meiofaunal community during periods of seasonal increases in permanent meiofauna (Bell,
1979a; Coull, unpubl.); 2) the majority of juvenile macrofauna utilize the upper 0.5 to 1.0 cm of sediment where meiofauna are also
most abundant (Bell, 1979a; Coull, unpubl.; Coull and Bell, 1979;
Yingst, 1978); and 3) juvenile macrofauna probably eat detritally-
associated microbes as do the meiofauna (Fenchel, 1970; Fauchauld
and Jumars, 1979). Thorson (1966) offered a somewhat complementary
suggestion that predation by permanent meiofauna on temporary meiofauna may have a significant effect on newly settled larvae. Woodin
(1976) has discussed interrelationships between adult macrofauna and

Table 1.  Summary of studies that quantify temporary meiofauna (juvenile macrofauna) as part of the total meiobenthos.  Columns 3 (% abundance) and 4 (% biomass) indicate the contribution of these temporary forms to the total meiofauna.

| Juvenile Macrofauna Taxa | Location | % Abundance | % Biomass | Reference |
|---|---|---|---|---|
| polychaetes bivalves | Scotland 101, 146 m | 1.4 - 3.0 | 22.0 | McIntyre 1964 |
| polychaetes bivalves | Rhode Island, USA Estuarine, 2-3 m | 0.1 - 4.6 | 4.5 - 11.4 | Tietjen 1969 |
| polychaetes bivalves | Bermuda, 2 - 5 m | 1.0 - 5.0 $\bar{x}$ = 1.3 | up to 26.4 | Coull 1970 |
| annelids lamellibranchs gastropods | South France 15 - 91 m | 7.0 - 15.8 | 10.0 - 25.0 | Guille and Soyer 1971 |
| molluscs | Chupinsky Inlet, USSR subtidal, mud-sand | 33.0 | 69.0 | Galtsova 1971 |
| "temporary meiofauna" | Finland shallow, soft bottom with macrophytes | up to 15.0 $\bar{x}$ = 5.0 | 60.0 | Elmgren and Ganning 1974 |
| *Pontoporeia affinis* (amphipod) bivalves polychaetes | Bothnian Gulf, Baltic subtidal, brackish and Arctic characteristics | < 1.0 | 35.0 | Elmgren et al. 1975 |
| mostly *Macoma baltica* | North Baltic archipelago, subtidal | 1.8 | 17.0 | Ankar and Elmgren 1975 |
| polychaetes oligochaetes | Tees Estuary, England subtidal, fine sand | 11.0 | 98.0 | Gray 1976 |
| polychaetes | Georgia Bight, USA 13 - 44 m | 1.6 | 15.0 | Tenore et al. 1978 |
| *Yoldia limatula* *Nucula annulata* (bivalve) polychaetes | Long Island Sound, USA 14 m silt-clay | up to 40.0 | up to 60.0% total meiofaunal ATP | Yingst 1978 |
| *Streblospio benedicti* (polychaete) | North Inlet, South Carolina, USA high marsh site | up to 9.0 | up to 14.0% of total meio-faunal ATP* | Bell 1979a |
| polychaetes bivalves | North Inlet, South Carolina, USA subtidal (1m), mud and sand | $\bar{x}$ = 3.4 (7 years) | 2.6 - 44.8 (7 years) | Coull unpublished |

*Based on values from Yingst 1978.

newly settling/settled larvae and Bell (1979b) suggested that inter-actions between juvenile macrofauna and meiofauna may also exist. As an extension of these ideas, we hypothesize that permanent meio-fauna-juvenile macrofauna interactions influence the abundance of both these groups.  Furthermore, we initiated a series of experiments to test if there were associations between juvenile macrofauna and meiofauna.

To provide background for our discussion of juvenile macrofauna-meiofauna interactions, a conceptual diagram of the different stages of a macrofaunal organism is outlined in Figure 1a and the potential mortality experienced by juvenile macrofauna in the "meiofauna stage" in Figure 1b.  As a general case, our model argues that significant mortality may be experienced by juvenile macrofaunal forms during their "temporary meiofauna" stage, and our hypothesis contends that such mortality may be due to interrelationships with permanent meio-fauna.  In specific instances, the curve may have a somewhat differ-

ent and as yet unknown shape (i.e., brooded vs. planktonic larvae;
opportunistic, short-lived macrofauna vs. long-lived forms).  We
argue that meiofauna-juvenile macrofauna interactions are potentially
important regardless of the life history features of the macrofauna.
We do acknowledge, however, that the relative importance of these interactions may vary depending on both abundance of juvenile macrofauna
and meiofauna, and duration spent in the meiofaunal size range by juvenile macrofauna.  Our study was designed to focus on the "meiofaunal bottleneck" (Fig. 1b), and the importance of such relationships between the juveniles of large-sized adults and adults of
smaller forms is not without precedent (Lynch, 1978; Neill, 1975;
Werner and Hall, 1979; Werner *et al.*, 1977).

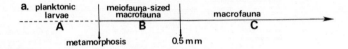

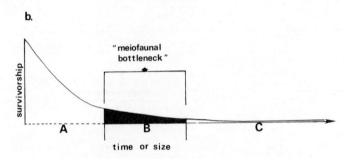

Figure 1.  (a)  Schematic diagram showing
different life history stages of a macrobenthic organism.  (b)  Theoretical survivorship
curve for macrofaunal organism for each of
the stages outlined in (a).  The "meiofaunal
bottleneck" is specifically addressed in this
study.

METHODS AND MATERIALS

To test our hypothesis of juvenile macrofauna-meiofauna interactions we used manipulative caging techniques which are known to
elicit density increases in permanent meiofauna inside cages vs. non-
caged areas (Bell, 1979b).  If juvenile macrofauna-meiofauna interactions were important, one would predict an increase in juvenile
macrofauna mortality (or decreased abundance) in areas with higher
densities of permanent meiofauna.  Likewise, if juvenile macrofauna
had a negative effect on permanent meiofauna, one would predict juvenile macrofauna increases would lead to permanent meiofauna decreases.

Our study was conducted in the high salt marsh of North Inlet,
Georgetown, South Carolina ($30^{\circ}21.0'$N, $79^{\circ}11.5'$W).  The area is covered by the tide for 3-5 h per tidal cycle and dominated by *Spartina*

*alterniflora* Loisel approximately 0.8 m in height.  The site is
uniquely suited for our study:  1) this area is amenable to manipulation (Bell, 1979b); 2) there is a dense root and rhizome mat which
stabilizes the sediment and this dense plant assemblage minimizes environmental alterations associated with the physical presence of a
cage compared to open mudflat or subtidal areas (Virnstein, 1978);
3) tidal velocities are reduced and therefore sediment movement and
resuspension minimal (Richard, 1978); 4) juvenile *Streblospio benedicti* Webster (Polychaeta:  Spionidae) settle during the fall (density range 1978, 4-12·10 cm$^{-2}$; Bell, 1979a).  Additionally, we
have extensive background information on this meiofaunal community
(Bell, 1979a).  Nematodes comprise approximately 70% of the community by number and copepods ($\bar{x}$ = 18.0%) and polychaetes ($\bar{x}$ = 8.8%)
rank second and third in abundance, respectively.  Four species comprised up to 80% of the copepod community:  *Stenhelia (D.) bifidia*
Coull, *Microarthridion littorale* (Poppe), *Enhydrosoma propinquum*
(Brady), and *Schizopera knabeni* Lang.  Distinct fall and spring peaks
of copepod densities were observed in each of three consecutive years
(1976-79).  *Manayunkia aestuarina* (Bourne), a minute tube-building
sabellid polychaete which is a permanent member of the meiofauna
(Bell and Coull, 1978a), comprised 90% of the polychaete assemblage
throughout the year, except in the fall when juveniles of *Streblospio benedicti* appeared.  Further information on the community structure of this high marsh assemblage is available in Bell (1979a).

     Cages (1 m x 1 m) of 2 mm galvanized steel mesh screening were
utilized to exclude natant macroepifauna from the high marsh since
previously we observed increased meiofauna densities in the absence
of macroepifauna (see Bell, 1979b).  Each cage had four sides approximately 76 cm in height stapled to wooden stakes and recessed 15 cm
into the sediment.  No tops were necessary because tidal waters rarely exceeded 50 cm in depth.  Two cages were erected on 2 September
1978 and equivalent marsh control areas (uncaged) were staked adjacent to cage treatments.  Two "two sides only" (1 m in length) cage
controls of identical mesh size and oriented perpendicular to tidal
flow were also included in the experimental design.  Such controls
were included to assess the possible effects of cage structure while
allowing macroepifauna access to the partially-caged area.

     A 5 x 5 (0.2 m x 0.2 m) grid was established within the cages,
cage controls (two sides only), and open marsh sites.  Within each
treatment area samples were taken randomly in areas free of biogenic
structures.  Those sectors touching the cage edge were excluded from
sampling.  Eight samples were taken initially from a marsh control
site and four from each cage or treatment on three subsequent dates
(11 September, 2 October, 12 November).  All samples were taken with
a hand-held plastic corer (2.54 cm inner diameter) to a sediment
depth of 3.0 cm.  A total of 80 cores were taken.

     In the laboratory all samples were washed live through 500 μm
and 63 μm mesh sieves.  The 500 μm sieve retained the root material
and these contents were subsequently saved for examination of large
polychaetes (e.g. *Streblospio benedicti*).  The contents of the 63 μm
mesh sieve were subjected to Ludox$^{Tm}$ centrifugation (de Jonge and
Bouwman, 1977) to extract the animals.  Samples were preserved in
10% buffered seawater formalin and stained with Rose Bengal.  Meio-

fauna were enumerated under a dissecting microscope and identified
to major taxon or species when possible.

Student-Newman-Keuls (SNK) multiple comparison tests (Miller,
1966) were used to test for density differences between treatments
and within treatments over time. All values were subjected to the
$F_{max}$ test (Sokal and Rholf, 1969) to test for homogeneity of variance prior to multiple comparison. A $\log_{10}$ (x + 1) transformation
was used if homogeneity of variance was rejected by the $F_{max}$ test.

RESULTS

Figure 2 illustrates density changes of the dominant, permanent
meiofauna (i.e. nematodes, copepods, *Manayunkia aestuarina*), and juvenile *Streblospio benedicti* (10-27 setigers in both cage and control
treatments) from our experiments. A SNK multiple comparison test of
the densities of both temporary and permanent meiofauna from all
treatments (Table 2) indicates the following: 1) densities of both
permanent and temporary meiofauna were significantly different in
cage and open marsh treatments; 2) densities of taxa from cage controls (two sides only) and open marsh areas were not significantly
different. All differences between cages and open controls were also significantly different for cages and cage controls; 3) abundance
of all taxa in both controls changed over time. *Streblospio benedicti* juveniles were the only organisms that did not change over
time in the cages.

A summary of the important differences between cage and open
marsh treatments is presented in Table 3. Copepods increased quickly inside cages and then returned to background levels. This rapid
increase and subsequent decline in copepod abundance was due primarily to the dominant, *Stenhelia* (*D.*) *bifidia*. Such patterns of abundance are similar to these reported and discussed by Bell (1979b).
Densities of *Streblospio benedicti* juveniles were very low throughout September in both open and cage sites but in October and November, 1978, densities increased significantly in open marsh areas
compared to cages (Fig. 2). In contrast both nematodes and the sabellid polychaete, *Manayunkia aestuarina*, increased inside cages compared to open marsh areas during this time.

DISCUSSION

The results of these experiments support our hypothesis that
permanent meiofauna influence juvenile macrofauna abundance. The
significant decrease in juvenile *Streblospio benedicti* density associated with increased numbers of permanent meiofauna suggest that increases
in permanent meiofauna may negatively affect recruitment/survival of
juvenile macrofauna. However, we cannot definitively discern whether the reduced density in *Manayunkia aestuarina* or nematodes in open
control sites was due to increased abundance of *S. benedicti* or mac-
roepifaunal effects (see Bell, 1979b).

Although our experimental findings support our hypothesis, our
data do not identify the exact nature of the interactions. We propose a number of possible mechanisms which may contribute to the ob-

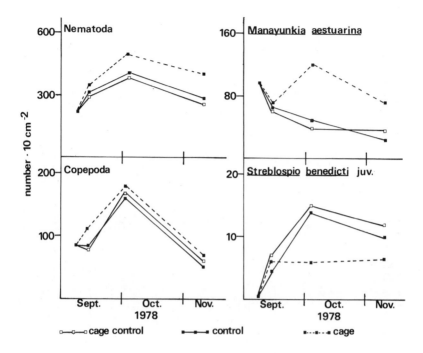

Figure 2. Mean densities of meiofauna taxa from cages, control and cage control sites after initiation of caging (September 2, 1978). Statistical comparisons of treatments are presented in Table 2.

served negative relationship between permanent meiofauna and juvenile *Streblospio benedicti*:

1) Permanent meiofauna may be better able to utilize or exploit food sources compared to the deposit-feeder, *Streblospio benedicti*, because there are many different types of nematode feeding groups represented in the high marsh (Bell, unpubl.) and *M. aestuarina* is known to be able to feed both as a deposit and secondary-suspension feeder (Lewis, 1968).

2) *Manayunkia* feeding processes (described by Lewis, 1968) as well as nematode feeding (e.g., Riemann and Schrage, 1978; Tietjen and Lee, 1977) and burrowing activities (Cullen, 1973; Frankel and Mead, 1973; Neumann *et al.*, 1970) may alter sediment properties and associated microbial assemblages. These changes may differentially affect settlement of macrofauna larvae and/or subsequent survival of juveniles.

3) *Manayunkia aestuarina*, a discretely mobile tube-builder, if disturbed from its tube will search for another, or build a new one. Under laboratory conditions we have observed *M. aestuarina* displacing *S. benedicti* from its tube and such aggressive behavior may adversely affect establishment of *S. benedicti*.

4) Increased copepod density may change sediment properties and/or food resources during recruitment. Such changes, however, are probably short-lived.

Table 2. A SNK multiple comparison test comparing meiofauna densities from cage, control and cage control (2 sides only) treatments over time. Since there were no differences (p < 0.05) between replicate cages, controls or cage controls, samples from replicate treatments were combined in the analyses. Those dates that are underlined represent densities that are not significantly different (p < 0.05).

| Nematodes | 11-12 cage control | 11-12 control | 9-11 cage control | 9-11 control | 9-11 cage control | 10-2 cage control | 10-2 control | 11-12 cage | 10-2 cage |
|---|---|---|---|---|---|---|---|---|---|
| Copepods | 11-12 control | 11-12 cage control | 9-11 cage control | 9-11 control | 9-11 cage | 10-2 cage | 10-2 control | 10-2 cage control | 10-2 cage |
| Manayunkia aestuarina | 11-12 control | 11-12 cage control | 10-2 cage control | 10-2 control | 9-11 cage control | 9-11 control | 9-11 cage control | 9-11 cage | 11-12 cage | 10-2 cage |
| Streblospio benedicti | 9-11 control | 9-11 cage | 10-2 cage | 11-12 cage | 11-12 cage control | 9-11 cage control | 11-12 control | 11-12 cage control | 10-2 control control | 10-2 cage control |

Table 3.  Summary of significant (p ≤ 0.05)[*] density differences in meiofauna taxa from cage and control treatments after initiation of caging (September 2, 1978).

|  | September 11, 1978 | October 2, 1978 | November 2, 1978 |
|---|---|---|---|
| Nematodes | n.s. | [*]higher inside cage | [*]higher inside cage |
| Copepods | [*]higher inside cage | n.s. | n.s. |
| *Manayunkia aestuarina* | n.s. | [*]higher inside cage | [*]higher inside cage |
| *Streblospio benedicti* (juveniles) | n.s. | [*]higher in control sites | [*]higher in control sites |

The decrease in *Streblospio benedicti* juveniles is probably not due to predation by meiofauna since known meiofaunal predators (i.e., *Protohydra leukarti*, proseriate turbellarians) were not abundant in our cages.  Macroinfauna predation is also unlikely since macroinfauna densities are extremely low in the high marsh ($1-10 \cdot m^{-2}$; Bell, 1979a), and we encountered no macroinfauna in any of our samples. Additionally, because there was no difference in juvenile *S. Benedicti* density between cage control areas and open marsh sites, differential recruitment of polychaete larvae into cages is not suggested.  In fact, the opposite was observed (higher densities outside cages).  Furthermore, we reject an alternative interpretation that macroepifauna may have a direct positive effect on juvenile polychaetes (*S. benedicti*).  We cannot imagine how macroepifauna would enhance juvenile *S. benedicti* survival and/or recruitment except through its influence on other meiofaunal taxa (Bell, 1979b). Therefore, we feel that our interpretation of changes in juvenile macrofauna abundance as a function of juvenile macrofauna-meiofauna interactions is justified.

Large fluctuations in densities of naturally occurring populations of nematodes, copepods and *Manayunkia aestuarina* have been observed over years and within seasons in the high marsh habitat (e.g. *Manayunkia aestuarina* densities increased 200% in control sites in Fall, 1979, compared to Fall, 1978; Bell, 1979a).  Accordingly, such variations may play an important role in influencing

juvenile macrofauna patterns of abundance and subsequent adult density levels.  Although juvenile macrofauna-meiofauna interactions
are important in our high marsh site, their relative importance in
other systems and for other juvenile forms, such as bivalves, remains to be confirmed.

The strong possibility of the existence of such interactions,
however, challenges us to reconsider the role of meiofauna in benthic dynamics.  We envision three ways potential meiofauna-juvenile
macrofauna interactions may augment our present concepts of benthic
processes:  1) Meiofauna recolonization and meiofauna-juvenile macrofauna interrelationships may be previously unrecognized factors
influencing successful establishment of colonizing macrofauna and
community development.  2) In macrofaunal studies that have excluded suspected predators from a given site, the observed changes in
macrofauna abundance may not necessarily represent a direct result
of predator exclusion but rather may be due to associated changes in
sediment properties, meiofauna, and subsequently, juvenile macrofauna.  Our predicted effects provide an alternative explanation
that should be considered or evaluated in experimental interpretation.  3) Juvenile macrofauna-meiofauna associations may play an
ancillary role to adult-larval interactions (Woodin, 1976) in influencing community composition.  For instance, brooding suspension-
feeding bivalves release larvae directly into the "meiofaunal size
category" and such juveniles may not only be susceptible to predation by adults but also to interactions with permanent meiofauna as
well.  Likewise, because permanent meiofauna are known to be positively associated with burrows or tubes of adult macrofauna (Aller
and Yingst, 1978; Bell *et al.*, 1978), meiofauna-juvenile macrofauna interactions may be important in influencing distribution and
abundance of juvenile macrofauna in dense assemblages of such tube-
building or burrowing macrofauna.

Figure 3 summarizes reported or proposed biological interactions between three operationally defined groups:  1) permanent
meiofauna; 2) adult macrofauna; 3) temporary meiofauna (juvenile
macrofauna) within a soft bottom assemblage.  Such interactions,
which may vary in time and space, or even between taxa within
groups, may act in complementary or inverse directions.  Likewise,
it is possible that observed associations between any two groups
may be modified or mediated by a third group.  Thus, any of these interrelationships may be either directly or indirectly sensitive to
changes in abundance of another group.  It is in this context that
the importance of meiofauna-juvenile macrofauna interactions becomes evident in marine benthic dynamics.

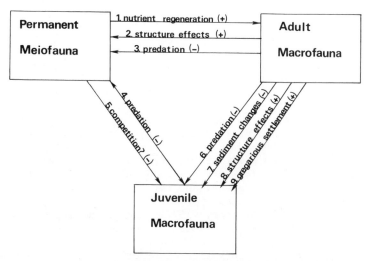

Figure 3. Model of documented biological interactions between three categories of soft-bottom assemblages. (+) or (-) refers to the type of association
observed or predicted between groups as a result of
interaction. Arrows point from the category producing
the effect. Numbers represent literature source: 1.
Tenore *et al.*, 1977; 2. Aller and Yingst, 1978; Bell
*et al.*, 1978; Lee *et al.*, 1977; 3. Bell and Coull,
1978b; Buzas, 1978; 4. Thorson, 1966; Sarvala, 1971;
5. Bell and Coull, this paper; 6. Woodin, 1976 (and
references within); 7. Woodin, 1976; 8. Woodin, 1976;
9. Woodin, 1979.

ACKNOWLEDGMENTS

This research was partially supported by the National Science
Foundation, Grant No. OCE 78-09450.

REFERENCES

Aller, R. C. and J. Y. Yingst. 1978. Biogeochemistry of tube-
    dwellings: A study of the sedentary polychaete *Amphitrite or-
    nata*. (Leidy). J. Mar. Res. 36: 201-254.
Ankar, S. and R. Elmgren. 1975. A survey of the benthic macro-
    and meiofauna of the Asko-Landsort area. Merentutkimuslait.
    Julk./Havsforskninginst. 239: 257-264.
Bell, S. S. 1979a. Short- and long-term variation in a high marsh
    meiofauna community. Est. Cstl. Mar. Sci. (in press).
Bell, S. S. 1979b. Meiofauna-macrofauna interactions in a high
    salt marsh habitat. Ecological Monogr. (in press).
    Columbia. 63 pp.
Bell, S. S. and B. C. Coull. 1978a. The meiobenthic polychaete
    *Manayunkia aestuarina* in South Carolina salt marshes. Amer.
    Zool. 18: 643 (Abstract).

Bell, S. S. and B. C. Coull. 1978b. Field evidence that shrimp predation regulates meiofauna. Oecologia (Berl.). 35: 141–148.

Bell, S. S., M. C. Watzin and B. C. Coull. 1978. Biogenic structure and its effect on the spatial heterogeneity of meiofauna in a salt marsh. J. Exp. Mar. Biol. Ecol. 35: 99–107.

Brousseau, D. J. 1978. Population dynamics of the soft-shell clam *Mya arenaria*. Mar. Biol. 50: 63–71.

Buzas, M. A. 1978. Foraminifera as prey for benthic deposit feeders: results of predator exclusion experiments. J. mar. Res., 36: 617–625.

Coull, B. C. and S. S. Bell. 1979. Perspectives of marine meiofaunal ecology, pp. 189–216. *In* R. J. Livingston (ed.), Ecological Processes in Coastal and Marine Ecosystems, Plenum Press, New York.

Coull, B. C. and J. W. Fleeger. 1977. Long-term temporal variation and community dynamics of meiobenthic copepods. Ecology 58: 1136–1143.

Cullen, D. J. 1973. Bioturbation of superficial marine sediments by interstitial meiobenthos. Nature 242: 323–324.

Elmgren, R. and B. Ganning. 1974. Ecological studies of two shallow, brackish water ecosystems. Contr. Asko Lab, Univ. Stockholm. 6: 1–56 pp.

Elmgren, R., R. Rosenberg, A. B. Andersin, S. Evans, P. Kangas, J. Lassig, E. Lappakoski, and R. Varmo. 1975. Benthic macro- and meiofauna in the Gulf of Bothnia (Northern Baltic). Unpubl. manus. 32 pp.

Fauchauld, K. and P. A. Jumars. 1979. The diet of worms: a study of polychaete feeding guilds. Ann. Rev. Mar. Biol. Oceanogr. 17: 193–284.

Fenchel, T. 1970. Studies on the decomposition of organic detritus derived from the turtle grass *Thalassia testudinum*. Limnol. Oceanogr. 15: 14–20.

Frankel, L. and D. J. Mead. 1973. Mucilaginous matrix of some estuarine sands in Connecticut. J. Sed. Petrol. 43: 1090–1105.

Galtsova, V. V. 1971. A quantitative characteristic of meiobenthos in the Chupinsky Inlet of the White Sea. Zool. Zhur. 50: 641–647.

Gray, J. S. 1976. The fauna of the polluted River Tees Estuary. Est. Cstl. Mar. Sci. 4: 653–676.

Guille, A. and J. Soyer. 1971. Contribution a l'etude comparee des biomass du macrobenthos et du meiobenthos de substrat meuble au large de Banyuls-sur-Mer II. Vie et Milieu Suppl. 22: 15–29.

De Jonge, V. N. and L. A. Bouwman. 1977. A simple density separation technique for quantitative isolation of meiobenthos using the colloidal silica Ludox-TM. Mar. Biol. 42: 143–148.

Lee, J. J., J. H. Tietjen, C. Mastropaolo and H. Rubin. 1977. Food quality and the heterogeneous spatial distribution of meiofauna. Helgoland. wiss. Meeresunters. 30: 272–282.

Lewis, D. B. 1968. Feeding and tube-building in the Fabriciinae (Annelida: Polychaeta). Proc. Linn. Soc. Lond. 179: 37–49.

Lynch, M. 1978. Complex interactions between natural coexploiters-*Daphnia* and *Ceriodaphnia*. Ecology. 59: 552–564.

McCall, P. L.   1977.   Community patterns and adaptive strategies of
    the infaunal benthos of Long Island Sound.   J. Mar. Res. 35:
    221–266.
McIntyre, A. D.   1964.   Meiobenthos of sublittoral muds.   J. Mar.
    Biol. Ass. U. K. 44:   665–674.
Miller, R. G., Jr.   1966.   Simultaneous statistical inference.
    McGraw–Hill, New York.
Muus, K.   1966.   A quantitative 3–year survey on the meiofauna of
    known macrofauna communities in the Oresund.   Veroff. Inst.
    Meresforch. (Bremerhaven) 2:   289–292.
Muus, K.   1973.   Settling, growth and mortality of young bivalves
    in the Oresund.   Ophelia. 12:   79–116.
Neill, W. E.   1975.   Experimental studies of microcrustacean compe-
    tition, community composition, and efficiency of resource utili-
    zation.   Ecology. 56:   809–826.
Neumann, A. C., C. D. Gebelein and T. P. Scoffin.   1970.   The com-
    position, structure and erodability of subtidal mats, Abaco,
    Bahamas.   J. Sed. Petrol. 49:   249–273.
Richard, G. A.   1978.   Seasonal and environmental variations in
    sediment accretion in a Long Island salt marsh.   Estuaries. 1:
    29–35.
Riemann, F. and M. Schrage.   1978.   The mucus-trap hypothesis on
    feeding of aquatic nematodes and implications for biodegradation
    and sediment texture.   Oecologia (Berl.). 34:   75–88.
Sarvala, J.   1971.   Ecology of *Harmothöe sarsi* (Malgrem) (Polychae-
    ta, Polynoidae) in the northern Baltic area.   Ann. Zool. Fennici.
    8:   231–309.
Shaffer, P. L.   1978.   Population dynamics of the capitellid poly-
    chaete *Heteromastus filiformis*.   Amer. Zool. 18:   663
    (Abstract).
Sokal, R. R. and R. J. Rholf.   1969.   Biometry.   W. H. Freeman, San
    Francisco.   775 pp.
Tenore, K. R., C. F. Chamberlain, W. M. Dunstan, R. B. Hanson, B.
    Sherr and J. H. Tietjen.   1978.   Possible effects of Gulf Stream
    intrusions and coastal runoff on the benthos of the continental
    shelf of the Georgia Bight, pp. 557–598.   *In* M. L. Wiley (ed.)
    Estuarine Interactions, Academic Press, New York.
Tenore, K. R., J. H. Tietjen and J. J. Lee.   1977.   Effect of meio-
    fauna on incorporation of aged eelgrass, *Zostera marina*, detritus
    by the polychaete *Nepthys incisa*.   J. Fish. Res. Bd. Can. 34:
    563–567.
Thorson, G.   1966.   Some factors influencing the recruitment and
    establishment of marine benthic communities.   Nethl. J. Sea Res.
    3:   267–293.
Tietjen, J. H.   1969.   The ecology of shallow water meiobenthos in
    two New England estuaries.   Oecologia (Berl.). 2:   251–291.
Tietjen, J. H. and J. J. Lee.   1977.   Feeding behavior of marine
    nematodes, pp. 21–36.   *In* B. C. Coull (ed.) Ecology of Marine
    Benthos, Univ. South Carolina Press, Columbia.
Virnstein, R. W.   1978.   Predator caging experiments in soft-sedi-
    ments:   caution advised, pp. 261–274.   *In* M. L. Wiley (ed.) Es-
    tuarine Interactions, Academic Press, New York.
Werner, E. E. and D. J. Hall.   1979.   Foraging efficiency and habi-

tat switching in competing sunfishes.  Ecology. 60:  256-264
Werner, E. E., D. J. Hall, D. R. Laughlin, D. J. Wagner, L. A.
    Wilsmann, and F. C. Funk.  1977.  Habitat partitioning in a
    freshwater fish community.  J. Fish. Res. Bd. Can. 34:  360-370.
Woodin, S. A.  1976.  Adult-larval interactions in dense in-
    faunal assemblages: Patterns of Abundance.  J. Mar. Res. 34:
    25-41.
Woodin, S. A.  1979.  Settlement phenomena:  the significance of
    functional groups, pp. 99-106.  *In* S. E. Stancyk (ed.) Repro-
    ductive Ecology of Marine Invertebrates, Univ. South Carolina
    Press, Columbia.
Yingst, J. Y.  1978.  Patterns of micro- and meiofaunal abundance
    in marine sediments, measured with adenosine triphosphate assay.
    Mar. Biol. 47:  41-54.

# NUTRIENT CYCLING

# Sediment-Water Interactions in Nutrient Dynamics

Bernt Zeitzschel

ABSTRACT: In shallow water ecosytems there is good evidence that sediments play an important role in supplying up to 100% of the nutrient requirements for phytoplankton primary productivity.

There are two different approaches to obtain data on the nutrient flux from sediments: 1) calculation of the flux using concentration gradients of constituents of the interstitial water or the water column and 2) measuring the nutrient flux directly by *in situ* or laboratory experiments. Both methods similarly indicate a correlation of total oxygen consumption and nutrient release from sediments with high rates characteristic of shallow water areas and rates two to four orders of magnitude less in the deep ocean.

The cycling of matter in shallow water ecosystems includes: 1) phytoplankton blooms, 2) sedimentation of these blooms (directly or via fecal pellets of herbivores), 3) decomposition of organic material (mainly at the water-sediment

interface), and 4) release of nutrients from the
bottom and recycling of these nutrients to the
euphotic zone by turbulent mixing.

These four events follow one another with
a certain time lag. The disturbance of this
pattern causes the observed variability of phytoplankton primary production in space and time.


INTRODUCTION

There are two important boundaries in the ocean where governing processes dominate: 1) the boundary between the air and the sea
and 2) the sediment-water interface. At the former, important processes dominate which influence the climate; at the latter, numerous physical, chemical, and biological reactions occur which are
important for life in the sea.

Phytoplankton plays a dominant role in the coupling of the
pelagic and benthic regime in the open ocean (In shallow areas seaweeds are also important). Phytoplankton are of great ecological
significance because they comprise the major portion of primary
producers in the open sea. According to Koblentz-Mishke $et$ $al$.
(1970) the total annual net primary production by phytoplankton of
the world ocean amount to 15–18 x $10^9$ metric tons of carbon.
In addition, primary production of phytobenthos adds at least 0.65
x $10^9$ metric tons more of carbon (Bunt, 1975). Primary production
by phytoplankton contributes substantially to the energy required
both by pelagic zooplankton and in shallow water ecosystems by
benthic animals. There is a controversy, however, as to whether
or not there is a direct input of organic matter produced in the
euphotic zone to the bottom of the open ocean. The variation of
plankton production in time and space in the pelagic environment
is very remarkable: primary production in the open ocean, which
accounts for about 90% of the total surface area, averages 50 g C
$m^{-2}y^{-1}$. Production in the continental shelf regions, which make
up about 10% of the whole ocean, and where the water depth is <
200 m is of the order of 100–150 g C $m^{-2}y^{-1}$. The highest of these
production values (300–500 g C $m^{-2}y^{-1}$) are recorded in upwelling
areas and estuarine environments, but the areas concerned make up
only a fraction of 1% of the whole surface of the ocean (Ryther,
1969).

Factors governing phytoplankton production in the sea are
light, nutrients, and water column stability. In this paper I
shall concentrate mainly on the role of nutrients and specifically
nutrient dynamics of the sediment-water interface. There is a
tremendous amount of literature in chemical oceanography describing the distribution of nutrients in the oceans. A short summary
may be as follows: nutrient concentration in the deeper parts of
the ocean, below a few hundred meters, is usually high during the
entire year, whereas nutrients are used up by phytoplankton in
the euphotic zone (e.g., when light conditions are favorable

for growth).  If there is a mechanism to transport nutrient-rich
deeper water to the surface, plankton blooms might develop (e.g.,
in upwelling areas of the west side of the continents or at the
equator).  If no such supply of nutrients occurs, especially in
low latitudes, productivity is rather poor (e.g., in the oceanic
gyres).  The supply of nutrients for phytoplankton growth in these
areas is limited.  The available nutrients are recycled in the
upper water column by decomposition of organic matter and exuda-
tion and excretion of organisms.  According to Bishop *et al*. (1978),
in the deep oceans more than 90% of the organic matter produced in
the euphotic zone is recycled in the upper 400 m.  The recycling
efficiency is nearly 99% in areas of low productivity.

In shallow-water systems, there is good evidence that the
sediments at the bottom of the sea play an important role in
nutrient supply to the euphotic zone.  A schematic diagram of a
shallow water ecosystem is depicted in Figure 1.  In the euphotic
zone (about 10 m water depth), organic matter is produced by photosynthesis.  Part of the standing stock of phytoplankton is eaten
by herbivorous zooplankton; part sinks directly to the bottom
depending on biological (physiological) and physical conditions.
Sinking particles may be decomposed partly in the water column;
most of the decomposition takes place at the sediment-water interface.  On the left side of the figure the primary productivity
versus depth is given; on the right side are shown the exponential decrease of light with depth and the concentration of nutrients in early summer.

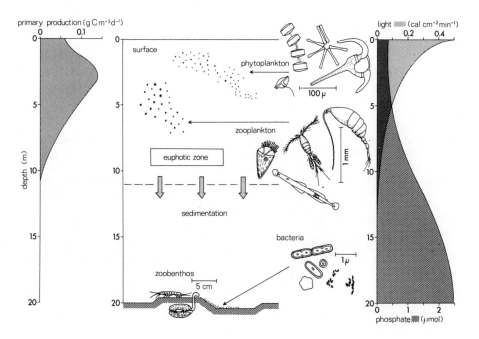

Figure 1.  Schematic diagram of a shallow water ecosystem.

On the shelf and in other shallow parts of the ocean, regeneration of nutrients involves the sediment surface because the settling time for phytoplankton and organic detritus is relatively short and the supply of organic matter is large (Parsons *et al.*, 1977).  One indication that supports this idea is the remarkably high concentration of inorganic nutrients in the pore water of the sediments (Sholkovitz, 1973; Berner, 1974; Gieskes, 1975; Manheim, 1976; Suess, 1976; and Lerman, 1977).

Input to the Sediment

There are two main sources of material that may be deposited on the sea floor: 1) terrigenous material and 2) organic matter and biogenic hard parts (e.g., $CaCo_3$, $SiO_2$) produced in the sea by living organisms.

Terrigenous material is transported to the ocean by rivers, by wind, or by wave action via erosion.  The input of man-made substances (e.g., by dumping) is, at least up to now, only of local importance.  With increasing distance from land the input of terrigenous material decreases.

Representative values of the annual carbon deposition and the percentage it represents of primary production in the euphotic zone are given in Table 1.  In summary, approximately 1-10% of the primary production reaches the deep ocean, whereas between 25 and 60% sinks down to the bottom in shallow water.  Consequently, the accumulation of particulate matter in the deep sea is low (few mm per 1,000 years).  The content of organic carbon in the settled material is less than 1% of the dry weight.  In contrast, the accumulation of organic carbon in shallow water is of the order of a few mm per year, and the organic carbon concentration is in the range of 5-10%.  The sedimentation of particles in the ocean is a very complex subject (for further information see McCave, 1975; Hargrave *et al.*, 1976; and Smetacek *et al.*, 1978).  The different factors affecting the flux of organic matter in marine environments is discussed by Hargrave (1980).

Inorganic detritus and biogenic particles reaching the ocean floor and pore water exchange are the main avenues of transport to the sediment (Fig. 2).  Fluxes within the sediment of the benthic boundary layer may be grouped into two categories: 1) the flux downward due to deposition and burial of sediment and 2) diffusional fluxes that may go in any direction depending on the chemical and physical conditions of the reacting system.  In general, the magnitudes and nature of the fluxes are poorly known (Berner *et al.*, 1976).

Chemical Reactions at the Sea Floor

Chemical reactions taking place in the benthic boundary layer are fundamentally influenced by the amount and composition of organic matter.  This component in many sediments is the single most important source of chemical energy.  Because this material is far from equilibrium, it is highly reactive and undergoes alterations mainly at the benthic boundary layer.  It is an energy source and

Table 1. Annual carbon deposition and percentage of primary production in different environments (after Parsons *et al.*, 1977, supplemented and updated)

| Location | Bottom depth (m) | Trap Depth (m) | Sedimentation $(g\ C\ m^{-2}y^{-1})$ | % of Primary production | Reference |
|---|---|---|---|---|---|
| Baltic (Kiel Bight) | 25 | 24 | 40 | 25 | Zeitzschel, 1965 |
| Departure Bay | 32 | 30 | 200 | 50 | Stephens, *et al.*, 1967 |
| Southampton Water | | | | | |
| Calshot | 2,5 | 2 | 376 | | |
| Marchwood | 5,7 | 5 | 1683 | | Trevallion, 1967 |
| Netley | 5,7 | 5 | 339 | | |
| Loch Ewe | 25 | 18 | 30 | 30 | Steele and Baird, 1972 |
| Shallow Water Venezuela | 4 | 4 | – | 40 | Edwards, 1973 |
| Loch Etive | | | | | |
| E 6 | 40 | 36 | 247 | | |
| E 24 | 20 | 18 | 82 | 8 | Ansell, 1974 |
| Loch Creran C 3 | 40 | 24 | 262 | | |
| Loch Thurnaig | 30 | 17 | 28 | 27 | Davies, 1975 |
| St. Margaret's Bay | 15 | 13 | 118 | 36 | Webster, *et al.*, 1975 |
| | 70 | 65 | 134 | | |
| | 62 | 20 | 57 | | |
| | 62 | 30 | 69 | | |
| Bedford Basin | 62 | 40 | 77 | 44 | Hargrave, *et al.*, 1976 |
| | 62 | 50 | 85 | | |
| | 62 | 60 | 91 | | |
| La Jolla Bight | 18,3 | 16,5 | 1214 | | Hartwig, 1976 |
| Baltic (Kiel Bight) | 20 | 19 | 94 | 60 | Bröckel, v., 1978 |
| Baltic (Bornholm Basin) | 64–70 | 60–65 | + | >14 | Smetacek *et al.*, 1978 |
| Bedford Basin | 70 | 20 | 55 | 40 | Hargrave and Taguchi, 1978 |
| | | 30 | 66 | | |
| | | 40 | 76 | | |
| | | 60 | 76 | | |
| Western North | 2200 | 30 above bottom | 6,3 | 6 | Rowe and Gardner, 1978 |
| Atlantic | 2800 | 500 | 2,3 | | |
| | 3650 | 518 " | 4,2 | 4 | |
| Upwelling Peru | 62–1010 | 30 | ++ | 6,7–18,9 | Bröckel, v. 1979 |

+   short-term measurements with moored traps $(0.18\ g\ C\ m^{-2}d^{-1})$
++  short-term measurements with drifting sediment traps $(0,24-0,51\ g\ C\ m^{-2}d^{-1})$

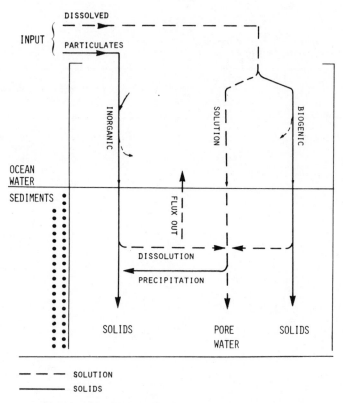

Figure 2. Major transport and reaction paths
in ocean water and sediments. Dashed-line arrows
indicate dissolved material (after Lerman, 1977).

substrate for microorganisms that mediate many physical-chemical
reactions (Berner *et al.*, 1976). Some of the most important diagenetic reactions directly resulting from the bacterial decomposition
of organic matter are the removal of dissolved oxygen, the reduction
of sulfate and the production of ammonia, phosphate, hydrogen sulfide and methane (Berner, 1976). Representative reactions are
given in Table 2. According to Suess (1976), organic decomposition at the benthic boundary layer often changes from oxygen consumption to sulfate reduction. The processes can be described by
reactions based on chemical models, e.g., by Almgren *et al.* (1975),
using model substances like peptides for nitrogen compounds and
phosphate-ester for phosphorus. The chemical reaction described
by Almgren *et al.*(1975) is as follows:

$$(CH_2O)_{89}(NHCO)_{16}C(H_2PO_4)^- + 45SO_4^{2-}$$
$$106CO_2 + 45S^{2-} + 106H_2O + 16NH_4^+ + H_2PO_4^-$$

The stability of the interface and its susceptibility to mixing through turbulence also directly influence the reaction and

Table 2.  Some representative overall biogenic chemical reactions in sediments (After Berner, 1976)

---

Oxygen utilization; $CO_2$ production

$$CH_2O + O_2 \longrightarrow CO_2 + H_2O$$

Nitrate reduction

$$5CH_2O + 4NO_3^- \longrightarrow 2N_2 + 4HCO_3^- + CO_2 + 3H_2O$$

Sulfate reduction

$$2CH_2O + SO_4^{2-} \longrightarrow H_2S + 2HCO_3^-$$

Ammonium formation

$$CH_2NH_2COOH + 2(H) \longrightarrow CH_3COOH + NH_3$$

Methane formation

$$CO_2 + 8(H) \longrightarrow CH_4 + 2H_2O$$

---

chemical migration through accelerated mixing in the uppermost sediments.  Such mixing enhances this mode of transport which exceeds that occurring through molecular diffusion.  Reaction at the benthic boundary layer also affects the physical properties of the sediments.  Breakdown of the organic matter that binds aggregates of detrital material leads to disaggregation and a reduction in grain size.  Precipitation of inorganic salts such as carbonates, phosphates, and Fe-Mn compounds can drastically alter physical properties.

Reactions in the sediment release a number of compounds and yield concentrations that far exceed those found in sea water.  In Figure 3, examples of interstitial water concentrations for phosphate, nitrate and ammonia are given for three different marine environments:  Kiel Bight in the Baltic, the Santa Barbara Basin, and deep Pacific Ocean.

Causes and Mechanisms of Migration of Nutrients at the Water-Sediment Interface

According to Berner *et al.* (1976), the migration of reactants and products in the benthic boundary layer is caused by: 1) physical processes in the water column immediately above the benthic boundary layer and 2) biological, physical and chemical processes within the sediment.  These processes are:

1) Turbulence and resuspension of sediments – mechanical energy transferred from the ocean to sediments may result in resuspension of sediments and dispension in the pore water within the upper part

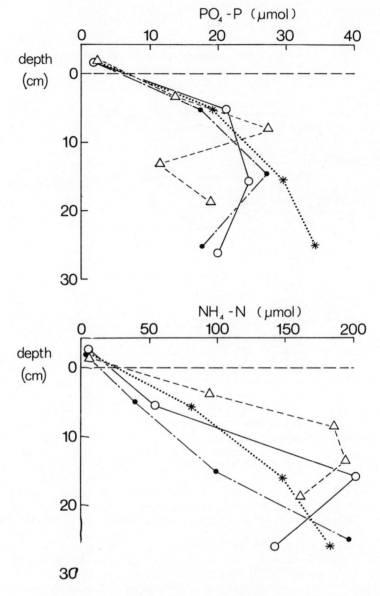

Figure 3a. Interstitial water phosphate and ammonia concentrations including the supernatant water at station Boknis Eck, Kiel Bight. Water depth 22 m. Symbols indicate different samples at the same station. (After Balzer, 1978)

of the layer. Turbulent stresses affect the physical structure of the sediment-water interface and may enhance the transfer of dissolved constituents from the upper parts of the sediment column (Rhoads *et al.*, 1975).

2) Bioturbation of the sediment by deposit-feeders and other moving organisms may result in homogenization of the chemical com-

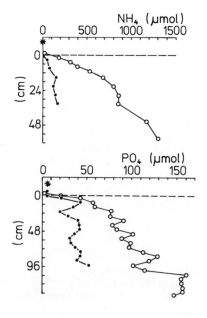

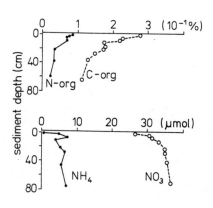

Figure 3c.  Organic carbon and organic nitrogen from sediments of the central Pacific Ocean.  Water depth is 5,700 m.  Interstitial water ammonia and nitrate concentration is from the same area. (After Suess, 1976)

Figure 3b.  Interstitial water ammonia and phosphate concentrations including the supernatant water (circles) and the bottom water (asterisks) from the Santa Barbara Basin.  Water depth left profile is 430 m, right profile, 585 m.  (After Sholkowitz, 1973)

position of the upper layers of sediment and flushing of interstitial water by irrigation of benthos.  The rates of mixing and irrigation are functions of organism type and community structure. These may vary with the type of sediment and the morphology of the ocean floor (Rhoads, 1974).

3) Diffusion of dissolved species, both reactants and products, is a migrational mechanism of primary importance.  The estimation of the diffusion coefficients of the reacting species is rather difficult.  Diffusive fluxes of dissolved material across the sea water-interface depend upon the gradient at the interface.  Evaluation of fluxes across the interface requires accurate knowledge of these gradients (Lerman, 1977).

4) Advection, or bulk flow, is a migrational mechanism caus-
ing net displacement of mass relative to the sediment-water interface.

5) Vertical gravity displacement of interstitial water (flushing), due to changes in the density of the bottom water - This
mechanism of nutrient release from the sediments is presumably of
great ecological importance, both for phytoplankton and the benthos,
mainly in coastal waters and estuaries.   The microbenthos would
be provided with oxygen and the phytoplankton with nutrients (Smetacek *et al.*, 1976).

Oppenheimer and Ward (1963) described an additional mechanism
of nutrient liberation from sediment occurring during exposure of
intertidal mud flats.   Interstitial water is drawn to the surface
by capillary action caused by evaporation.   The load of nutrient
salts precipitates at the surface and is dissolved by the incoming
tide.

Nutrient Release from Sediments

There are two fundamentally different approaches to obtain
data on the nutrient flux from sediments: 1) calculation of the flux
using concentration gradients of constituents of the interstitial
water or the water column and 2) measurement of flux directly
by *in situ* or laboratory experiments.

Berner (1976) described an approximate expression to calculate the flux of dissolved substances from sediments from pore
water concentration gradients:

$$J_i = \frac{C_o \, J_s}{P_s} \, \frac{(\phi_x)}{(1-\phi_x)} \; - \; \phi_o \, D_o \, \frac{\partial C}{\partial x} \qquad x = o$$

where $J_i$ is the flux of the dissolved constituent i between sediment and supernatant water, measured positively downward in terms
of mass per unit area of sediment per unit time; $J_s$ is the flux of
solid particles to the sediment due to deposition; $P_s$ is the mean
density of solids; $C_o$ is the concentration of the sediment-water
interface in terms of mass per unit volume of water; $D_o$ is the
diffusion coefficient at the sediment-water interface; $\phi$ is the
porosity at the sediment-water interface; and $\phi_x$ is the porosity
at a depth x below which either porosity remains constant or the
continuous upward flow of water due to compaction is interrupted
by the presence of basement, permeable sand layers, etc.   The
first term on the right represents burial of pore water by deposition, while the latter term represents diffusion between sediment
and overlying water.   $\phi_x$ and not $\phi_o$ is used to describe burial
(Berner, 1976).   According to this author, derivation of the equation assumes steady-state porosity and concentration gradients near
the sediment surface.   A qualitative summary of the direction of
fluxes to be expected is given in Figure 4.

In shallow water sediments where bioturbation and wave plus
current mixing are important and where depositional rates are high
and many chemical reactions take place practically at the sediment-
water interface, it is perhaps better not to make any conclusions
based on simple diffusive calculations until better data are ob-

tained and/or until more *in situ* flux measurements have been made (Berner, 1976). Rowe and Smith (1977) calculated the nutrient flux from ammonia profiles in the water column from the Mid-Atlantic Bight. Depending on the choice of vertical eddy diffusion coefficients, they derived values between 70 and 350 $\mu$moles $NH_4^+$ $m^{-2}$ $h^{-1}$.

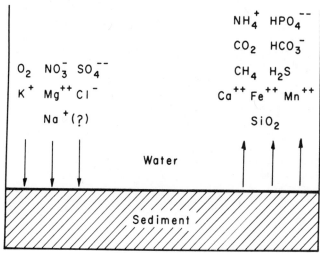

Figure 4. Direction of flux (shown by arrows) expected for dissolved constituents between sea water and sediment pore water. (After Berner, 1976)

For the second approach – the direct measurement of the flux of dissolved constituents – a variety of experimental *in situ* and laboratory apparatus have been designed and described in the literature. Some of these apparatus are depicted schematically in Figure 5. According to Pamatmat (1973, 1975, 1977), Hargrave and Connolly (1978) and Zeitzschel and Davies (1978), two different techniques have been used successfully to assess the relationship between bottom sediments and dissolved substances like oxygen and inorganic nutrients in aquatic systems. Sediment water interactions can be measured by *in situ* enclosure of water over natural sediment in bell jars, tunnels, boxes or other chambers or by removal of sediment and overlying water in cores or grabs for laboratory incubation.

*In situ* studies in shallow water down to approximately 30 m are mostly SCUBA-operated. In deep water, deep-sea submersibles (DSRV 'Alvin') or remote underwater manipulators (RUM) were used to conduct specific experiments in depths of several thousand meters. Smith *et al.* (1976) described a free vehicle respirometer (FVR) which measures the oxygen consumption of benthic communities *in situ* to abyssal depth. The advantage of these *in situ* devices is that most relevant environmental factors can be simulated. The only factor where serious difficulties arise is turbulence. The alternative to *in situ* measurements of community metabolism or nutrient release is the use of sediment corers or specially designed grabs.

With these devices, a small portion of the sea floor is cut out with some overlying water and is used on board ship or back in a land laboratory to measure the oxygen demand and/or nutrient release under controlled conditions (Pamatmat, 1971, 1973; Pilson *et al.*, 1979). Pamatmat has shown that there is a close agreement between *in situ* and core measurements of community metabolism from samples obtained in Puget Sound (Washington) down to 185 m depth. For deep sea cores, however, the effect of decomposition and temperature changes is noticeable. Smith and Pamatmat (Smith, 1978) compared data of deep sea samples measured with a grab respirometer and core samples. The data obtained indicate that for samples at comparable depths and environmental conditions, the shipboard method yielded values which were generally an order of magnitude higher than the *in situ* measurements. A new, improved technique which combines the advantages of both previously discussed methods, has been described recently by Smith *et al.* (1978). With this grab respirometer the *in situ* measurements can be related directly to the organisms of the enclosed patch of sediment, because the sediment can be retrieved after the *in situ* measurement.

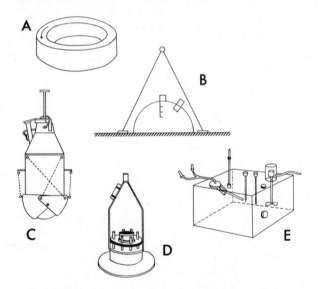

Figure 5. Schematic diagram of devices to measure the release of nutrients at the water-sediment interface: (A) annular closed tunnel (Davies, 1975), (B) 2 m$^3$ bell jar with sampling automatic and sensor package (Balzer, 1978), (C) grab respirometer (Smith *et al.*, 1978), (D) bell jar to collect supernatant water automatically (Hargrave and Connolly, 1978), and (E) Self-contained bell jar with instruments and sampling ports (Hallberg *et al.*, 1973).

According to Hargrave and Connolly (1978), the flux of dissolved material into or out of undisturbed sediments can be calculated ($m^{-2}$) as

$$\frac{V\,(C_o - C_t)}{A} \quad x \quad \frac{10^4}{T}$$

where V is the volume (liters) of water over the sediment, $C_o$ and $C_t$ are the dissolved concentrations ($liter^{-1}$) before and after time T, and A is the sediment area ($cm^2$) enclosed.  The calculation requires that the water is homogeneously mixed, that changes in concentration are known or assumed to be linear over time, and that dissolved material is only exchanged at the sediment surface.

The general problem with direct measurements is that by introducing an experimental set up at the sea bottom or by taking samples into the laboratory, the system to be measured undergoes unavoidable perturbations.  For instance, a bell jar placed over the bottom to measure outfluxing nutrients must be closed to flow from surrounding sea water, and, thus, flow conditions at the bottom and chemical conditions sensitive to flow (e.g., oxygenation) are unavoidably altered (Davies, 1975; Berner, 1976).  Artificial simulation of flow with a stirrer or a flow through pumping system can, as a compromise, overcome this problem.

Actual data on nutrient release from the marine environment are relatively scarce.  A selection of published data is summarized in Table 3.  In Figure 6, nutrient release data for ammonia are plotted versus depth and compared with measurements of the total oxygen uptake from *in situ* and core sample measurements from a variety of locations (Zeitzschel and Davies, 1978).  These nutrient release data have been obtained by *in situ* measurements with bell jars (Rowe *et al.*, 1975; Nixon *et al.*, 1976; Rowe *et al.*, 1977; Balzer, 1978, and Balzer and Keller, 1978).  Balzer (1978) and Hargrave and Connolly (1978) used a bell jar with a clock-driven release mechanism to collect water samples trapped over undisturbed sediments, whereas Smith *et al.* (1978) employed a grab respirometer which was manipulated by SCUBA or by DSRV 'Alvin'.  Pilson *et al.* (1979) used a box corer for sampling and a flow-through system for incubation in the laboratory, whereas Davies (1975) measured the nutrient flux with an annular closed tunnel at different flow rates (Fig. 5).

The release of different inorganic nutrients is by no means uniform.  According to Nixon *et al.* (1976), almost all of the inorganic nitrogen released from the bottom was in the form of ammonia. Nitrite fluxes were, in this study, virtually always below the level of detection.  All the nitrate in the water was derived from pelagic processes and the benthic communites did not participate directly in the seasonal nitrate cycle in the water.

According to Smith *et al.* (1978), the basic pattern of nutrient exchange at two deep water stations (2,200 and 2,750 m) are regeneration of ammonia and uptake of nitrate and phosphate by the benthic community.  The authors pointed out that the evolution of ammonia suggests denitrification processes, but, at least to the maximum depth of grab penetration, no reducing conditons were observed visually.  The ammonia might be a normal excretory product of the fauna, but the uptake of nitrate and nitrite was not as easily explained because dissolved oxygen was present as a hydrogen acceptor (Smith *et al*. 1978).

As mentioned before, chemical reactions fundamentally influence, and are influenced by, biological and physical processes in and above the sediments. The four major factors affecting the nutrient release across the water-bottom interface which will be discussed in this paper are the influence of the input of organic matter, bioturbation by benthos and the effects of temperature and waterflow at the bottom.

Table 3.  Nutrient release from sediments in different marine environments

| Location | Sediment Type | Depth (m) | Nutrient release from sediment ($\mu mol\ m^{-2}h^{-1}$) | | | | | | Reference |
|---|---|---|---|---|---|---|---|---|---|
| | | | $NH_4^+$ | $NO_2^-$ | $NO_3^-$ | $\sum N$ | $PO_4^{3-}$ | $\sum P$ | |
| Buzzards Bay (Massachusetts) [3] | | 17 | 68,8 | 0,05 | 0,15 | 69 | | | Rowe *et al.*, 1975 |
| Eel Pond (Massachusetts) | | 2 | 84,59 | | | 85,29 | | | |
| Loch Thurnaig | mud | 30 | | | | 23[6] | | | Davies, 1975 |
| Narragansett Bay (Rhode Island) | silt/ clay | 6-7 | Max. 400 | | 0,87[2] | | | | Nixon *et al.*, 1976 |
| Northwest African upwelling[1] | silt/ sand | 25 | 235 | | | 410 | 50 | | Rowe *et al.*, 1977 |
| Baltic | mud | 20 | 88,7[4] 27,8[5] | | | 16,2[5] | 29,6[4] 2,3[5] | | Balzer, 1978 |
| Harrington Sound (Bermuda) | | 9 18 24 | | | | 28 18 20 | | | Balzer and Keller, 1978 |
| Eastern Passage (Nova Scotia) | | 1-2 | | | | | | 7,9-8,9 | Hargrave and Connolly, 1978 |
| Atlantic | clayed silt clayed silt very fine sand | 2200 2750 17 | 0,95 0,61 33,8 | | | | | | Smith *et al.*, 1978 |
| Marine Microsm (MERL) | mud | 5 | 38,8 | | | 65 | 8,5 | 0,062[2] | Pilson *et al.*, 1979 |
| Narragansett Bay (Rhode Island | silt/clay | 6-7 | 99,3 | | | | | 0,121[2] | |

[1] average for two stations
[2] mol $m^{-2}y^{-1}$
[3] approximate average per year
[4] anoxic
[5] oxic
[6] average April - September

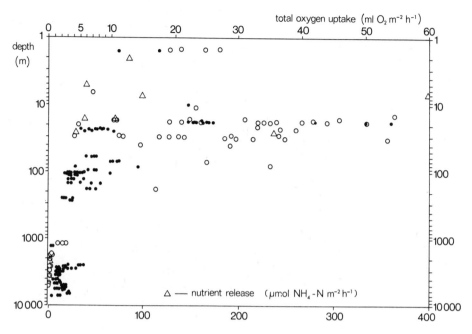

Figure 6. Rate of total oxygen consumption of the sediment versus depth (n = 206) and flux of ammonia at the sediment-water interface (n = 10). Open circles are $O_2$ values from *in situ* experiments, filled circles are $O_2$ values from grab and core samples, triangles are flux of ammonia. (Data from literature up to 1978)

The input of organic matter to the sediment is not uniform, but takes place in irregular events, depending on physical (turbulence, mixed layer) conditions in the water column and the physiological state of the phytoplankton. Time scales of settling events are in the order of a few days (Hargrave, 1975; Smetacek *et al.*, 1978). There is good evidence from experiments that the decomposition of organic matter and the regeneration of nutrients is directly correlated with the input of organic matter to the bottom (Hallberg, *et al.*, 1973). High rates of deposition of organic matter are favored by high primary production in the overlying water and quick settling and burial to avoid decomposition in the water column. The decomposition in this zone is a function of water depth, turbulence at the sediment-water interface, dissolved oxygen content, and rate of burial of enclosing sediment particles (Berner, 1976).

Benthic organisms influence the chemical processes at the sediment-water interface. This phenomenon is called bioturbation. Especially in shallow water ecosystems, where macrobenthos is abundant, intensive biogenic mixing and irrigation of the bottom takes place in the upper few centimeters of the sediment. Mixing activities of benthic organisms are important in accelerating vertical diffusion and transport of ions or compounds absorbed on particles or in solution in pore water (Rhoads, 1974).  Mechanisms of biological transport by various organisms in sediment are illustrated in Figure 7.

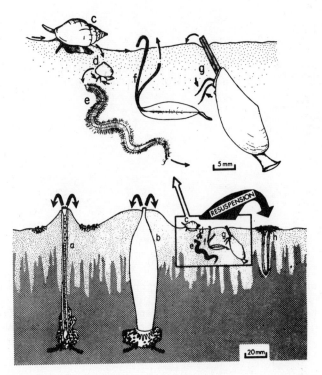

Figure 7.  Methods of mixing and recycling of sediment by deposit-feeders:  a. maldanid polychaete; b. holothurian; c. gastropod (*Nassarius*); d. nuculid bivalve (*Nucula* sp.); e. errant polychaete; f. tellinid bivalve (*Macoma* sp.); g. nuculanid bivalve (*Yoldia* sp.); h. anemone (*Cerianthus* sp.); oxidized mud lightly stippled; reduced mud densely stippled.  Species a and b are conveyor belt species, pumping reduced sediment from below the RPD to the oxidized surface; cycling within the oxidized surface is done by species c-g; arrows show routes of sediment ingestion and egestion of feces.  The coprophagous relationships are purely speculative.  Re-suspended fecal pellets may be utilized by suspension-feeders. (After Rhoads, 1974)

In Figure 8 the theoretically derived relationships between temperature and release of three nutrients are shown (Kremer and Nixon, 1978).  These relationships are supported by data from Nixon *et al.* (1976) who studied the seasonal aspects of nutrient flux of coastal marine bottom communities at three stations in the West Passage of Narragansett Bay, Rhode Island.  Examples of these studies are given by Nixon *et al.* (1980).

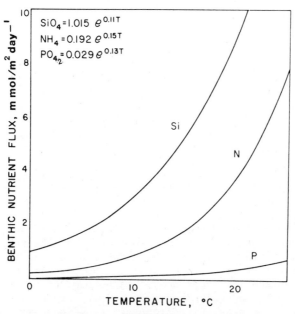

Figure 8. Benthic regeneration of
nutrients represented in the model as
a function of temperature. Fluxes from
three major bottom communities in Narragansett Bay over an annual cycle were
used to fit the empirical equations.
(After Kremer and Nixon, 1978)

The fourth factor - the effect of flow rate on oxygen consumption and nutrient release - was studied by Rhoads *et al*. (1975),
Davies (1975), and Vanderborght *et al*. (1977). Davies concluded
that high flow rates (10-20 cm sec$^{-1}$) cause resuspension of the
flocculant material, resulting in a notable increase in the rate of
oxygen consumption. This seems to be related to the amount of
reduced products being formed in the sediment which is dependent
on the available food supply and the diffusion into the sediment
of aerated water. Ammonia release from sediment at two waterflow rates of 3 and 11 cm sec$^{-1}$ was rather similar. Davies (1975)
suspected that there are probably two processes of nutrient release,
a slow, steady state of diffusion of nutrients into the water column,
and a more sporadic release of nutrients occurring when wind-driven
turbulence causes resuspension of the bottom sediment.

The Ecological Significance of Sediment-Water Interactions

The ecological significance of sediment-water interactions in
respect to nutrient dynamics can be summarized as follows: there
is good evidence that total oxygen consumption and nutrient release
follow the same trend, which means rates are high in shallow water
areas, whereas deep water fluxes are 2-4 orders of magnitude less
(Fig. 6). This difference is caused: 1) by the different input of
organic material to the bottom (25-60% of primary production in
shallow water ecosystems compared to 1-10% in deep water systems),

2) by the relatively low metabolic rates and/or activity of deep water organisms, and 3) by relatively slow chemical reactions due to low temperatures (about 1°C) in deep water.

Released nutrients from the bottom are only of direct ecological importance if they are supplied at a time when there is a direct need for nutrients. For instance, a high nutrient supply in boreal areas in winter will not affect the primary productivity in the euphotic zone because light is the major limiting factor at this period.

There is, however, good evidence, at least for some areas like Bedford Basin, Kiel Bight, the Mid-Atlantic Bight, or Narragansett Bay, that released nutrients from the sediment transported into the euphotic zone in summer play an important role in inducing phytoplankton blooms. Rowe *et al.* (1975), Davies (1975), Rowe and Smith (1977), and Hargrave and Connolly (1978) calculated that the release of nutrients from the sediments may make up between 30 and 100% of nutrient requirements of the phytoplankton in the euphotic zone. This implies that in shallow water ecosytems bottom regeneration keeps pace with primary production and that primary producers depend very little on pelagic regenerative processes (Rowe and Smith, 1977).

Figure 9 is a schematic diagram of components and interrelationships in a shallow water ecosystem. This diagram is the basis for the formulation of a system of differential equations to simulate a system like this under a variety of assumptions (Bölter *et al.*, 1977). Figure 10 shows a simple example of the influence of nutrient uptake by phytoplankton and nutrient recycling at the sediment-water interface on the phytoplankton. In this model eight variables are included, but only two are depicted. It is obvious from this simulation that the flux of nutrients from the sediment drastically influences the standing stock of phytoplankton.

In conclusion, there is enough evidence to state that in shallow water ecosystems, a succession of four major events take place:

1) phytoplankton blooms (e.g., a spring bloom in boreal regions)

2) sedimentation of this bloom (directly or via fecal pellets of herbivores)

3) decomposition of organic material (mainly at the water-sediment interface)

4) release of nutrients from the bottom and recycling of these nutrients to the euphotic zone by turbulent mixing.

The time scales for the duration and the succession of these processes are of the order of days to weeks, and the depth range where these events occur, may be limited to a maximum of 200 m. These four events occur regularly on the shelf, in coastal waters, and in estuaries. I feel that the different events follow one another with a certain time lag and that the disturbance of this pattern causes the observed variability of phytoplankton primary productivity in space and time.

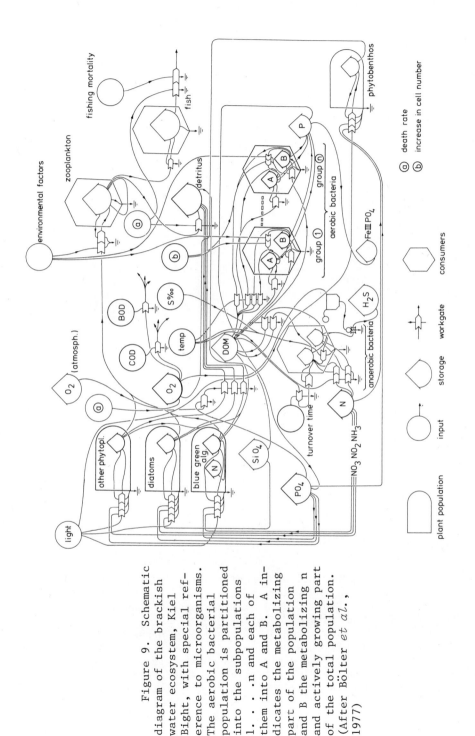

Figure 9. Schematic diagram of the brackish water ecosystem, Kiel Bight, with special reference to microorganisms. The aerobic bacterial population is partitioned into the subpopulations 1 . . .n and each of them into A and B. A indicates the metabolizing part of the population and B the metabolizing n and actively growing part of the total population. (After Bölter *et al.*, 1977)

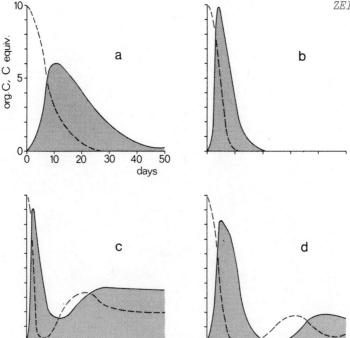

Figure 10. Simulation of the interaction between nutrient supply and phytoplankton stand-ing stock as a function of uptake rates and re-cycling flux (subcomponent of an eight-compart-ment model): a. Initial nutrient addition and low uptake coefficient; b. Initial nutrient ad-dition and maximum uptake coefficient; c. As in b. with the addition of high nutrient recycling (turnover rate of phosphate about 20 per year); d. As in b. with the addition of a best estimate of recycling rate (turnover rate of phosphate eight per year). (After Probst, pers. comm.)

REFERENCES

Aller, R. C. 1980. Relationships of tube-dwelling benthos with sediment and overlying water chemistry. pp. 000-000. *In* K. R. Tenore and B. C. Coull (eds.), Marine Benthic Dynamics. Univ. of South Carolina Press, Columbia.

Almgren, T., L. G. Danielsson, D. Dyrrsen, T. Johanson, and G. Nyquist. 1975. Release of inorganic matter from sediments in a stagnant basin. Thhalassia jugos 1. 11: 19-23.

Ansell, A. D. 1974. Sedimentation of organic detritus in Loch Etive and Creran, Argyll, Scotland. Mar. Biol. 27: 263-273.

Balzer, W. 1978. Untersuchungen über Abbau organischer Materie und Nährstoff-Freisetzung am Boden der Kieler Bucht beim über-gang vom oxischen zum anoxischen Milieu. Rep. Sonderforschungs-bereich 95 - Wechselwirkung Meer-Meeresboden, Kiel University. 36: 1-137.

Balzer, W., and I. Keller. 1978. Freisetzung von Nährstoffen, pp. 24-28. *In* G. Wefer (ed.), Das Harrington Sound Projekt des SFB 95 - Bericht über die zweite Meß-phase (März/April 1978). Reports Sonderforschungsbereich 95 - Wechselwirkung Meer-Meeresboden - Kiel University.

Berner, R. A. 1974. Kinetic models for the early diagenesis of nitrogen, sulfur, phosphorus, and silicon in anoxic marine sediments, pp. 427-449. *In* E. D. Goldberg (ed.), The Sea, V. 5. Wiley.

Berner, R. A. 1976. The benthic boundary layer from the viewpoint of a geochemist, pp. 33-35. *In* I. N. McCave (ed.), The Benthic Boundar Layer, Plenum.

Berner, R. A., S. E. Calvert, R. Chesselet, R. C. Cooke, A. J. de Groot, J. C. Duinker, A. Lermann, J. M. Martin, N. B. Priee, F. L. Sayles, E. Suess, and R. Wollast. 1976. Solution-sediment chemical interactions, pp. 261-272. *In* I. N. McCave (ed.), The Benthic Boundary Layer, Plenum.

Bishop, J. K. B., D. R. Ketten, and J. M. Edmond. 1978. The chemistry, biology and vertical flux of particulate matter from the upper 400 m of the Cape Basin in the southeast Atlantic Ocean. Deep-Sea Res. 25: 1121-1161.

Bölter, M., L. -A. Meyer-Reil, and B. Probst. 1977. Comparative analysis of data measured in the brackish water of the Kiel Fjord and the Kiel Bight, pp. 249-280. *In* G. Rheinheimer (ed.), Microbial Ecology of a Brackish Water Environment. Ecological Studies. 25. Springer.

Bröckel, K. V. 1978. Energy flow in a shallow water ecosystem - an approach to quantify the energy flow through the pelagic part of the shallow water exosystem off Boknis Eck (Eckernförde Bay). Kieler Meeresforsch. Sonderheft. 4: 233-243.

Bröckel, K. V. 1979. A note on short-term production and sedimentation in the upwelling region off Peru. (submitted).

Bunt, J. S. 1975. Primary productivity of marine ecosystems, pp. 169-183. *In* H. Lieth and R. H. Whittaker (eds.), Primary productivity of the biosphere, Ecological Studies 14. Springer.

Davies, J. M. 1975. Energy flow through the benthos in a Scottish sea loch. Mar. Biol. 31: 353-362.

Edwards, R. R. C. 1973. Production ecology of two Caribbean marine ecosystems II. Metabolism and energy flow. Estuarine and Coastal Marine Science. 1: 319-353.

Gieskes, J. M. 1975. Chemistry of interstitial waters of marine sediments. Ann. Rev. Earth Planet. Sci. 3: 433-453.

Hallberg, R. O., L. -E. Bågander, A. G. Engvall, M. Lindström, S. Oden, and F. A. Schippel. 1973. The chemical microbiological dynamics of the sediment-water interface. Contr. from Askö Lab., Univ. Stockholm. 2: 1, 1-117.

Hargrave, B. T. 1975. The importance of total and mixed-layer depths in the supply of organic matter to bottom communities. Symp. Biol. Hung. 15: 157-165.

Hargrave, B. T. 1980. Factors affecting the flux of organic matter to sediments in a marine bay, pp. 000-000. *In* K. R. Tenore and B. C. Coull (eds.), Marine Benthic Dynamics. Univ. of Souty Carolina Press, Columbia.

Hargrave, B. T., G. A. Phillips, and S. Taguchi. 1976. Sedimentation measurements in Bedford Basin, 1973-74. Fish. Mar. Serv.

Res. Dev. Tech. Rep. 608:  129 pp.

Hargrave, B. T., and S. Taguchi.  1978.  Origin of deposited
material sedimented in a marine bay.  J. Fish. Res. Bd. Can.
35:  1604-1613.

Hargrave, B. T., and G. F. Connolly.  1978.  A device to collect
supernatant water for measurement of the flux of dissolved
compounds across sediment surfaces.  Limnol. Oceanogr. 23:
1005-1010.

Hartwig, E. O.  1976.  The impact of nitrogen and phosphorus release from a siliceous sediment on the overlying water, pp.
103-117.  *In* M. Wiley (ed.), Estuarine Processes.  Academic
Press.

Koblentz-Mishke, O. J., V. V. Volkovinsky, and J. G. Kabanova.
1970.  Plankton primary production of the world ocean, pp.
183-193.  *In* W. S. Wooster (ed.), Scientific Exploration of
the South Pacific.  Nat. Acad. Sci.

Kremer, J. N., and S. W. Nixon.  1978.  A coastal marine ecosystem,
simulation and analysis.  Ecological Studies 24, Springer.

Lerman, A.  1977.  Migrational processes and chemical reactions
in interstitial water, pp. 695-738.  *In* E. D. Goldberg, I. N.
McCave, J. J. O'Brien, J. H. Steele (eds.), The Sea, 6.  Marine
Modeling.  Interscience.

Manheim, F. F.  1976.  Interstitial water of marine sediments.
Chem. Oceanography. 6:  115-186.

McCave, I. N.  1975.  Vertical flux of particles in the ocean.
Deep-Sea Res. 22:  491-502.

Nixon, S. W., C. A. Oviatt, and S. S. Hale.  1976.  Nitrogen regeneration and the metabolism of coastal marine bottom communities, pp. 269-283.  *In* J. M. Anderson and A. Mac Fayden (eds.),
The Role of Terrestrial and Aquatic Organisms in Decomposition
processes.  Blackwell.

Nixon, S. W., J. R. Kelly, B. N. Furnas, C. A. Oviatt, and S. S.
Hale.  1980.  Phosphorus regeneration and the metabolism of
coastal marine bottom communities, pp. 000-000.  *In* K. R.
Tenore and B. C. Coull (eds.),  Marine Benthic Dynamics.  Univ.
of South Carolina Press, Columbia.

Oppenheimer, C. H., and R. A. Ward.  1963.  Release and capillary
movement of phosphorus in exposed tidal sediments, pp. 664-
673.  *In* C. H. Oppenheimer (ed.),  Symposium on Marine Microbiology.  Thomas.

Pamatmat, M. M.  1971.  Oxygen consumption by the seabed.  IV.
Shipboard and laboratory experiments.  Limnol. Oceanogr. 16:
536-550.

Pamatmat, M. M.  1973.  Benthic community metabolism on the continental terrace and in the deep-sea in the North Pacific.
Int. Rev. Gesamten Nydrobiol. 58:  345-368.

Pamatmat, M. M.  1975.  *In situ* metabolism of benthic communities.
Cha. Biol. Mar. 16:  613-633.

Pamatmat, M. M.  1977.  Benthic community metabolism: a review and
assessment of present status and outlook, pp. 89-111.  *In* B. C.
Coull (ed.), Ecology of Marine Benthos.  Univ. of South Carolina
Press, Columbia.

Parsons, T. R., M. Takahaski, and B. Hargrave.  1977.  Biological
Oceanographic Processes.  Pergamon.

Pilson, M. E. Q., C. A. Oviatt, and S. W. Dixon. 1979. Annual nutrient cycles in a marine microcosm. Symposium on Microcosms
in Ecological Research, Savannah River Ecological Laboratory,
Nov. 8-10, 1978, Augusta, Ga. (submitted).

Rhoads, D. C. 1974. Organism-sediment relations on the muddy sea
floor, pp. 263-300. *In* H. Barnes (ed.), Oceanogr. Mar. Biol.
Ann. Rev. 12, Allan and Unwin.

Rhoads, D. C., K. Tenore, and M. Browne. 1975. The role of resuspended bottom mud in nutrient cycles of shallow embayments, pp.
563-579. *In* Chemistry, Biology and the Estuarine System.
Estuar. Res. 1.

Rowe, G. T., C. H. Clifford, K. L. Smith, Jr., and P. L. Hamilton.
1975. Benthic nutrient regeneration and its coupling to primary
productivity in coastal waters. Nature. 255: 215-217.

Rowe, G. T., C. H. Clifford, and K. L. Smith, Jr. 1977. Nutrient
regeneration in sediments off Cape Blanc, Spanish Sahara. Deep-
Sea Res. 24: 57-63.

Rowe, G. T., and K. L. Smith, Jr. 1977. Benthopelagic coupling
in the Mid-Atlantic Bight, pp. 55-65. *In* B. C. Coull (ed.),
Ecology of Marine Benthos. Univ. of South Carolina Press,
Columbia.

Rowe, G. T., and W. D. Gardner. 1978. Sedimentation rates in the
slope water of the Northwest Atlantic Ocean measured directly
with sediment traps. Contr. No. 4087 from the Woods Hole
Oceanographic Institution.

Ryther, J. H. 1969. Photosynthesis and fish production in the sea.
Science. 166: 72-76.

Sholkovitz, E. 1973. Interstitial water chemistry of Santa Barbara basin sediments. Geochim. Cosmochim. Acta. 37: 2043-
2073.

Smetacek, V., B. v. Bodungen, K. v. Bröckel, and B. Zeitzschel.
1976. The plankton tower. II. Release of nutrients from
sediments due to changes in the density of bottom water. Mar.
Biol. 34: 373-378.

Smetacek, V., K. v. Brockel, B. Zeitzschel, and W. Zenk. 1978.
Sedimentation of particulate matter during a phytoplankton spring
bloom in relation to the hydrographical regime. Mar. Biol. 47:
211-226.

Smith, K. L., Jr. 1978. Benthic community respiration in the N.W.
Atlantic Ocean: *in situ* measurements from 40-5200 m. Mar. Biol.
47: 337-347.

Smith, K. L., Jr., G. A. White, M. B. Laver, and J. A. Haugsness.
1978. Nutrient exchange and oxygen consumption by deep-sea
benthic communities: Preliminary *in situ* measurements. Limnol.
Oceanogr. 23: 997-1005.

Steele, J. H., and I. E. Baird. 1972. Sedimentation of organic
matter in a Scottish sea loch. Mem. Ist. Ital. Idrobiol. 29
(Suppl.): 73-88.

Stephens, K., R. W. Sheldon, and T. R. Parsons. 1967. Seasonal
variations in the availability of food for benthos in coastal
environments. Ecology. 48: 852-855.

Suess, E. 1976. Nutrients near the depositional interface,
pp. 57-79. *In* I. N. McCave (ed.), The Benthic Boundary Layer,
Plenum.

Trevallion, A. 1967. An investigation of detritus in Southampton
    water. J. mar. biol. Ass. U.K. 47: 523-532.
Vanderborght, J. -P., R. Wollast and G. Billen. 1977. Kinetic
    models of diagenesis in disturbed sediments. Part 1. Mass
    transfer properties and silica diagenesis. Limnol. Oceanogr.
    22: 787-793.
Webster, T. J. M., M. A. Paranjape, and K. H. Mann. 1975. Sedi-
    mentation of organic matter in St. Margaret's Bay, Nova Scotia.
    J. Fish. Res. Bd. Can. 32: 1399-1407.
Zeitzschel, B. 1965. Zur Sedimentation von Seston, eine produk-
    tionsbiologische Untersuchung von Sinkstoffen und Sedimenten
    der Westlichen und Mittleren Ostsee. Kieler Meeresforsch.
    21: 55-80.
Zeitzschel, B., and J. M. Davies. 1978. Benthic growth chambers.
    Rapp. R. -v. Réun. Cons. int. Explor. Mer. 173: 31-42.

# Phosphorus Regeneration and the Metabolism of Coastal Marine Bottom Communities

S. W. Nixon
J. R. Kelly
B. N. Furnas
C. A. Oviatt
S. S. Hale

ABSTRACT: *In situ* measurements of the exchange of phosphate between sediment and the overlying water at three stations in Narragansett Bay, Rhode Island, showed that there was almost always a net flux out of the sediments. The magnitude of the flux ranged from near zero in winter to almost 60 $\mu$moles $m^{-2}$ $hr^{-1}$ in summer. The flux was strongly correlated with temperature during the spring warming and did not decrease with increasing phosphate concentrations in the overlying water. Calculations indicate that the phosphate cycle in coastal waters such as Narragansett Bay is dominated by sediment-water exchanges. Some 120 mg-at of inorganic P $m^{-2}yr^{-1}$ were released annually to the overlying water, or enough phosphorus to support about 50% of the annual phytoplankton primary production. The flux of dissolved organic phosphorus was erratic and lower, and appreciable uptake as well as release was often observed.

Laboratory experiments using replicate cores collected from the bay at different times of year showed that oxygen uptake, carbon dioxide release,

and phosphorus exchange by the sediments were also influenced by the availability of fresh organic matter.  It was clear from the laboratory and
field measurements that the regeneration of organic matter by the benthos results in a return
of inorganic nutrients to the water column that
is anomalously low in fixed nitrogen relative to
phosphorus.  This remarkable feature of benthic
regeneration, along with the fact that a large
amount of organic matter is decomposed on the
bottom in shallow areas compared with the open
sea, appears to be responsible for the characteristically low N/P ratio of coastal marine waters
and for the importance of nitrogen rather than
phosphorus as a major limiting nutrient in these
areas.

INTRODUCTION

     The nutrient dynamics of estuarine and coastal marine waters
differ from those of the open sea in a number of ways.  At least
two of these differences are striking and of considerable ecological importance, and involve phosphorus.  While the amount of phosphate in the ocean varies from place to place and from time to time,
it has been found repeatedly that the ratio of dissolved nitrogen
to phosphorus remains remarkably constant at about 16 N to 1 P by
atoms (Redfield, 1934; Fleming, 1940; Alvarez-Borrego *et al.*, 1975).
In nearshore waters, however, the ratio of inorganic nitrogen to
phosphorus is usually much lower (Riley, 1941; Jeffries, 1962), with
the result that nitrogen is generally the most limiting nutrient
for primary production in coastal areas (Ryther and Dunstan, 1971;
Goldman *et al.*, 1973).  In Narragansett Bay, for example, the N/P
ratio is about 10 to 15 during the winter but declines sharply in
spring and remains below 5 throughout the summer (Kremer and Nixon,
1978).
     In addition to the great abundance of phosphorus with respect
to nitrogen in coastal waters, the seasonal cycle of abundance of
phosphorus differs from that found offshore.  Instead of a winter
maximum in phosphate with low concentrations in summer as found in
offshore waters, the seasonal pattern in areas like Narragansett
Bay is reversed, with high concentrations during summer following
a rapid increase in the spring (Fig. 1) (Smayda, 1957; Taft and
Taylor, 1976).
     The processes responsible for the anomalously low N/P ratio
and characteristic seasonal cycle of phosphorus in coastal waters
are not well understood.  While coastal waters are subject to influences from fresh water and from anthropogenic inputs, they also
differ from the open sea in the much greater potential impact that
the sediments may exert on the overlying, relatively shallow water.
In an earlier paper (Nixon *et al.*, 1976), we reported oxygen uptake
rates and net inorganic fixed nitrogen fluxes over an annual cycle

between sediments and the waters in Narragansett Bay, Rhode Island. On the basis of those measurements, it appeared that the total annual benthic oxygen uptake was on the order of 370 g $O_2$ m$^{-2}$yr$^{-1}$ or, using an empirically determined RQ of 1 (see Fig. 5), a remineralization of 140 g C m$^{-2}$yr$^{-1}$. Recent primary production estimates for the bay show an annual fixation of some 310 g C m$^{-2}$yr$^{-1}$ (Furnas *et al.*, 1976), so that some 45% of the organic matter produced in the bay is remineralized on the bottom. However, the inorganic nitrogen returned to the water column only amounted to about 0.87 g-at N m$^{-2}$yr$^{-1}$ or some 22% of the nitrogen presumably required to support the annual primary production. It seemed likely from these data that the sediments might play a dominant role in estuarine phosphorus cycles as well, and that the passage of a large fraction of the organic matter through benthic rather than pelagic regeneration might be responsible for at least a part of the characteristic pattern of phosphorus dynamics in coastal waters.

This paper reports the results of *in situ* sediment-water phosphorus flux measurements over an annual cycle from three stations in Narragansett Bay, as well as a number of laboratory experiments using large diameter cores to determine the response of sediment-water phosphorus fluxes to temperature, dissolved oxygen, the addition of organic matter, and macrofauna.

## METHODS

### Field Measurements

*In situ* measurements of benthic oxygen uptake and the net flux of phosphate and dissolved organic phosphorus (DOP) between the sediment and overlying water were made over an annual cycle at three stations in Narragansett Bay using opaque PVC chambers. Simultaneous nitrogen flux measurements, characteristics of the stations, and details of the chamber design and sampling procedure were reported earlier (Hale, 1974; Nixon *et al.*, 1976). Briefly, the chambers covered 0.24 m$^2$ of bottom, contained 27 L of water, and were incubated for 3-4 h with control bottles of bottom water alongside. Three or four chambers were used at each station. Water depths were about 5.8 m at Station 1 in the upper West Passage, 7.3 m at Station 2 near mid bay, and 6.1 m at Station 3 in the lower West Passage. The sediments at Station 1 were largely sand, while silt-clay fractions dominated the lower 2 stations.

### Laboratory Experiments

On a number of occasions, three or four box cores were collected by SCUBA divers at Station 2 and brought back to the laboratory for various experiments. The cores were similar to those used by Aller (1977) and contained about 0.022 m$^2$ of sediment to a depth of 10-15 cm with 2 L of overlying water. Great care was taken in obtaining and transporting the cores to disturb the sediment as little as possible. The cores were maintained in the dark in a bay water bath at field temperature until they reached the laboratory where they were kept in the dark in a constant temperature incuba-

tor.  Water over the cores was changed about every two days and
gently aerated whenever flux measurements were not being made.  Unless otherwise noted, all laboratory flux measurements took from
3-6 h, during which oxygen levels seldom fell below 75% of saturation.  In some cases, long-term flux measurements (12-24 h) were
made, in which case air was blown over the surface of the water in
the cores to prevent oxygen concentrations from dropping below 75%
of saturation.  Oxygen uptake was not measured during the long-term
incubations.  As with the field incubations, fluxes were calculated
as the difference between initial and final concentrations of the
material of interest, multiplied by the volume of water and divided
by the area of sediment and the duration of the incubation.  This
procedure assumes that the rate of flux is linear over time, at
least for the relatively short incubations usually used in our
field and laboratory work.  Time series measurements on cores from
Long Island Sound (Aller, 1977) and from Narragansett Bay
(Elderfield *et al.*, in press) have shown linear behavior for ammonia, phosphate and silicate fluxes during incubations of 24-100 h
when the water over the cores was aerated.


### Analytical Methods

Dissolved oxygen was measured using the Winkler method as modified by Carritt and Carpenter (1966) in Strickland and Parsons
(1968).  Changes in carbon dioxide were determined from changes in
pH using a Corning Digital 112 meter and the $CO_2$ titration method
of Beyers *et al.* (1963).  Water samples for phosphorus analysis
were filtered immediately through Reeve Angle Type 934AH glass fiber filters and preserved (Gilmartin, 1967) until freezing (field
samples) or frozen immediately.  Reactive phosphate (Strickland and
Parsons, 1968) and DOP (Menzel and Corwin, 1965) were measured in
thawed samples.

The analytical precisions of the determinations and the resulting effects on the calculated fluxes for a 3-h incubation are given
below, assuming that the errors for chambers and controls are additive.  Differences between field and laboratory are due to differences between chamber areas and volumes.

|  | Field | Laboratory |
|---|---|---|
| $O_2 \pm 0.05$ mg $l^{-1}$ | $\pm 3.71$ mg m$^{-2}$hr$^{-1}$ | $\pm 4.62$ mg m$^{-2}$hr$^{-1}$ |
| $PO_4 \pm 0.03$ μmoles $l^{-1}$ | $\pm 2.25$ μmoles m$^{-2}$hr$^{-1}$ | $\pm 2.76$ μmoles m$^{-2}$hr$^{-1}$ |
| DOP $\pm 0.03$ μg-atp $l^{-1}$ | $\pm 2.25$ μg-atp m$^{-2}$hr$^{-1}$ | $\pm 2.76$ μg-atp m$^{-2}$hr$^{-1}$ |

The phosphorus content of sediments and of particulate organic
matter used in regeneration experiments was measured using a modification of the Williams *et al.* (1967) technique developed by R.
Beach and M. E. Q. Pilson at the University of Rhode Island.  Samples for total phosphorus were oven dried at 60°C, ground in a Wiley Mill (mesh #40) and combusted at 450°C.  After combustion, 1 N
HCl was added to the sediment, the samples sonicated for 15 min and
allowed to stand overnight.  The molybdate-blue method of Strickland
and Parsons (1968) was then used to measure phosphate after the
sample pH was adjusted with distilled water.

RESULTS AND DISCUSSION

Seasonal Cycle of Sediment-Water Exchange
*Magnitude and Timing of the Flux*

The flux of phosphate from the sediments to the overlying water in Narragansett Bay varied over the annual cycle from essentially zero during winter to 50 or 60 μmoles $m^{-2}h^{-1}$ during summer (Fig. 1). Only two or three individual measurements showed any significant uptake of phosphate by the sediments at any time during the year. This pattern is similar to the only other annual cycle of net phosphate flux measurements yet reported, where Hartwig (1976) found only occasional phosphate uptake in the sandy sediment of the La Jolla Bight. Moreover, there is only one uptake of appreciable magnitude in other published measurements of sediment-water phosphate fluxes (Table 1). It seems likely that marine sediments in general act as a source of phosphate to the overlying water. There was also an exchange of dissolved organic phosphorus across the sediment-water interface, at least at two of the stations, but the fluxes were erratic and usually much lower than for phosphate (Fig. 2). The DOP fluxes found here appear to be somewhat lower than those measured by Hartwig (1976) at La Jolla.

Figure 1. (Top) The seasonal cycle of phosphate in near-surface and near-bottom water near the middle of Narragansett Bay in 1972-73. The heavy broken line is the computed phosphate concentration using the phosphate flux regression in Figure 3, measured bottom water temperatures, and a numerical hydrodynamic mixing model (Kremer and Nixon, 1978). (Bottom) Net sediment-water phosphate fluxes measured *in situ* over an annual cycle in the upper, mid, and lower West Passage of Narragansett Bay. Data points are individual chamber measurements, positive values are fluxes out of the sediment. The smooth curve was drawn by eye to estimate the annual flux.

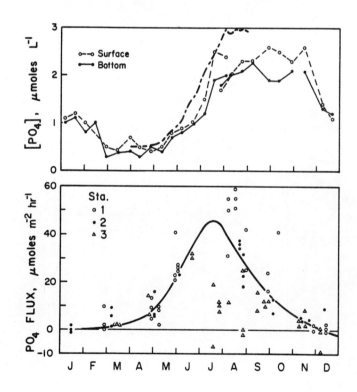

Table 1. Measured net flux of phosphate between marine sediments and overlying waters under aerobic conditions.

| Location | Sediment | Depth, m | Temp, °C | Incubation, h | PO$_4$ Flux $\mu$moles m$^{-2}$ h$^{-1}$ |
|---|---|---|---|---|---|
| **LOW TEMPERATURE (<10°C)** | | | | | |
| Matsushima Bay, Japan[1] | mud | ~1 | ~10 | ? | 16 |
| Great Bay, N.H.[2] | muddy-silt | ~1.4 | 4-11 | 24 | -0.6 (-) -0.8 |
| Vostok Bay, U.S.S.R.[11] | sand | 3 | 0.4-2.1 | 12 | 0.5-3 |
| Narragansett Bay, R.I.[3] | silt-clay | 6.5 | 10 | 3-4 | 4.7 |
| Vostok Bay, U.S.S.R.[11] | mud | 7 | 2.1-3.8 | 12 | 1.8-8.8 |
| L.I. Sound, N.Y.[4] | silt-clay | 15 | 4 | 50 | 0.9 |
| Buzzard's Bay, Mass.[5] | ? | 17 | 1.6 | 3.6 | 5.2 |
| L.I. Sound, N.Y.[4] | silt-clay | 34 | 4 | 73 | 0.5 |
| N.Y. Bight[6] | sand | 35.5 | 7.8 | 3-4 | 4-9 |
| N.W. Atlantic[7] | clayey-silt | 2200-2750 | 3 | 125(?) | -0.2 (-) -0.3 |
| **HIGH TEMPERATURE (>10°C)** | | | | | |
| Eel Pond, Mass.[6] | ? | 2 | 20 | 4.5-13 | 6.6-21.6 |
| Vostok Bay, U.S.S.R.[11] | mud | 5-7 | 20 | 10 | 14.8-26.4 |
| Narragansett Bay, R.I.[3] | silty-clay | 6.5 | 20 | 3-4 | 28.8 |
| L.I. Sound, N.Y.[4] | silty-clay | 8 | 22 | 54 | 5.4 |
| Kaneohe Bay, R.I.[8] | calcium-carbonate | 10 | 24 | ? | 2-5.6 |
| Buzzard's Bay, Mass.[5] | ? | 17 | 16 | 5.25 | -14.9 |
| La Jolla Bight, Calif.[6] | sand | 17 | 15 | 3(?) | -13.6 (-) -45.5 |
| La Jolla Bight, Calif.[9] | sand | 18 | 12-18 | 5-6 | 0.3 |
| Cap Blanc, Africa[10] | sandy-silt | 25 | ? | 3.5-12 | 36-50 |

[1] Okuda (1955), disturbed core in the laboratory
[2] Gilbert (1976), *in situ*
[3] This study, regression for multiple *in situ* measurements from three stations
[4] Aller (1977), undisturbed core in the laboratory, time series regression
[5] Rowe *et al.* (1975a) *in situ*
[6] Rowe *et al.* (1975) *in situ*
[7] Smith *et al.* (1978) *in situ*
[8] Harrison (in prep.), annual mean from three stations, highly bioturbated sediments
[9] Hartwig (1976), *in situ* dark box replication study
[10] Rowe *et al.* (1977) *in situ*
[11] Propp *et al.*, this volume, *in situ* dark measurement

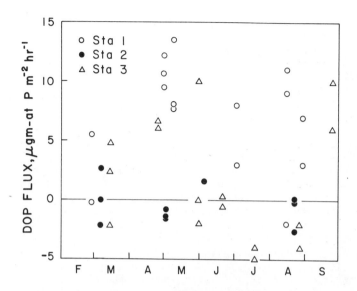

Figure 2. Net sediment-water fluxes of
dissolved organic phosphorus measured *in situ* in
the upper, mid, and lower West Passage of Narragansett Bay.

For Narragansett Bay, the net annual flux of phosphate from
the bottom was about 120 mg-at P $m^{-2}yr^{-1}$, or enough to provide some
50% of the phosphorus required to support the primary production of
310 g C $m^{-2}yr^{-1}$ (Furnas *et al.*, 1976). For comparison, the estimate of phosphorus excretion by zooplankton in the bay is about 30
mg-at P $m^{-2}yr^{-1}$ (Vargo, 1976). A very rough estimate of the annual
input of DOP appears to be about 50 mg-at P $m^{-2}yr^{-1}$.

### The Effect of Temperature

The flux of phosphate was strongly correlated with water temperature, at least through the winter-spring warming (Fig. 3),
though it appeared that exchanges in the fall were lower at a given
temperature than they were during spring. The temperature dependence found in Narragansett Bay may only apply to sediments in
which the top few cm or at least mm are oxidized, since Holm (1978)
and his colleagues have found no temperature effect on phosphate
release from anoxic sediments of the Baltic above 10°C.
By examining the effect of the benthic exchange on phosphate
concentrations in the overlying water, one can appreciate the importance of the temperature dependence and magnitude of the sediment-water phosphate flux. If the measured bottom water tempera-

ture is used to drive a simple numerical model consisting of the temperature-flux regression (Fig. 3) and the depth and tidal exchanges for Narragansett Bay (Kremer and Nixon, 1978), the computed concentration of phosphate in the water is close to that observed from April through August in the field both in timing and rate of increase (Fig. 1). The omission of phytoplankton uptake of phosphate from the calculation is probably responsible for the overestimate in the model. Additional evidence that sediment-water exchanges are responsible for the seasonal cycle in near-shore phosphate concentrations can be found in microcosm experiments at the Marine Ecosystems Research Laboratory (MERL). The MERL microcosms are analogues of Narragansett Bay and contain 13 $m^3$ of bay water. The 5 m deep water column in each microcosm overlies a pan of sediment collected from the bay measuring 2 m in diameter and about 30 cm deep. The seasonal phosphate cycle in the microcosm tanks is virtually identical to that observed in the bay even when the tanks are maintained in batch mode with no inputs from the bay (no river influence, no sewage, no offshore exchanges) (Pilson *et al.*, in press). It seems clear that, at least in Narragansett Bay, the characteristic and dramatic spring-summer phosphate increase is driven by outputs from the sediments.

Figure 3. The relationship between net sediment-water phosphate flux and bottom water temperature in Narragansett Bay. Only data from December-June were included in the regression.

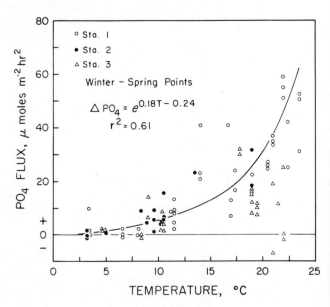

*Concentration Effects and Phosphate Cycle Buffering*

The flux of phosphate over the year from the sediments was not correlated in any obvious way with the concentration in the overlying water, at least up to 4 $\mu$moles $L^{-1}$ (Fig. 4). A similar lack of correlation between sediment-water flux and phosphate concentrations for individual cores maintained at various temperatures in the laboratory is also evident in the data given by Aller (1977), where phosphate fluxes out of Long Island Sound sediments were constant, at least up to 8 μmoles $PO_4$ $L^{-1}$, and in the measurements by Elderfield *et al.* (in press), where constant fluxes were maintained from Narragansett Bay sediments with phosphate concentrations in the overlying water varying from 5-17 μmoles $L^{-1}$. This behavior does not seem consistent with the widespread impression that sediments "buffer" the phosphate concentration in the overlying water (Rochford, 1951; Carritt and Goodgal, 1954; Pomeroy *et al.*, 1965), or with observations that pulses of phosphate added to marine systems do not usually result in permanently elevated concentrations in the water column, even when phosphate is presumably not limiting in the system (Pilson *et al.*, in press). The solution to this problem is not yet clear, but the measured fluxes in aerobic waters may be driven by biological activities in the top few mm of oxidized sediment, where there is a large amount of organic matter and where the exchange properties of the sediments may not influence the exchange dynamics as strongly as they do in deeper sediment-pore water interactions or in laboratory sediment slurry exchange experiments.

Figure 4. The relationship between net sediment-water phosphate flux and the mean phosphate concentration in the overlying water during an incubation.

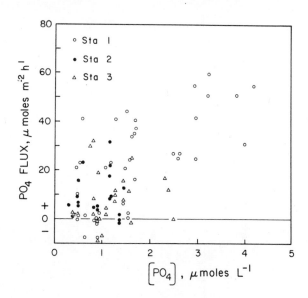

*Anomalous Nutrient Ratios and Benthic Stoichiometry*

In examining the ratios of phosphate flux to oxygen uptake and the release of $CO_2$ and nitrogen, it became clear that the stoichiometry of nutrient regeneration in the sediments as reflected in net sediment-water fluxes differed markedly in some respects from that of open ocean pelagic systems. While the ratio of phosphorus released to oxygen taken up by the sediment varied considerably in

individual measurements, the integrated value over the annual cycle
gave an O/P of 193 by atoms, or only slightly less than the 212
given by the Redfield (1934) model.  We assume that this also implies a Redfield type ratio for carbon and phosphorus fluxes, since
a number of simultaneous $CO_2$ release and oxygen uptake measurements
on sediment cores in the laboratory at various times during the
year gave an RQ of about 1 (Fig. 5).  Of course, the stoichiometry
of organic remineralization itself might be quite different from
that indicated by net sediment-water fluxes.  However, the ratio of
nitrogen released to phosphorus released by the sediments was much
lower at all three stations than the 16/1 by atoms expected from
the Redfield model (Fig. 6).  As we suggested in an earlier report
on sediment-water nitrogen exchanges (Nixon *et al.*, 1976), this
anomalously low N/P ratio in sediment-water fluxes appears to be
responsible for the characteristically low value of N/P in shallow
coastal waters and, at least to some degree, for the prominence of
nitrogen as the major limiting nutrient in near-shore primary production.

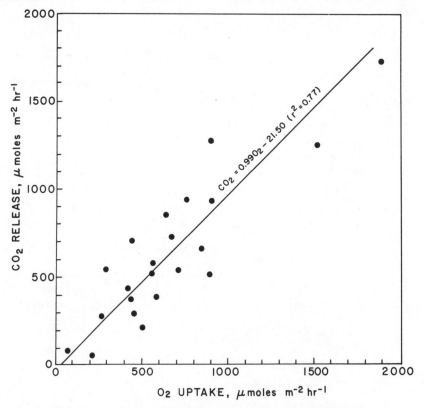

Figure 5.  The relationship between oxygen uptake and
carbon dioxide release by sediment cores collected near the
middle of Narragansett Bay at various times during the year.
A functional regression is given for the molar ratio of the
fluxes.

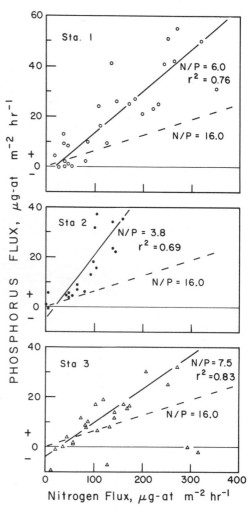

Figure 6. The relationship between net fluxes of fixed inorganic nitrogen (largely ammonia) and phosphate over an annual cycle from three stations in Narragansett Bay. Functional regressions are given for the atomic ratios. Broken lines represent the classic Redfield ratio.

The mechanism responsible for the low N/P ratio in sediment fluxes is not clear. We have measured the oxygen uptake and nutrient excretion of the macrofauna (> 0.5 mm) from the cores on numerous occasions and the ammonia to phosphate ratio is about 12 (functional regression, N=8, $r^2$ = 0.69), so it is not likely that the metabolism of the larger benthos is directly involved. We have also carried out decomposition experiments in which particulate organic matter was collected from Narragansett Bay (60-μ net), dried at 80°C, and added to sediment cores (∿85 g m$^{-2}$) or to bay water controls (∿0.5 g L$^{-1}$). The concentrations of carbon, nitrogen and phosphorus were then followed in solid and dissolved forms over time as the systems were kept at constant temperature in the dark. The results showed that phosphate was released very rapidly from the organic matter relative to nitrogen and carbon.

Estimated Amount of Particulate Matter Remaining
at 15°C in the Dark

| Day | Carbon[1] | Nitrogen[1,2] | Phosphorus[1] |
|-----|-----------|----------------|----------------|
| 0 | 100 (75 mg) | 100 (12.5 mg) | 100 (1.34 mg) |
| 2 | 97 | 94 | 37 |
| 7 | 91 | 82 | 29 |
| 15 | 84 | 72 | 26 |

[1]does not include DOM; [2]calculated from initial particulate nitrogen and ammonia increases

The same finding was reported earlier by Grill and Richards (1964) in their studies of diatom decomposition. While they confirmed Redfield's (1934) N/P ratio of 16 from decomposition, they also showed an initial phase of breakdown in which there was "a rapid release of phosphate with little or no increase in ammonia..."

It is difficult to see how this behavior can explain the overall, long-term deficit in nitrogen from the sediments. While the ratio of organic nitrogen to organic phosphorus in the surface sediments of Narragansett Bay (0-2 cm) is indeed very high, as might be expected if phosphorus were being remineralized more rapidly than nitrogen, the ratio of total nitrogen to total phosphorus is low (Fig. 7), and there is about 5-10 times more inorganic than organic phosphorus in the surface sediments of the bay (Sheith, 1974). Much of this inorganic phosphorus is probably tied up in ferric hydroxide gels and/or sorbed onto clay minerals. In fact, a comparison of the decomposition of organic matter in bay water and in bay water over sediment cores indicates that the sediments may take up an appreciable portion of the phosphate and DOP that is regenerated (Fig. 8). The results of this experiment, in which the overlying water was aerated, contrast with the organic enrichment experiments described by Holm (1978), who found that the release of phosphate from Baltic Sea sediments increased linearly with organic input. While the Narragansett Bay sediments appeared to take up perhaps 30 or 40% of the phosphorus regenerated from the organic input, the phosphorus released by the anoxic Baltic sediments was far greater than the phosphorus contained in the organic input. Under the anoxic conditions brought about by metabolism of the organic matter, the large store of inorganic phosphorus held by the iron became soluble and was exchanged with the overlying water.

If the water overlying cores from Narragansett Bay is allowed to go anoxic, or very nearly so (< 0.1 mg $O_2$ $L^{-1}$), the flux of phosphate also appears to increase dramatically. For example, in one experiment using cores collected from mid-bay (Sta. 2) in November and kept in the laboratory at 15°C, the flux of phosphate increased from less than 5 $\mu$moles $m^{-2}hr^{-1}$ when the oxygen in the overlying water was above 1.5 mg $O_2$ $L^{-1}$ to 25-65 $\mu$moles $m^{-2}hr^{-1}$ under anoxic conditions. It is not clear how long such a flux may last, however, since one core maintained under anoxic conditions for a week showed no phosphate flux at the end of that time.

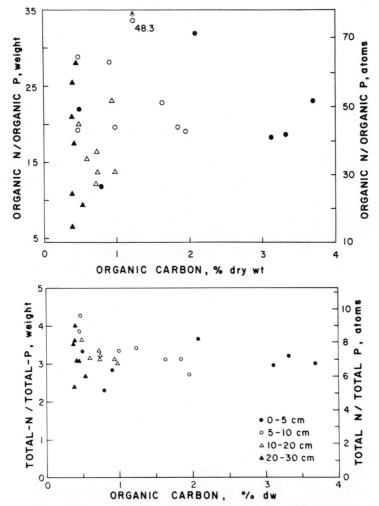

Figure 7. The ratio of organic nitrogen/organic phosphorus (top) and total nitrogen/total phosphorus (bottom) as a function of organic carbon content in sediments from Narragansett Bay and the Providence River. Data are from Sheith (1974).

It is clear that the processes governing phosphorus exchange at the sediment-water interface are complex, at least in sediments such as those in Narragansett Bay, where both chemical and biological transformations appear to be active. The situation is further complicated because the amount of phosphorus available is so large relative to the fluxes involved. Even at the highest exchange rates of 50 $\mu$moles m$^{-2}$hr$^{-1}$, the amount of phosphate in the upper 1 cm of sediment could support the flux for some 20 years if it was all "available". But on the basis of our present knowledge of the phosphorus dynamics in these sediments, it does not seem likely that an enhanced or enriched phosphorus exchange is responsible for the low N/P ratio of the sediment-water nutrient fluxes. In fact, the alternative of impoverished nitrogen return from the sediments seems far more important. Recent measurements in our laboratory of the production of $N_2$ gas from undisturbed sediment cores from the bay have shown fluxes of 100-150 µg-at N $m^{-2}hr^{-1}$ at 23°C (Seitzinger *et al.*, in prep.). Denitrification of this magnitude could essentially account for the anomalously low N/P ratios and hence for the overriding importance of fixed nitrogen in near-shore nutrient dynamics.

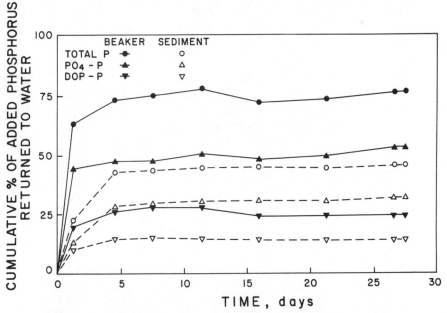

Figure 8. The amount of phosphorus regenerated in the dark at 15°C from organic matter collected in Narragansett Bay and placed in beakers with bay water or on sediment cores taken from the bay.


Laboratory Experiments

*Organic Inputs and the Seasonal Cycle*

Because the field measurements indicated that net phosphate fluxes showed a seasonal cycle that was not entirely governed by temperature, we carried out a number of experiments to see if there was also an effect of organic input from the water column. While a number of us have attempted to show the influence of benthic nutrient fluxes on the production of the overlying water (Thorstenson and MacKenzie, 1974; Davies, 1975; Rowe *et al.*, 1975a; Nixon *et al.*,

1976; Aller, 1977 and others), it has been more difficult to determine if, and how quickly the bottom might respond to variations in organic input.

To begin to answer this question, we collected cores from the mid-bay station before and after the intense winter-spring phytoplankton bloom that characterizes Narragansett Bay (Pratt, 1965; Smayda, 1973). The cores were brought into the laboratory and warmed 1-2°C day⁻¹ in an incubator and the oxygen uptake and nutrient fluxes measured at 5°C intervals. The results show that the potential benthic metabolism was clearly enhanced in cores collected shortly after the end of the plankton bloom (Fig. 9). The effect of temperature on the flux of phosphate from the sediment was some 40% greater after the bloom than it was before (Fig. 10). These data support the view that a large portion of benthic remineralization involves organic matter that has recently been deposited on the bottom, and that the exchange of nutrients between the sediments and the overlying water is responsive on a relatively short scale to events in the water column.

Figure 9. The uptake of oxygen and the net flux of ammonia and phosphate from sediment cores collected near the middle of Narragansett Bay before and after the winter-spring phytoplankton bloom. Phytoplankton cell counts were made on near-surface water in the lower West Passage. The cores were kept in the dark in an incubator and gradually warmed to 10 and 20°C.

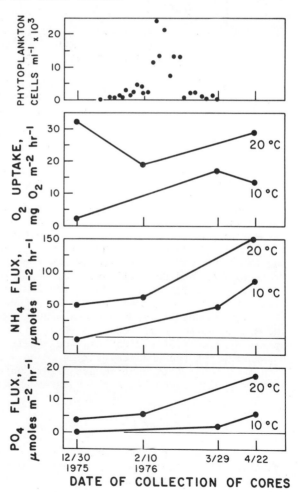

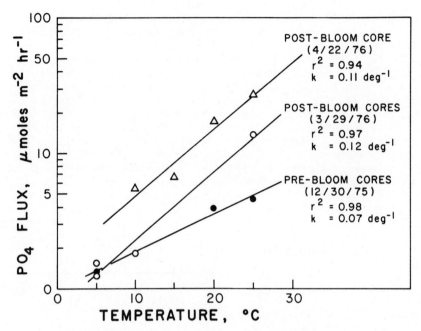

Figure 10. The net flux of phosphate as a function of temperature from Narragansett Bay sediments collected before and after the winter-spring phytoplankton bloom (see Fig. 9).

However, it is also true that the apparent metabolism of the sediments is much more stable than the overlying plankton system. For example, we have carried out "starvation experiments" at different times of year in which cores have been kept at constant temperature in the dark and given essentially no particulate organic input. The water over the cores was changed every 16-24 h, but the input water (1-2 liters) was first filtered through a 15μ Grainger Dirt and Rust filter, a 0.5μ ultrafine filter, and a 0.45 μ, 98% efficient Pall Ultrapore in-line cartridge filter system. Even with cores collected before the winter-spring plankton bloom, there was an uptake of oxygen and a net flux of phosphate out of the sediments for over 100 days (Fig. 11). The decline in oxygen uptake of some 0.5% per day was about half of that observed for phosphate, but the changeover time is so slow for all of the nutrients that laboratory experiments using "unfed" cores appear to give useful results even when carried out over several weeks.

*Macrofauna Excretion*

While a number of attempts have been made to estimate the contribution of the macrofauna to benthic oxygen uptake (Carey, 1967; Banse *et al.*, 1971; Smith *et al.*, 1972; Vernberg and Coull, 1974,

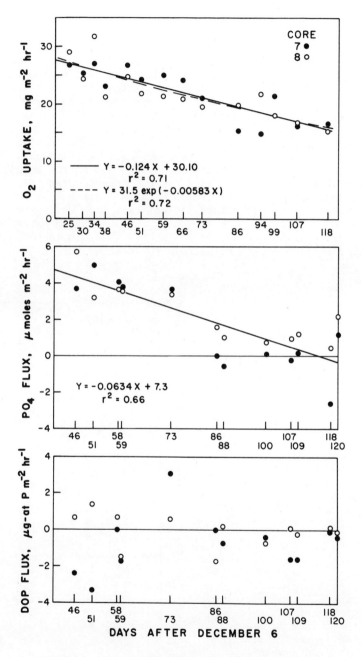

Figure 11. The uptake of oxygen and the net flux of phosphorus from sediments collected before the winter-spring phytoplankton bloom and maintained in the laboratory in the dark at 15°C with no particulate organic input.

and others), few, if any, measurements have been made of the contribution of the macrofauna to benthic nutrient fluxes. We have made
a series of preliminary measurements in which the uptake of oxygen
and the release of phosphate and DOP by sediment cores were obtained just before the cores were broken down and carefully screened to
remove the macrofauna (> 0.5 mm). All of the animals from each
core were then placed in bay water at the same temperature in the
dark and their oxygen uptake and phosphorus excretion measured.
When these data are all expressed per unit area of sediment, the
macrofauna appear to account for some 20-50% of the total oxygen
uptake, a somewhat high but not uncommon finding (Fig. 12). However, the excretion of phosphate and DOP by the macrofauna alone is
far higher than the observed flux for the intact cores (Fig. 12).
As mentioned earlier, the N/P ratio of animal metabolism appeared
normal as did the oxygen uptake per unit weight. While not defin-
itive, these data suggest that much of the phosphorus excretion by
the macrofauna living below the sediment surface is taken up by
sorption reactions or, especially in the case of DOP, by bacteria.

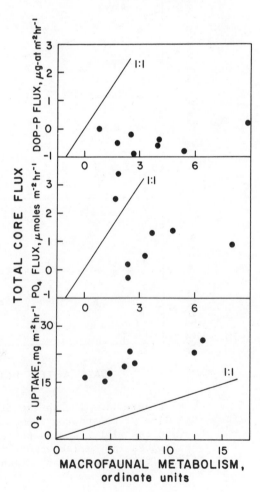

Figure 12. The contri-
bution of macrofauna (> 0.5
mm) to the oxygen uptake and
phosphorus flux of sediment
cores collected near the
middle of Narragansett Bay.

*Nutrient Fluxes and Community Metabolism*

To some degree the differences in phosphate fluxes from different areas reported in Table 1 arise from the fact that the very process of measurement itself can influence the findings. Most of the benthic metabolism work reported in the literature is based on concentration differences between initial and final samples, yet the assumption of linearity in flux rates over time is seldom tested. In the case of oxygen uptake, this clearly is not the pattern, and the apparent metabolic rate of Narragansett Bay sediments is strongly dependent on the oxygen concentrations of the overlying water (Fig. 13). While this also appears to be the case for $CO_2$ release, the flux of ammonia in short-term metabolic studies was much less influenced by the oxygen concentrations or by the apparent metabolic rate (Fig. 14). It is not clear why oxygen uptake and $CO_2$ release should be tightly coupled and responsive to oxygen concentration while ammonia release is not. Field data over the annual cycle showed that both ammonia and phosphate fluxes were strongly correlated with oxygen uptake (Nixon *et al.*, 1976). Unfortunately, phosphate fluxes were not measured as part of the low oxygen core experiments, but it is conceivable that as the redox discontinuity layer rose toward the surface sediment, more of the phosphate dissolved in the anoxic pore waters might be released into the overlying water, thus enhancing the apparent phosphate release.

*Variation and Evaluation of Benthic Fluxes*

Without being pessimistic, we emphasize that our present understanding of the metabolism and chemical dynamics of marine bottom communities is still primitive and that our efforts to understand the processes involved remain, for the most part, relatively crude. While we may be anxious to find out if the fluxes from different stations are different with different assemblages of animals, sediment types, or various organic inputs, and while we may wish to use net flux measurements to validate elaborate models based on sediment pore water profiles, chemical gradients, and animal activities, we need to be aware that the metabolic patchiness and variance of the sediments seems to be as troublesome as their well-known taxonomic patchiness. The errors involved in the chemical analysis of oxygen, phosphorus, nitrogen, etc. are not trivial when carried through the calculations of area-based fluxes, but they are usually a small problem compared with the large variance in fluxes measured at stations within a few meters of each other on the bottom or even from one station or core on successive days. This is usually a condition of nature rather than an error of measurement, and it is easy for us to forget that as benthic ecology moves from identifying, counting and weighing animals to the apparent sophistication of measuring chemical fluxes, the ever-frustrating cloud of uncertainty is still with us. We have compiled a summary of the variance in replicate flux measurements reported by various authors, along with some community structure data for com-

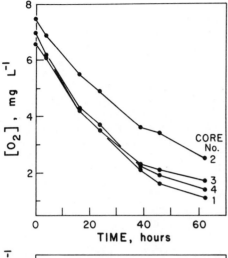

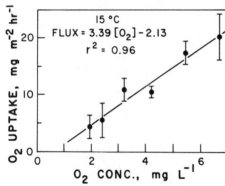

Figure 13. (Top) The concentration of dissolved oxygen over time in the water over four cores collected near the middle of Narragansett Bay and incubated in the dark at 15°C. (Bottom) The relationship between oxygen uptake and the mean oxygen concentration in the overlying water during the incubation. Data points are the mean ± 1 S.D. for the four cores shown.

Figure 14. (Top) The uptake of oxygen and the release of carbon dioxide as a function of oxygen concentration in the overlying water for sediment cores incubated in the dark at 15°C. Data points are the mean ± 1 S.D. for replicate cores. (Middle) The molar ratio of carbon dioxide released to oxygen taken up as a function of oxygen concentration. (Bottom) The net sediment-water flux of ammonia as a function of oxygen concentration for the four individual cores.

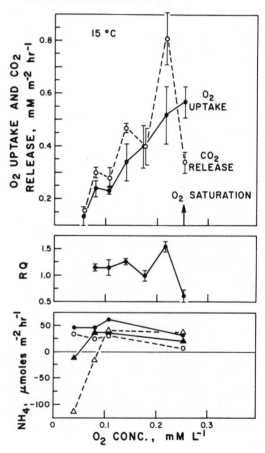

parison (Table 2). The results are humbling and should help to provide a useful if frustrating perspective on many of the conclusions we have drawn.

Table 2. Average coefficient of variation in replicate benthic flux measurements reported by various authors.

| Author | N | $O_2$ | $PO_4$ | DOP | $Si(OH)_4$ | $NH_4$ | $NO_3$ |
|--------|---|-------|--------|-----|------------|--------|--------|
| | | | | FLUXES, Coefficient of Variation, % | | | |
| Hale (1974) field chambers | 4 | 23 | 85 | --- | 27 | 25 | 410 |
| Rowe et al. (1975b) field chambers | 3 | 60 | 51 | --- | -- | 79 | 2000 |
| Hartwig (1976) field chambers | 3 | -- | 38 | 110 | -- | 37 | 150 |
| Gilbert (1976) field chambers | 4 | 40 | 400 | --- | 421 | -- | 23 |
| Elderfield et al. (in press) lab. cores | 4 | -- | 40 | --- | 92 | 42 | -- |
| | 4 | -- | 67 | --- | 63 | 45 | -- |
| This study lab. cores | 3 | 45 | 60 | 142 | 38 | 45 | 63 |

MACROFAUNA, Coefficient of Variation, %

| | | Numbers | Biomass |
|--|--|---------|---------|
| Hale (1974) replicate Smith-McIntyre grab samples from mid-bay (>0.75 mm) | 5 (November) | 127 | 50 |
| | 5 (March) | 47 | 63 |
| | 5 (July) | 12 | 35 |

## ACKNOWLEDGMENTS

We are grateful to Jim O'Reilly, Sharon Northby, Beth Evans, Ann Durbin, Eric Klos, Stephen Hobbs, Fred Short and Jim Kremer for their help with the extensive field work and laboratory analysis. Discussions of benthic nutrient dynamics with Michael Pilson, Michaèl Bender, Flip Froelich, Sybil Seitzinger, Jonathan Garber and Virginia Lee have been stimulating and helpful as well. Support for the work was provided by the Office of Sea Grant, NOAA, U.S. Dept. of Commerce and from the U.S. Environmental Protection Agency.

REFERENCES

Aller, Robert C.  1977.  The influence of macrobenthos on chemical
    diagenesis of marine sediments.  Ph.D. Thesis, Yale University,
    New Haven, CT, 600 pp.

Alvarez-Borrego, S., D. Guthrie, C. H. Culberson, and P. K. Park.
    1975.  Test of Redfield's model for oxygen-nutrient relation-
    ships using regression analysis.  Limnol. Oceanogr. 20:  795-805.

Banse, K., F. H. Nichols, and D. R. May.  1971.  Oxygen consumption
    by the seabed.  III.  On the role of the macrofauna at three
    stations.  Vie Milieu Suppl. 22:  31-52.

Beyers, R. J., J. L. Latimer, H. T. Odum, R. B. Parker, and N. E.
    Armstrong.  1963.  Directions for the determination of changes
    in carbon dioxide concentration from changes in pH.  Publ. Inst.
    Mar. Sci., Univ. Texas. 9:  454-489.

Carey, A. G.  1967.  Energetics of the benthos of Long Island Sound.
    I.  Oxygen utilization of sediment.  Bull. Bingham Oceanogr.
    Coll. 19:  136-144.

Carritt, D. E., and J. H. Carpenter.  1966.  Comparison and evalua-
    tion of currently employed modifications of the Winkler Method
    for determining dissolved oxygen in seawater, A NASCO Report.
    J. Marine Res. 24(3):  286-318.

Carritt, D. E., and S. Goodgal.  1954.  Sorption reactions and some
    ecological implications.  Deep-Sea Res. 1:  224-243.

Davies, J. M.  1975.  Energy flow through the benthos in a Scottish
    sea loch.  Mar. Biol. 31:  353-362.

Elderfield, H., N. Luedtke, R. J. McCaffrey and M. Bender.  In
    press.  Benthic flux studies in Narragansett Bay.  Am. J. of Sci.

Fleming, R. H.  1940.  The composition of plankton and units for
    reporting populations and production.  Proc. Sixth Pacific Sci.
    Congr. 3:  535-540.

Furnas, M. J., G. L. Hitchcock, and T. J. Smayda.  1976.  Nutrient-
    phytoplankton relationships in Narragansett Bay during the 1974
    summer bloom.  *In* Estuarine Processes, Vol. 1, Uses, Stresses
    and Adaptation to the Estuary, M. L. Wiley (ed.), Academic Press,
    N.Y., pp. 118-134.

Gilmartin, M.  1967.  Changes in inorganic phosphate concentration
    occurring during seawater sample storage.  Limnol. Oceanogr. 12:
    325-328.

Gilbert, P. M.  1976.  Nutrient flux studies in the Great Bay Estu-
    ary, New Hampshire.  M.S. Thesis, University of New Hampshire,
    Durham, N.H.

Goldman, J. C., Tenore, K. R., and H. I. Stanley.  1973.  Inorganic
    nitrogen removal from wastewater:  effect on phytoplankton
    growth in coastal marine waters.  Science. 180:  955-956.

Grill, Edwin V., and Francis A. Richards.  1964.  Nutrient regener-
    ation from phytoplankton decomposing in seawater.  J. of Mar.
    Res. 22(1-3):  51-69.

Hale, S. S.  1974.  The role of benthic communities in the nutrient
    cycles of Narragansett Bay.  M.S. Thesis, University of Rhode
    Island, Kingston, R.I.

Harrison, J. T.  In prep.  Biological mediation of benthic nutrient
    flux in Kaneohe Bay, Hawaii.  Ph.D. Thesis, Univ. of Hawaii.

Hartwig, Eric O. 1976. The impact of nitrogen and phosphorus release from a siliceous sediment on the overlying water. *In* M.
Wiley (ed.) Estuarine Processes, pp. 103-117.

Hartwig, Eric O. 1978. Factors affecting respiration and photosynthesis by the benthic community of a subtidal siliceous sediment.
Mar. Biol. 46: 283-293.

Holm, N. G. 1978. Phosphorus exchange through the sediment-water
interface. Mechanism studies of dynamic processes in the Baltic
Sea. Contrib. in Microbial Geochemistry, Dept. of Geology, Univ.
of Stockholm, No. 3, 149 pp.

Jeffries, H. P. 1962. Environmental characteristics of Raritan
Bay, a polluted estuary. Limnol. Oceanogr. 7: 21-31.

Kremer, J. N., and S. W. Nixon. 1978. A Coastal Marine Ecosystem,
Simulation and Analysis, Ecological Studies 24, Springer-Verlag,
N.Y., 217 pp.

Menzel, D. W., and N. Corwin. 1965. The measurement of total phosphorus in seawater based on the liberation of organically bound
fractions by persulfate oxidation. Limnol. Oceanogr. 10: 280-
282.

Nixon, S. W., C. A. Oviatt, and S. S. Hale. 1976. Nitrogen regen-
eration and the metabolism of coastal marine bottom communities.
*In* The Role of Terrestrial and Aquatic Organisms in Decomposition
Processes, pp. 269-283, J. M. Anderson and A. Macfadyed (eds.),
Proc. 17th Symposium, British Ecological Soc., Blackwell Scien-
tific Publication, April 1975.

Okuda, J. 1955. On the soluble nutrients in bay deposits. III.
Examinations on the diffusion of soluble nutrients to sea water
from mud. Bull. Tohoku Regional Fish. Res. Lab. 2: 215-242.

Pilson, M. E. Q., S. W. Nixon, and C. A. Oviatt. In press. Annual
nutrient cycles in a marine microcosm. Proceedings of the
Savannah River Ecology Laboratory Symposium on Microcosms in
Ecological Research, Augusta, Ga., November 8-11, 1978.

Pomeroy, L. R., E. E. Smith and C. M. Grant. 1965. The exchange
of phosphate between estuarine water and sediments. Limnol.
Oceanogr. 10(2): 167-172.

Pratt, D. M. 1965. The winter-spring diatom flowering in Narragansett Bay. Limnol. Oceanogr. 40: 173-184.

Propp, M. V., V. G. Tarasoff, I. I. Gherbadgi, and N. V. Lootzik.
1979. This volume.

Redfield, A. C. 1934. On the proportion of organic derivatives in
sea water and their relation to the composition of plankton, pp.
176-192. *In* James Johnson Memorial Volume, University Press,
Liverpool, 348 pp.

Riley, G. A. 1941. Plankton studies III. Long Island Sound.
Bull. Bingham Oceanogr. Coll. 7(3): 1-93.

Rochford, D. J. 1951. Studies in Australian estuarine hydrology.
I. Introductory and comparative features. Australian J. Mar.
Freshwater Res. 2: 1-116.

Rowe, G. T., C. H. Clifford, K. L. Smith, Jr., and P. L. Hamilton.
1975a. Benthic nutrient regeneration and its coupling to primary productivity in coastal waters. Nature. 255: 215-217.

Rowe, G. T., K. L. Smith, Jr., and C. H. Clifford. 1975b. Benthic
pelagic coupling in the New York Bight, pp. 370-376. *In* M. Grant

Gross (ed.) Middle Atlantic Continental Shelf and the New York Bight. ASLO Special Symposium, Vol. 2, pp. 441.

Rowe, G. T., C. H. Clifford, and K. L. Smith, Jr. 1977. Nutrient regeneration in sediments off Cap Blanc, Spanish Sahara. Deep-Sea Res. 24: 57–63.

Ryther, J. H., and Dunstan, W. M. 1971. Nitrogen, phosphorus and eutrophication in the coastal marine environment. Science. 171: 1008–1013.

Seitzinger, S., S. Nixon, M. Pilson, and S. Heffernan. In prep. Denitrification and $N_2O$ production in near-shore marine sediments.

Sheith, M-S., J. 1974. Nutrients in Narragansett Bay sediments. M.S. Thesis, University of Rhode Island, Kingston, R.I.

Smayda, T. J. 1957. Phytoplankton studies in lower Narragansett Bay. Limnol. Oceanogr. 2: 342–359.

Smayda, T. J. 1973. The growth of *Skeletonema costatum* during a winter-spring bloom in Narragansett Bay, Rhode Island. Norw. J. Bot. 20: 219–247.

Smith, K. L., H. A. Burns, and J. M. Teal. 1972. *In situ* respiration of benthic communities in Castle Harbor, Bermuda. Mar. Biol. 12: 196–199.

Smith, K. L., Jr., G. A. White, M. B. Laver, and J. A. Haugsness. 1978. Nutrient exchange and oxygen consumption by deep-sea benthic communities: Preliminary *in situ* measurements. Limnol. Oceanogr. 23(5): 997–1005.

Solorzano, L. 1969. Determination of ammonia in natural waters by the phenolhypochlorite method. Limnol. Oceanogr. 14: 799–801.

Stainton, M. P., Capel, M. J., and F. A. J. Armstrong. 1974. The chemical analysis of fresh water. Dept. of the Environment (Canada) Fisheries and Marine Service, Miscellaneous Special Publication, No. 25.

Strickland, J. D. H., and T. R. Parsons. 1968. A practical handbook of sea-water analysis. Fish. Res. Bd. Can. Bull. 167: 1–311.

Taft, J. L., and W. R. Taylor. 1976. Phosphorus dynamics in some coastal plain estuaries, 79–89. *In* M. Wiley (ed), Estuarine Processes, Vol. 1. Academic Press, New York.

Thorstenson, Donald C., and Fred T. Mackenzie. 1974. Time variability of pore water chemistry in recent carbonate sediments, Devil's Hole, Harrington Sound, Bermuda. Geo. et Cosmochimica Acta. 38: 1–19.

Vargo, G. A. 1976. The influence of grazing and nutrient excretion by zooplankton on the growth and production of the marine diatom *Skeletonema costatum* (Greville) Cleve in Narragansett Bay, Ph.D. Thesis, University of Rhode Island, Kingston, Rhode Island.

Vernberg, W. B., and B. C. Coull. 1974. Respiration of an interstitial ciliate and benthic energy relationships. Oecologia (Berl.) 16: 259–264.

Williams, J. D. H., J. K. Syers, and T. W. Walker. 1967. Proc. Soil. Sci. Soc. Amer., 31: 736.

# Factors Affecting the Flux of Organic Matter to Sediments in a Marine Bay

B. T. Hargrave

ABSTRACT: This paper extends analysis of data collected in Bedford Basin to consider relations between seasonal changes in mixed-layer depth, the establishment of planktonic populations, and the amount and nature of deposited material and its oxidation in sediments. The observations show that annual cycles of organic matter sedimentation and benthic metabolism are coupled through seasonal changes in phytoplankton production and water column stratification. Breakdown of the thermocline during autumn leads to an apparent increase in sedimentation rate, partly as a result of sediment resuspension. Increases in phytoplankton production during spring and summer lead to higher rates of organic matter sedimentation and a corresponding increase in benthic respiration. During the fall, benthic respiration decreases even though there is an apparent increase in sedimentation rate. The seasonal change in the organic composition of settled material, especially nitrogen and plant pigment content, is an important factor which regulates benthic metabolism. The result is a highly dynamic response by the benthic

community to changing conditions of organic supply
from the water column.

INTRODUCTION

Recent studies on the flux of dissolved nutrients between
marine sediments and water have renewed interest in processes that
link pelagic and benthic populations.  The idea that metabolic processes, organic matter degradation, or nutrient remineralization
in sediments can be quantitatively related to biological production
and hydrographic conditions in the water column is not new, however.
The concept forms the basis of Thienemann's (1927) and Hutchinson's
(1938) trophic classification of lakes.  Ohle's (1956) proposal of
a bioactivity index for stratified water bodies based on carbon
dioxide accumulation relative to mean depth was an extension of
these ideas.  He suggested that the amount of organic material sedimented in stratified lakes was inversely related to the mean depth
and directly related to the concentration of material per unit volume.
These early studies were based on a holistic view of material cycling
between water and sediments.  The observations imply that in areas
where allochthonous input of organic matter is not important, mineralization of deposited material is directly related to supply through
sedimentation, which is proportional to autotrophic production in
surface waters.

Although sedimentation has often been measured, data have usually not been used to quantify relationships identified in these early
studies.  Measurements of material deposition in collectors suspended at fixed depths have been used to consider the origin of settled
material (Stephens *et al.*, 1967; Steele and Baird, 1972; Smetacek
*et al.*, 1978) or compared with estimates of benthic metabolism, nutrient release or secondary production (Johnson and Brinkhurst, 1971;
Davies, 1975; Hartwig, 1976).  All of these studies have shown seasonal changes in sedimentation which demonstrate that rates of metabolism and nutrient remineralization by benthic populations reflect variations in organic supply from surface waters.  In addition, annual organic carbon sedimentation in various areas was shown to be linearly
related to the ratio of carbon supply:mixed layer (seasonal thermocline)
depth (Hargrave, 1975).  A proportionality (after logarithmic transformation) between this ratio and annual sediment oxygen uptake in
different areas also exists (Hargrave, 1973).  Annual studies of sedimentation in Bedford Basin, a coastal embayment in Nova Scotia, were
also used to calculate that daily loss of particulate organic carbon
through sedimentation was equivalent to 1-6% of suspended concentrations integrated per unit area above the depth of trap exposure (Taguchi and Hargrave, 1978).  This confirms one of Ohle's suggestions
and corresponds with the general idea that there is a small daily
loss of suspended material through sedimentation in most natural
waters.  What has not been considered in these studies, however, is how
physical and biological pathways interact to determine organic matter
supply to sediments.

A Conceptual Model Linking Water Column Physical Structure,
Sedimentation and Benthic Metabolism

Mechanisms of Sedimentation

Past studies of sedimentation have shown that different pathways
of supply transport particulate matter from the water column to sediments. Direct settling of phytoplankton cells during and following
blooms can occur (Hargrave and Taguchi, 1978; Smetacek *et al.*, 1978).
This observation supplements the conventional view that zooplankton
fecal pellets are a dominant vector for particulate matter deposition
(Steele and Baird, 1972; Ferrante and Parker, 1977; Honjo and Roman,
1978). Differences in sedimentation measured simultaneously at various depths have also been interpreted as reflecting resuspension of
previously settled material which is transported horizontally to
deeper depths during periods of mixing (Wetzel *et al.*, 1972; Davis,
1973; Hargrave and Taguchi, 1978). Alternatively, vertical differences in current speed could cause scouring and loss of material settled
in traps (Smetacek *et al.*, 1978). The relative importance of these
effects is likely to be different in different water bodies and to
change seasonally to alter the nature of particulate matter deposited
in collectors. However, differences in quantity and quality of
settled material actually settled to the bottom will affect the rate
and nature of metabolic processes in sediments.

Effects of Mixing on Exchange Processes Between Water and Sediment

Possible interactions between water column physical structure and
biological production processes observed to affect particulate sedimentation and benthic metabolism, based on previous observations, are summarized in Figure 1. The inverse relation between mixed-layer depth
during stratification, the proportion of organic matter supply sedimented, and total oxygen uptake by sediments in different areas, discussed by Hargrave (1973, 1975), appears to be due to the recognized
importance of mineralization of organic matter in well-mixed surface
waters. The deeper the thermocline (or halocline), the longer the particulate matter remains in suspension and is susceptible to consumption
and decomposition. Deep mixing also distributes phytoplankton below
the compensation depth, and when cellular respiration exceeds production, decreases in biomass occur. This factor, coupled with the low
production per unit volume that accompanies deep vertical mixing, is
generally associated with low biomass of all sizes of grazing organisms.

Lack of stratification in any water body also usually results in
increased supply and availablility of dissolved inorganic nutrients in
surface water layers. Phytoplankton blooms during fall and winter may
occur in response to low grazing pressure and high nutrient supply as
reported for Bedford Basin (Taguchi and Platt, 1977). Deposition of
phytoplankton cells during and following these periods of phytoplankton growth (Hargrave and Taguchi, 1978) shows that nutrient depletion
rather than grazing, terminated these intervals of high production.

Observations by Walsh *et al.*(1978) also imply that blooms which occur
following wind-induced periods of mixing are not consumed by pelagic
herbivores, but may be transferred directly to benthic populations.

Horizontal and vertical mixing, as well as gravitational settling, transports suspended material to the bottom and also resuspends
sediment into the water column.  These conditions minimize gradients
in dissolved gases and nutrients across the sediment water boundary.
They may also account for seasonal changes in chemical constituents
in interstitial water, if sediments are porous enough to allow water
exchange (Thorstenson and Mackenzie, 1974).  Increased oxygen supply
to the interface means that aerobic respiration in sediments should
predominate over anaerobic metabolism during these periods.  Exceptions would occur if deposition of freshly produced organic matter
increased rates of oxygen consumption such that mixing could not
replenish oxygen deficiencies formed at the sediment surface (Davies,
1975).

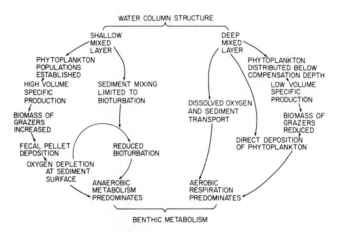

Figure 1.  Inter-relationships of processes
affecting the supply and metabolism of particulate matter in sediments beneath a water column
which undergoes seasonal stratification.  Another
important mechanism, which has not been quantitatively compared to depositional flux, is the transport of organic matter by vertically migrating
organisms or sinking of carcasses.  This additional flux has not been estimated in previous comparisons between sedimentation and benthic metabolism.


Effects of Stratification on Exchange Processes Between Water
and Sediment

Stratification and the formation of a shallow mixed-layer
dramatically alters conditions for sedimentation.  Only when the
mixed-layer depth becomes less than the critical depth (the depth
where integrated photosynthetic production and respiration are equal)
can net production by phytoplankton occur (Sverdrup, 1953).  The

general importance of physical processes of mixing and water column
stabilization and the interaction of these processes with light and
nutrient supply in permitting phytoplankton populations to become
established is now clearly documented (Pingree *et al.*, 1978).
Once these populations become established, volume-specific
production in the mixed-layer increases.  Phytoplankton may be
transported horizontally if advective exchange occurs (tidal
flushing of coastal embayments or long shore currents, for
example).  However, as Walsh *et al.* (1978) observed, once near-
surface vertical gradients in phytoplankton productivity are
formed and per sist, increases in biomass of herbivorous grazing
populations follow.  Lags in the occurrence of peak herbivore
biomass should be inversely related to generation time (Vinogradov *et al.*, 1970).

High concentrations of plant pigment degradation products in
suspended material at certain times have often been assumed to arise
from seasonal changes in grazing pressure by planktonic herbivores
(Glooschenko *et al.*, 1972).  Certainly studies cited above have
shown that seasonal differences in abundance of zooplankton fecal
pellets and pheopigment content in material collected in sediment
traps and material resulting from grazing may predominate in material
deposited at certain times.  These intermittent or extended periods of
fecal pellet deposition represent times when particulate matter is
rapidly transported to deeper water.  Even if fragmentation, dissolution or consumption of pellets occurs during sinking (Ferrante and
Parker, 1977), this material forms a supply of organic matter susceptible to degradation.  This may reduce dissolved oxygen concentrations at
any depth where the material accumulates or in bottom water and at
the sediment surface if pellets settle to the bottom (Seki *et al.*,
1974).

Effects of oxygen depletion on the movement of dissolved inorganic ions between sediments and water were first recognized and
described in lakes that undergo periodic stratification (Mortimer,
1971).  Oxygen consumption by suspended and sedimented material in
any water body where oxygen supply is restricted by stratification
will reduce dissolved oxygen concentration and oxygen deficits will
form most rapidly at sites of greatest uptake.  Such conditions at a
sediment surface result in an accumulation of reduced compounds
through the combined effects of altered oxidation-reduction potentials
and accumulation of by-products of anaerobic metabolism (respiration,
fermentation, sulphate reduction).  Oxidation of organic matter added to anoxic sediments with low redox potentials and lacking oxygen
supply can occur rapidly through these and other anaerobic processes
(Rich and Wetzel, 1978; Jørgensen, 1977).

Vertical mixing at the sediment surface should also be minimized during periods of stratification.  Sediment resuspension in
lakes, for example, appears to be most intense during periods of
overturn with physical transport reduced during periods of water
column stabilization (Davis, 1973; Pennington, 1974).  It also seems
likely that sediment mixing and transport of pore water through bioturbation, at least that due to large burrowing benthos dependent
on adequate supplies of dissolved oxygen (Aller, 1978), will be
reduced during periods of low dissolved oxygen concentration.  Ces-

sation of growth and lower respiration in profundal lake benthos occur during summer stratification (Jónasson, 1972). Also, while a high abundance of tolerant species of fauna may occur in sediments that become temporarily anoxic, numbers are generally reduced (Fenchel, 1969; Coull, 1969; Rowe, 1971). Thus, stratified conditions should enhance anaerobic metabolic processes in sediment through reduced rates of physical and biological mixing (Fig. 1).

### Annual Cycles of Organic Supply, Sedimentation and Sediment Oxygen Uptake in a Coastal Marine Bay

Seasonal observations in Bedford Basin have provided data which can be used to test these general ideas of interactions between physical and biological factors that link pelagic and benthic populations. Measurements of phytoplankton production, sedimentation and sediment oxygen uptake in this coastal marine bay are described elsewhere (Hargrave and Taguchi, 1978; Hargrave, 1978).

The period between December and January is a time of minimum phytoplankton production in Bedford Basin (Fig. 2). Maximum rates of chlorophyll a and pheopigment sedimentation and moderate levels of organic carbon and nitrogen sedimentation occurred at this time (Figs. 2 and 3). Phytoplankton (*Ceratium* and *Skeletonema*) blooms during the preceding two months could have contributed to material deposited in traps, but only if settling of cells produced 4-6 weeks earlier was delayed. Increased river discharge during November and December caused marked density stratification during December (Fig.4) (Hargrave and Taguchi, 1978), but increased mixing after this time could have transported material previously settled above 20 m to deeper depths. Increases in microzooplankton volume during December followed by higher net zooplankton biomass approximately two weeks later (Fig. 4) also imply that some of the material originating from these phytoplankton blooms could have been consumed by pelagic herbivores. High rates of pheopigment deposition (Fig. 2) probably reflect the combined effects of grazing and settling of senescent algae.

The importance of the deepening in mixed-layer depth as a causal mechanism for increased sedimentation during December and January can also be inferred by comparison of depositon at shallow and greater depths. Hargrave and Taguchi (1978) observed that differences in sedimentation over depth in Bedford Basin were greatest during periods of least stratification. Resuspension, vertical and horizontal transport could have contributed material to increase sedimentation with depth during these periods. In addition, advective flow and turbulence should decrease with increasing depth in this enclosed bay. If loss of settled material from traps by resuspension occurred, this would be reduced with depth to yield apparently higher sedimentation rates. The magnitude of the effect differed for various substances, however (Table 1). If particulate matter deposition at the shallower depths was low only because of loss of settled material through scouring, similar decreases should have occurred for all substances. Increased deposition with increased depth is clearly seen in measures of pigment deposition (Fig. 2).

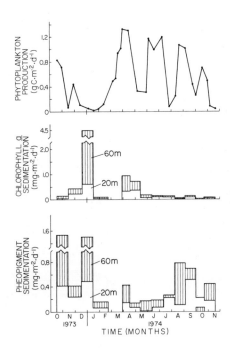

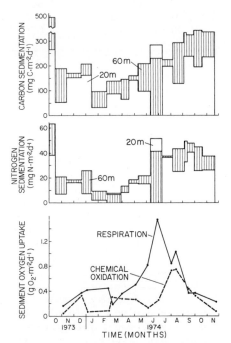

Figure 3. (Below) Particulate organic carbon and nitrogen sedimentation at two depths and oxygen uptake by undisturbed sediment cores as respiration (formalin-sensitive) and chemical oxidation (residual uptake after treatment with 1% formalin) at 60 m in Bedford during 1973 and 1974. Measurements of oxygen uptake by undisturbed sediment cores (with and without treatment with formalin to separate respiration and chemical oxidation) are presented in Hargrave (1978).

Figure 2. (Above) Phytoplankton primary production (integrated to the depth of 1% incident radiation) and chlorophyll a and pheopigment sedimentation at two depths at a central station in Bedford Basin during 1973 and 1974. Sediment traps (open-ended cylinders 7.5 x 31 cm) were clamped to a taut wire mooring at 10 m intervals between 20 m and 60 m. Collection and analysis of settled material is described by Hargrave and Taguchi (1978).

Whatever the cause, excess deposition at 60 m over that occurring at 20 m was greatest for chlorophyll a and pheopigments (eight and three-fold increase respectively) during December and April - times when rapid changes in mixed-layer depth were occurring (Fig. 4). Other periods of changes in stratification (October-November and August-September) were also times of increased pheopigment, but not chlorophyll a, sedimentation with depth. The accumulation of pigment degradation products arising from zooplankton feeding would be expected to be maximum during these periods. Although net zooplankton biomass declined during autumn (Fig. 4),

grazing activity could still have been high.  Low chlorophyll a
content observed in material deposited during the fall (Hargrave and
Taguchi, 1978) would be expected if zooplankton feeding pressure was
maintained.


Table 1.  Annual sedimentation (g $m^{-2}yr^{-1}$) of various
substances at two depths in Bedford Basin.  Seasonal data
are presented in Hargrave et al. (1976).

| Depth (m) | 20 | 60 | Percentage Increase |
|-----------|------|--------|---------------------|
| Dry Weight | 638.0 | 1105.9 | 73 |
| Organic Matter | 155.2 | 238.2 | 53 |
| Carbon | 55.1 | 75.7 | 37 |
| Nitrogen | 7.6 | 9.4 | 24 |
| Chlorophyll a | .056 | .183 | 227 |
| Pheopigments | .076 | .203 | 167 |

Figure 4.  Changes in density stratification (as the
gradient in $t$ between 0 and 20 m
and mixed-layer depth taken as
that depth with a density change
less than 0.2 $\sigma_t$ units, micro-
and net zooplankton biomass (integrated between 0 and 20 m and
per unit volume between 5 and
10 m, respectively), and sediment respiration as a percent of
total oxygen uptake at 60 m in
Bedford Basin during 1973 and
1974.  Data for microzooplankton
biomass are given in Taguchi and
Platt (1978) and for net zooplankton biomass in Conover and Mayzaud (1975).

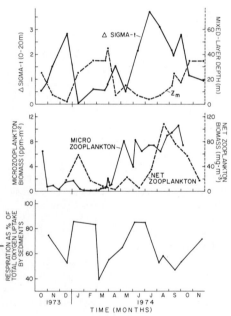

Periods of minimal stratification and deepest mixing in Bedford Basin corresponded to times when oxygen uptake by sediments
at 60 m was low (Figs. 3 and 4).  However, the proportion of total
oxygen uptake (with chemical oxidation and respiration partitioned
by formalin addition) due to respiration generally exceeded 60%
during these periods.  An exception to this trend occurred during

June, when density gradients and a shallow mixed-layer were forming
rapidly. Dissolved oxygen concentration in water above the sediment exceeded 6 mg $L^{-1}$ at this time (Hargrave, 1978). Thus, while
oxygen consumption due to respiratory demand increased sharply during May and June (Fig. 3) oxygen supply was apparently not limiting.

Phytoplankton production increased gradually during February
to maximum values during the spring bloom in early April (Fig. 2).
Although the depth of the mixed-layer exceeded 30 m, the water column was becoming increasingly stratified (Fig. 4). The abrupt termination of the bloom during late April, a time when the biomass of
micro- and net zooplankton was beginning to increase, shows that
grazing pressure alone was probably not the cause of decreased phytoplankton production. High sedimentation rates of chlorophyll a
during April (Fig. 2) show that undegraded plant pigments were deposited. Between June and August, however, when vertical gradients
in stratification were intensified, successive phytoplankton blooms
did not contribute to increased chlorophyll a sedimentation. Feeding by herbivorous zooplankton, with increased biomass throughout
these summer months (Fig. 4), could have contributed to progressive
increases in pheopigment, carbon and nitrogen deposition.

Maximum rates of oxygen uptake by sediments in Bedford Basin
occurred during July and August coincident with high rates of carbon sedimentation and maximum rates of nitrogen deposition at 20 m
(Fig. 3). Separation of total oxygen uptake into chemical and respiratory demand shows that maximum respiration coincided with the
peak in nitrogen sedimentation in late June and early July. Increased chemical oxidation four to six weeks later was associated
with a slight decrease in dissolved oxygen (to 5 mg $L^{-1}$) above the
sediment surface (Hargrave, 1978). High rates of sedimentation and
respiration in surface sediments could have contributed to the accumulation of reduced compounds during this period. Respiration,
as a proportion of total oxygen uptake, decreased continuously between July and mid-September (Fig. 4). The trend shows that aerobic metabolism decreased with increased sedimentation.

After September vertical stratification and density gradients
in Bedford Basin began to decrease as the mixed-layer deepened rapidly (Fig. 4). Although successive blooms of phytoplankton continued to occur, zooplankton biomass decreased (Figs. 2 and 4). High
daily rates of carbon and nitrogen sedimentation were maintained
(Fig. 3), probably due to the resuspension of previously settled
material. Measurements of sedimentation (at monthly intervals)
would have had to be made more frequently to document these events.
One indication that the particulate organic matter deposited during
this time was metabolically more refractory than that settled during the summer, was that both respiration and chemical oxidation in
sediments decreased continuously during the period (Fig. 3). Oxygen uptake by material collected in traps and stirred in aerated
water also decreased to stable and seasonal low rates after October
(Hargrave 1978). However, an increasing proportion of oxygen uptake by bottom sediments occurred as respiration (Fig. 4). Thus,
despite the continued high rates of deposition of particulate carbon and nitrogen during the fall, the refractory nature of this material allowed aerobic metabolism to predominate.

Quantitative Relations between Organic Supply, Sedimentation
and Oxidation of Organic Matter in Bedford Basin Sediments

The foregoing description of seasonal changes in phytoplankton
production, mixed-layer depth, rates of sedimentation and benthic metabolism in Bedford Basin conform to the conceptual model
presented in Figure 1. Stratification appears to play an important
role in determining pathways of both supply and oxidation of organic matter in sediments of this bay. The observations are of limited general significance, however, since they do not provide quantitative or predictive relationships which can be tested with data
from other areas. If the data are expressed as time-averaged or
time-integrated totals, however, correlations between variables
over time may be examined.

Monthly integral values show that while changes in phytoplankton production, sedimentation and sediment oxygen uptake followed
similar patterns through the year (with summer maxima), the relative magnitude of changes and timing of peak rates were different
(Table 2, Figs. 2 and 5). Phytophankton production was high from
March to October, but sedimentation of organic carbon increased
gradually. This contrasted nitrogen sedimentation (at 20 m) and
sediment respiration which rapidly increased to maximum values by
early July. This was not a time of maximum temperature, but stratification was pronounced and microzooplankton biomass had risen
sharply (Fig. 4, Table 2). Grazing by these herbivores could have
enhanced both carbon and nitrogen deposition. Monthly carbon sedimentation at 20 m was equivalent to between 20 and 30% of phytoplankton production throughout the period June to October (Fig. 5).

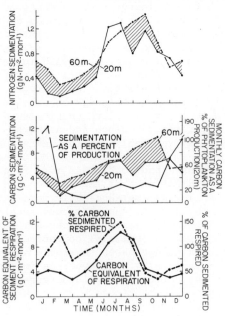

Figure 5. Seasonal
changes in carbon and ni-
trogen sedimentation, phytoplankton production and
sediment respiration (ex-
pressed as carbon by assuming an R.Q. value of 0.85)
calculated as monthly
totals throughout one year
in Bedford Basin (see Table
1).

Table 2. Monthly integrated or average values calculated for different variables measured over an annual period (October 1973 to November 1974) at a central station in Bedford Basin. Data for depth-integrated phytoplankton production (Fig. 2) are presented in Taguchi and Platt (1977). Conover and Mayzaud (1975) provided data for calculation of net zooplankton biomass (dry weight m⁻³ at 5-10 m depth)[a]. Average monthly values for bottom water (60 m) temperature, mixed-layer depth and microzooplankton volume were calculated from data presented by Taguchi and Platt (1978) in Figure 4. Measures of carbon and nitrogen sedimentation, sediment respiration and chemical oxidation (Fig. 3) were integrated to calculate monthly total values.

| Month | Temperature °C | Mixed-Layer Depth m | Phytoplankton Production g C·m⁻² | Carbon Sedimentation 20 m g C·m⁻² | Carbon Sedimentation 60 m | Nitrogen Sedimentation 20 m g N·m⁻² | Nitrogen Sedimentation 60 m | Net Micro-Zooplankton mg·m⁻³ | Net Micro-Zooplankton ppm | Sediment Oxygen Uptake Respiration g O₂·m⁻² | Chemical Oxidation |
|---|---|---|---|---|---|---|---|---|---|---|---|
| January | 2.0 | 15 | 1.5 | 2.9 | 4.6 | 0.2 | 0.5 | 59.3 | 0.8 | 12.9 | 1.1 |
| February | 1.5 | 32 | 5.6 | 1.2 | 3.1 | 0.1 | 0.3 | 17.4 | 0.3 | 12.2 | 2.0 |
| March | 1.0 | 35 | 22.6 | 2.6 | 4.0 | 0.2 | 0.4 | 8.6 | 0.6 | 9.1 | 4.0 |
| April | 2.0 | 17 | 34.5 | 3.1 | 4.6 | 0.3 | 0.5 | 2.8 | 2.2 | 12.6 | 4.6 |
| May | 3.0 | 12 | 12.4 | 3.7 | 5.8 | 0.4 | 0.6 | 23.4 | 6.1 | 18.3 | 3.1 |
| June | 4.5 | 7 | 28.5 | 6.4 | 6.7 | 1.2 | 1.0 | 5.7 | 7.2 | 27.0 | 7.7 |
| July | 6.0 | 5 | 23.3 | 6.9 | 6.9 | 1.3 | 1.2 | 36.8 | 7.5 | 32.1 | 19.4 |
| August | 7.5 | 8 | 18.6 | 4.4 | 8.6 | 0.8 | 1.3 | 114.5 | 7.9 | 27.2 | 15.0 |
| September | 8.2 | 17 | 21.0 | 6.5 | 9.6 | 1.2 | 1.4 | 77.0 | 9.9 | 13.9 | 7.4 |
| October | 9.5 | 27 | 24.8 | 6.5 | 10.5 | 0.8 | 1.0 | 57.2 | 5.0 | 11.8 | 7.3 |
| November | 5.5 | 35 | 9.8 | 7.0 | 5.5 | 0.7 | 0.6 | 17.4 | 0.5 | 9.4 | 2.2 |
| December | 4.5 | 5 | 4.7 | 4.9 | 5.8 | 0.4 | 0.6 | 18.7 | 1.3 | 11.3 | 3.9 |

[a] Observations between June 1974 and June 1975.

Between November and January, however, carbon deposited approximately equalled or exceeded that produced by phytoplankton. This reflects processes, like post-depositional resuspension and transport, discussed above, which deposit material in traps in addition to that settled as products from current phytoplankton production.

Differences in the origin of material deposited in traps throughout the year can be observed by comparison of monthly totals of sediment respiration (expressed in terms of carbon) with measured amounts of carbon sedimentation. The ratio carbon respired: carbon deposited as a percent is equivalent to an efficiency of oxidation for material settled in traps (Fig. 5). Carbon degraded in sediments calculated from measured oxygen uptake and an assumed conversion factor is underestimated since anaerobic decomposition has occurred. Also time lags between sedimentation of organic matter and release of inorganic carbon make comparisons on a monthly basis misleading. On an annual basis, however, the calculation does permit a comparison of the amount of organic carbon settled at a particular depth with an estimate of that which could have been respired aerobically. Also, if collectors used to trap material underestimate (or overestimate) sedimentation, the direction and magnitude of the bias can be assessed.

On an annual basis the carbon equivalent of sediment respiration at 60 m in Bedford Basin (63 g $m^{-2}$) equalled 83% of organic carbon supply (76 g $m^{-2}$) (Table 1). The ratio of carbon respired: carbon sedimented changed between months, however (Fig. 5). During February and between July and August, for example, calculated rates of carbon respired in sediments exceeded the rate of organic carbon supply. Anaerobic metabolism was not measured but this might be expected to be high during and following periods of rapid sedimentation. This occurred during December and between July and August coincident with peaks in chemical oxidation (Fig. 3). If reduced material had accumulated in sediments to cause chemical oxygen uptake, anaerobic metabolic processes would have been enhanced. Thus, although monthly estimates of carbon release calculated as respiration exceed carbon supply during these periods, additional organic matter was probably also consumed anaerobically. On the other hand, aerobic respiratory demand for organic carbon from September to December was substantially less (50-60%) than that sedimented. The organic matter deposited must have been refractory to rapid degradation or unavailable for oxidation because chemical oxygen uptake and respiration decreased continuously during this time (Fig. 3).

There are obvious seasonal differences in the proportions of aerobic and anaerobic metabolism in sediments in Bedford Basin (Fig. 4) which, until quantified by other measurements, preclude any conclusions concerning the total oxidation of sedimented organic matter. The calculations show, however, that organic carbon settled in traps suspended at 60 m could account for that estimated to be respired aerobically in sediments during one year. Carbon sedimentation measured at 20 m (55 g $m^{-2}$) would not be sufficient to meet this demand.

Simple linear regressions were calculated between monthly values of different variables presented in Table 2 to quantify the degree to which observations were correlated throughout the year. No effort was made to incorporate time lags into the comparisons, nor were transformations performed to improve the fit of data to a calculated line. The matrix of correlation coefficients (Table 3) shows that seasonal changes in some variables were closely related (temperature and sedimentation rates, nitrogen sedimentation rates and sediment respiration and chemical oxidation, for example). Changes in other variables (phytoplankton production, mixed-layer depth) were poorly correlated with most other measurements. Temperature and sediment oxygen uptake, compared as linear ($r = 0.41$) and power ($r = 0.51$) functions, showed that factors other than seasonal variations in temperature were involved in affecting benthic metabolism.

Table 3. Correlation coefficients ($r$) from linear regression analysis comparing monthly averaged or integrated values of phytoplankton primary production (g $C \cdot m^{-2}$) (PP), mixed-layer depth (m) ($Z_m$), microzooplankton volume (ppm, 0-20 m) (MZV), net zooplankton biomass (mg$\cdot m^{-3}$, 5-10 m) (NZB), particulate carbon and nitrogen sedimentation at 20 and 60 m (g$\cdot m^{-2}$) (CS, NS), and sediment oxygen uptake (g$\cdot m^{-2}$) as respiration (SRESP), chemical oxidation (SCMOX) and total uptake (SOX) derived from seasonal measurements in Bedford Basin (data presented in or calculated from Table 2).

| | $Z_m$ | PP | $PP/Z_m$ | MZV | NZB | $CS_{20}$ | $CS_{60}$ | $NS_{20}$ | $NS_{60}$ | SRESP | SCMOX | SOX |
|---|---|---|---|---|---|---|---|---|---|---|---|---|
| T | 0.20 | 0.24 | 0.22 | 0.67* | 0.42 | 0.79* | 0.95* | 0.75* | 0.84* | 0.28 | 0.55 | 0.41 |
| $Z_m$ | | 0.10 | – | 0.57 | 0.10 | 0.26 | 0.28 | 0.44 | 0.42 | 0.70* | 0.52 | 0.66* |
| PP | | | – | 0.47 | 0.10 | 0.30 | 0.55 | 0.45 | 0.50 | 0.32 | 0.45 | 0.39 |
| $PP/Z_m$ | | | | 0.63* | 0.005 | 0.47 | 0.28 | 0.73* | 0.57 | 0.89* | 0.81* | 0.89* |
| MZV | | | | | 0.54 | 0.56 | 0.77* | 0.81* | 0.91* | 0.70* | 0.69* | 0.81* |
| NZB | | | | | | 0.20 | 0.66* | 0.30 | 0.70* | 0.28 | 0.46 | 0.37 |
| $CS_{20}$ | | | | | | | 0.71* | 0.88* | 0.35 | 0.35 | 0.48 | 0.42 |
| $CS_{60}$ | | | | | | | | 0.73* | 0.88* | 0.32 | 0.53 | 0.41 |
| $NS_{20}$ | | | | | | | | | 0.87* | 0.66* | 0.72* | 0.71* |
| $NS_{60}$ | | | | | | | | | | 0.62* | 0.75* | 0.71* |
| SRESP | | | | | | | | | | | 0.84* | 0.97* |
| SCMOX | | | | | | | | | | | | 0.94* |

*$P > 0.05$, df = 10

Without the application of techniques like path analysis (Li, 1975) or loop analysis (Levins, 1975), which require that causal associations between coupled variables be known or assumed, regression analyses can only be used to infer causality. For example, there are many ways variables listed in Table 3 could be linked in a network of interactions without knowledge of cause and effect. One possible configuration (Fig. 6), consistent with the conceptual model presented in Fig. 1, demonstrates for example, the apparent central role of both net and microzooplankton in seasonal changes in carbon and nitrogen sedimentation in Bedford Basin. Another conclusion is that nitrogen deposition throughout the year is more closely linked to sediment respiration and chemical oxygen uptake

than is carbon deposition. What is valuable in such a representa-
tion, rather than the subjective choice of what seem to be reasonable pathways, is that similarities of changes in pairs of variables over time can be illustrated.

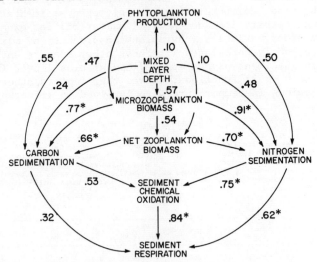

Figure 6. Simple linear regression coefficients derived by comparing monthly values
of pairs of variables (n = 12) derived from
seasonal observations in Bedford Basin (see
Tables 1 and 2). Asterisks indicate significant correlations (p < 0.05).

These correlations do not consider either interactions between
variables which determine a subsequent effect on another variable
or reciprocal (feedback) interactions where there could be a lag
in response. For example, relationships between seasonal changes
in water column stratification, increased phytoplankton production
and biomass of grazers which appeared to exist at certain times of
the year (Figs. 2 and 4), were not evident when data collected
throughout the year were compared (Table 3). Also, while seasonal
changes in mixed-layer depth and monthly total phytoplankton production were not significantly related to measures of zooplankton
biomass or sedimentation as single variables, correlations of production per unit mixing depth with microzooplankton volume and nitrogen sedimentation at 20 m were significant. All of these variables were closely correlated with measure of sediment respiration
and chemical oxygen uptake.

The use of only simple linear correlations to test relations
between variables may miss significant non-linear relationships.
For example, Johnson and Brinkhurst (1971) suggested that utilization of organic matter supplied to sediments might not increase in
direct proportion to input. They observed that at high rates of
sedimentation utilization by benthic consumers (estimated as the
fraction of organic supply utilized for macroinvertebrate production) decreased. Comparison of monthly total nitrogen sedimentation and sediment respiration in Bedford Basin illustrates

this effect (Fig. 7). There was an approximate linear increase in respiration with increased sedimentation from March through July. Even higher levels of deposition during August and September, however, coincided with a large decrease in respiration. No substantial changes in respiration occurred with subsequent reductions in nitrogen sedimentation.

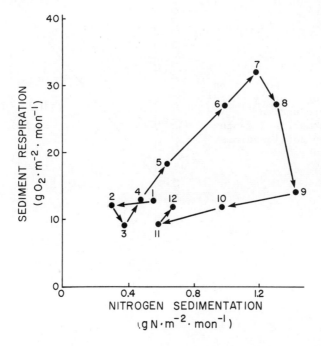

Figure 7. Changes in monthly nitrogen sedimentation with respect to sediment respiration at 60 m in Bedford Basin over an annual period.

The ratio of annual organic carbon supply or phytoplankton production: mixed-layer depth has been compared with annual carbon sedimentation and total sediment oxygen uptake by combining data from different locations (Hargrave, 1973; 1975). Correlations between these variables in Bedford Basin show that nitrogen, rather than carbon, sedimentation is more closely related to both phytoplankton production per unit mixing depth and oxidation of material in sediments over an annual period (Table 3). Also, while previous comparisons were used to derive a relation between these two variables measured on an annual basis, the variables are also closely correlated throughout the year (Fig. 8). Whatever the intermediary pathways or lags in response, these empirical relationships show that processes of supply and oxidation of organic matter in sediments are linked over time. These site-specific observations thus support conclusions derived from comparisons of data collected in different locations.

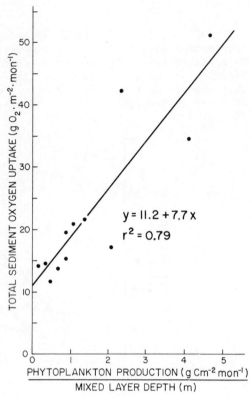

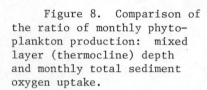

Figure 8. Comparison of the ratio of monthly phytoplankton production: mixed layer (thermocline) depth and monthly total sediment oxygen uptake.

GENERAL DISCUSSION

There are at least two distinct, but not mutually exclusive, pathways of particulate matter transport to sediments at 60 m in Bedford Basin. Sedimentation of material originating from herbivorous grazing, which increased during the summer, corresponds to previous observations in other locations which show that zooplankton fecal pellets predominate in settled material (Steele and Baird, 1972; Seki *et al.*, 1974; Davies, 1975; Knauer *et al.*, 1979). However, sedimentation of ungrazed products of phytoplankton blooms is also seasonally important, especially at times when vertical stratification is weak and the biomass of grazing organisms is low (Smetacek *et al.* 1978). Sedimentation of material originating during spring blooms could account for the positive correlation observed between benthic macrofauna biomass and spring maximum suspended chlorophyll concentrations in different areas (Parsons *et al.*, 1977). Direct consumption of algal cells, without intermediatry transfer to bacteria, would increase the trophic feeding efficiency of benthic deposit feeders. Input of freshly produced products of photosynthesis to sediments by this pathway over limited time periods could be a critical factor determining maximum levels of biomass of benthic fauna supported throughout the remainder of the year.

The relation between nitrogen supply and respiratory demand of sediments in Bedford Basin (Fig. 7) confirms Johnson and Brinkhurst's idea that organic matter input and utilization need not be linearly related. However, aerobic respiration increased in proportion to nitrogen (and carbon) deposition, between March and July. This was a time when organic material was efficiently utilized (Fig. 5) and accumulation of unoxidized metabolic end products was probably minimized. This interpretation is supported by the increasing proportion of total oxygen demand represented by respiration (from 40% in March to 85% by July) (Fig. 4). During this period, organic matter was consumed at a rate which equalled or exceeded the rate at which it was added to the sediments - a condition which Marshall (1970) and McIntyre *et al.* (1970) have suggested exists in many benthic communities.

Following July, however, either the settled material was not as susceptible to aerobic degradation and/or conditions in the sediments prevented respiratory oxidation. Higher temperatures, which continued to increase during and following this period, did not lead to increased sediment respiration. Hargrave (1978) presented evidence that material settled in traps at 60 m in Bedford Basin was more refractory to oxidation than that collected at shallower depths. Decreased respiration in particulate matter collected at 60 m occurred from September through the autumn and winter following peaks in May and August. Also, chlorophyll a as a percent of total pigment and numbers of fecal pellets in settled material decreased and carbon:chlorophyll ratios increased sharply during July (Hargrave and Taguchi, 1978). The absence of any substantial change in sediment respiration, despite apparent large changes (three to four fold) in sedimentation during this time (Fig. 7) illustrates that the quantity of material sedimented was not the only factor which determined benthic metabolism. For example, erosion of surface sediment may temporarily enhance chemical oxidation and lead to a reduction in respiration. This would not have been detected with the monthly measurements which were made.

The nature of settled organic matter, the relative proportion of carbon:nitrogen:phosphorus, for example, is one of the factors that can affect the susceptibility of organic matter to degradation during and following sedimentation. Increased carbon:nitrogen and carbon:phosphorus ratios in sedimented particulate organic matter are observed with depth in oceanic waters (Knauer *et al.*, 1979). Also high carbon:nitrogen ratios in sediments that accumulate at low rates are indicative of more rapid regeneration of nitrogen and phosphorus than carbon (Toth and Lerman, 1977). Seki *et al.* (1968) demonstrated this by comparing sedimentation of these three elements with assumed steady-state concentrations in sediments in a shallow marine bay to calculate that phosphorus and nitrogen turnover rates were greater than similar estimates for carbon. In the present study correlation coefficients between nitrogen sedimentation and all other variables (except phytoplankton production) were higher than those derived for carbon (Fig. 6). Thus production in pelagic and benthic populations in Bedford Basin may be coupled more by nitrogen than by carbon flux. This is consistent with observations of rapid and extensive nitrogen mineralization in epilimnetic waters of lakes (Bloesch *et al.*, 1977).

Despite limitations associated with using carbon alone as a basis for describing organic matter flux, current methods allow a measure of expression of phytoplankton production, sedimentation and sediment oxygen uptake in terms of this element. The inadequacy of estimating total benthic metabolism by measuring oxygen consumption alone has been discussed by Pamatmat (1977), as has the need for developing methods which simultaneously quantify aerobic and anaerobic processes by which particulate organic matter in sediments is decomposed (Jørgensen, 1977). Even without this information, however, sediment respiratory demand can be compared with the input of organic matter to estimate the fraction utilized. Johnson and Brinkhurst (1971) and Davies (1975) made this comparison in Lake Ontario and in a semi-enclosed marine bay and observed that, as in Bedford Basin, up to 90% of the organic matter sedimented annually was respired. Exceptions to this efficient aerobic degradation of settled material occurred at an inner bay station with a high input of allochthonous material (Johnson and Brinkhurst, 1971). Also, there were imbalances between supply and utilization in Bedford Basin over short periods of time during the year (Fig. 5). In contrast to these studies, Hartwig (1976) measured annual carbon deposition to shallow marine sediments which exceeded sediment respiratory demand by ten times. Resuspension and physical transport were assumed to be critical factors which determined both the supply and removal of organic matter because there was no significant variation in sediment organic content. Roman and Tenore (1978) also observed large changes in particulate matter resuspended from shallow marine sediments exposed to tidal currents. In these environments, physical transport and mixing must continuously play a central role in determining rates of benthic metabolism by affecting the net supply of particulate organic matter which remains to be oxidized at the sediment surface.

Consideration of the adequacy of measuring sedimentation by traps moored at fixed depths is beyond the scope of this discussion. Smetacek *et al.* (1978) have considered the effect of current flow on estimating sedimentation in fixed traps. One must assume that material settled in traps is similar in amount and organic composition to that actually deposited at the sediment surface. These problems need critical attention. However, measures of organic matter sedimentation and sediment oxygen uptake in this and other studies have shown that estimates of annual deposition in some environments may approximate those of utilization. If these types of observations were combined with measures of organic matter remaining in sediments, the magnitude of the flux of organic matter could be determined by independent methods. The idea that the metabolism of benthic populations in depositional sediments is generally limited by the supply of oxidizable organic matter then could be tested directly.

## REFERENCES

Aller, R. C. 1978. Experimental studies of changes produced by deposit feeders on pore water, sediment, and overlying water chemistry. Am. J. Sci. 278: 1185-1234.

Bloesch, J., R. Stadelmann, and H. Buhrer. 1977. Primary produc-
    tion, mineralization, and sedimentation in the euphotic zone of
    two Swiss lakes. Limnol. Oceanogr. 22: 511-526.

Conover, R. J., and P. Mayzaud. 1975. Respiration and nitrogen
    excretion of neritic zooplankton in relation to potential food
    supply, pp. 151-163. *In* Proc. 10th Eur. Symp. Mar. Biol.,
    Ostend, Belgium. Persoone, G. and E. Jaspers (eds.). Universa
    Press, Wetteren. Vol. 2.

Coull, B. C. 1969. Hydrographic control of meiobenthos in Bermu-
    da. Limnol. Oceanogr. 14: 953-957.

Davies, J. M. 1975. Energy flow through the benthos in a Scottish
    sea loch. Mar. Biol. 31: 353-362.

Davis, M. B. 1973. Redeposition of pollen grains in lake sediment.
    Limnol. Oceanogr. 18: 44-52.

Fenchel, T. 1969. The ecology of marine microbenthos IV. Struc-
    ture and function of the benthic ecosystem, its chemical and
    physical factors and the microfauna communities with special
    reference to the ciliated protozoa. Ophelia 6: 1-182.

Ferrante, J. G., and J. I. Parker. 1977. Transport of diatom frus-
    tules by copepod fecal pellets to the sediments of Lake Michigan.
    Limnol. Oceanogr. 22: 92-98.

Glooschenko, W. A., J. E. Moore, and R. A. Vollenweider. 1972.
    The seasonal cycle of pheo-pigments in Lake Ontario with partic-
    ular emphasis on the role of zooplankton grazing. Limnol.
    Oceanogr. 17: 597-605.

Hargrave, B. T. 1973. Coupling carbon flow through some pelagic
    and benthic communities. J. Fish. Res. Board Can. 30: 1317-
    1326.

Hargrave, B. T. 1975. The importance of total and mixed-layer
    depth in the supply of organic material to bottom communities.
    Symp. Biol. Hung. 15: 157-165.

Hargrave, B. T. 1978. Seasonal changes in oxygen uptake by settled
    particulate matter and sediments in a marine bay. J. Fish. Res.
    Board Can. 35: 1621-1628.

Hargrave, B. T., G. A. Phillips, and S. Taguchi. 1976. Sedimenta-
    tion measurements in Bedford Basin, 1973-1974. Fish. Mar. Ser.
    Rept. 608, 129 pp.

Hargrave, B. T., and S. Taguchi. 1978. Origin of deposited mate-
    rial sedimented in a marine bay. J. Fish. Res. Board Can. 35:
    1604-1613.

Hartwig, E. O. 1976. Nutrient cycling between the water column
    and a marine sediment. I. Organic carbon. Mar. Biol. 34:
    285-295.

Honjo, S., and M. R. Roman. 1978. Marine copepod fecal pellets:
    production, preservation and sedimentation. J. Mar. Res. 36:
    45-56.

Hutchinson, G. E. 1938. On the relation between the oxygen defi-
    cit and the productivity and typology of lakes. Int. Rev.
    Hydrobiol. 36: 336-355.

Jørgensen, B. B. 1977. The sulfur cycle of a coastal marine sed-
    iment (Limfjorden, Denmark). Limnol. Oceanogr. 22: 814-832.

Johnson, M. G., and R. O. Brinkhurst. 1971. Benthic community me-
    tabolism in Bay of Quinte and Lake Ontario. J. Fish. Res.
    Board Can. 28: 1715-1725.

Jónasson, P. M. 1972. Ecology and production of the profundal benthos in relation to phytoplankton in Lake Esrom. Oikos Suppl. 14: 1-148.

Knauer, G. A., J. H. Martin, and K. W. Bruland. 1979. Fluxes of particulate carbon nitrogen, and phosphorus in the upper water column of the northeast Pacific. Deep Sea Res. 26: 97-108.

Levins, R. 1975. Evolution in communities near equilibrium, pp. 16-50. *In* Ecology and Evolution of Communities, M. Cody and J. Diamond (eds.), Belknap.

Li, C. C. 1975. Path Analysis - a primer. Boxwood Press. Pacific Grove, Calif.

McIntyre, A. D., A. L. S. Munro, and J. H. Steele. 1970. Energy flow in a sand ecosystem, pp. 19-31. *In* Marine Food Chains, J. H. Steele (ed.), Oliver and Boyd, Edinburgh.

Marshall, N. 1970. Food transfer through the lower trophic levels of the benthic environment, pp. 52-66. *In* Marine Food Chains, J. H. Steele (ed.), Oliver and Boyd, Edinburgh.

Mortimer, C. H. 1971. Chemical exchanges between sediments and water in the Great Lakes - speculations on probable regulatory mechanisms. Limnol. Oceanogr. 16: 387-404.

Ohle, W. 1956. Bioactivity, production, and energy utilization of lakes. Limnol. Oceanogr. 1: 139-149.

Pamatmat, M. M. 1977. Benthic community metabolism: a review and assessment of present status and outlook, pp. 89-111. *In* Ecology of Marine Benthos, B. C. Coull (ed.), Univ. S. Carolina Press, Columbia.

Parsons, T. R., M. Takahashi, and B. T. Hargrave. 1977. Biological Oceanographic Processes. Pergamon Press, Toronto.

Pennington, W. 1974. Seston and sediment formation in five lake district lakes. J. Ecol. 62: 215-251.

Pingree, R. D., P. M. Holligan, and G. T. Mardell. 1978. The effects of vertical stability on phytoplankton distributions in the summer on the northwest European Shelf. Deep Sea Res. 25: 1011-1028.

Rich, P. H., and R. G. Wetzel. 1978. Detritus in the lake ecosystem. Am. Nat. 112: 57-71.

Roman, M. R., and K. R. Tenore. 1978. Tidal resuspension in Buzzard's Bay, I. Seasonal Changes in the resuspension of organic carbon and chlorophyll $\underline{a}$. Est. Cst. Mar. Sci. 6: 37-46. 37-46.

Rowe, G. T. 1971. Benthic biomass and surface productivity, pp. 441-454. *In* Fertility of the Sea, J. D. Costlow, (ed.) Gordon and Breach, New York.

Seki, H., J. Skelding, and T. R. Parsons. 1968. Observations on the decomposition of a marine sediment. Limnol. Oceanogr. 13 440-447.

Seki, H., T. Tsuji, and A. Hattori. 1974. Effect of zooplankton grazing on the formation of the anoxic layer in Tokyo Bay. Estuarine Coastal Mar. Sci. 2: 145-151.

Smetacek, V., K. von Brockel, B. Zeitzschel, and W. Zenk. 1978. Sedimentation of particulate matter during a phytoplankton spring bloom in relation to the hydrographic regime. Mar. Biol. 47: 211-226.

Steele, J. H., and I. E. Baird. 1972. Sedimentation of organic matter in a Scottish sea loch. Mem. Ist. Ital. Idrobiol. 29 (Suppl.): 73–88.

Stephens, K., R. H. Sheldon, and T. R. Parsons. 1967. Seasonal variations in the availability of food for benthos in a coastal environment. Ecology 48: 852–855.

Sverdrup, H. U. 1953. On conditions for the vernal blooming of phytoplankton. J. Conseil Exp. Mar., 18: 289–295.

Taguchi, S., and T. Platt. 1977. Assimilation of $14CO_2$ in the dark compared to phytoplankton production in a small coastal inlet. Est. Cstl. Mar. Sci. 5: 679–684.

Taguchi, S., and B. T. Hargrave. 1978. Loss rates of suspended material sedimented in a marine bay. J. Fish. Res. Board Can. 35: 1614–1620.

Taguchi, S., and T. Platt. 1978. Size distribution and chemical composition of particulate matter in Bedford Basin, 1973 and 1974. Fish. Mar. Ser. Tech. Rept. 56: 370 pp.

Thienemann, A. 1927. Der Bau des Seebeckens in seiner Bedeutung für das Leben im See. Ver. Zool. Bot. Ges. 77: 87–91.

Thorstenson, D. C., and F. T. Mackenzie. 1974. Time variability of pore water chemistry in recent carbonate sediments, Devil's Hole, Harrington Sound, Bermuda. Geochim. Cosmochim. Acta. 38: 1–19.

Toth, D. J., and A. Lerman. 1977. Organic matter reactivity and sedimentation rates in the ocean. Am. J. Sci. 277: 465–485.

Vinogradov, M. E., I. I. Gitelzon, and YU. I. Sorokin. 1970. The vertical structure of a pelagic community in the tropical ocean. Mar. Biol. 6: 187–194.

Walsh, J. J., T. E. Whitledge, F. W. Barvenik, C. D. Wirick, S. O. Howe, W. E. Esaias, and J. T. Scott. 1978. Wind events and food chain dynamics within the New York Bight. Limnol. Oceanogr. 23: 659–683.

Wetzel, R. G., P. H. Rich, M. C. Miller, and H. L. Allen. 1972. Metabolism of dissolved and particulate detrital carbon in a temperature hard-water lake. Mem. Ist. Ital. Idrobiol. 29 (Suppl.): 185–243.

# Benthic-Pelagic Oxygen and Nutrient Exchange in a Coastal Region of the Sea of Japan

M. V. Propp
V. G. Tarasoff
I. I. Cherbadgi
N. V. Lootzik

ABSTRACT: Measurements of benthic macrofauna biomass, sediment chlorophyll and ATP content, rates
of benthic photosynthesis, respiration, nutrient
release and uptake were made throughout the year
on sand (3 m) and mud (7 m) in Vostok Bay of the
Sea of Japan. Benthic photosynthesis was greater
than respiration only on sand during spring; the
rates of photosynthesis correlated with chlorophyll a content and underwater light radiation.
Benthic respiration rate was dependent primarily
on temperature, although chlorophyll content and
experiment duration were also significant. The
rates of orthophosphate, ammonia, nitrate and silicate release from the sediments confirmed that
nutrient regeneration from sediment is an important source of nutrients in this neritic ecosystem. The atomic ratios between respiration and
nutrient release varied with season and differed
from those typical for planktonic regeneration.
Oxygen uptake and silica release were significantly correlated with the fluxes of orthophosphate
and ammonia.

INTRODUCTION

Transformation of organic matter and oxygen consumption in marine sediments are linked with primary production in the photic layer and may be significant in the dynamics of marine benthic ecosystems (Hargrave, 1973; Pamatmat, 1977; Smith, 1978). It has often been assumed that release of nutrients from sediments is essential for processes of primary production, chiefly in the neritic part of the sea where the distance between sediments and surface is small. The important role of bottom regeneration in lakes, estuaries and stagnated parts of the sea is well known (Rittenberg et al., 1955; Suess, 1976). There have also been recent studies in the coastal regions of the sea to quantify the exchange of nutrients across the sediment-water interface (Hale, 1974; Hartwig, 1974, 1976a, 1976b; Davies, 1975; Rowe et al., 1975; Nixon et al., 1976a; Jansson and Wulf, 1977; Propp, 1977).

The objective of this study included measurements of the rates of photosynthesis, respiration, release and uptake of nitrogen, phosphorus and silicate in shallow benthic communities on sand and mud, and an assessment of the interdependence between these fluxes and environmental conditions and macrobenthos abundance. A method is described which permits the measurement of benthic metabolic rates in situ over a period of several days to allow statistical evaluation of relationships between measured variables.

MATERIALS AND METHODS

Measurements of benthic community metabolism were made in situ in the coastal part of the Sea of Japan in Vostok Bay (part of the Bay of Peter the Great) at the "Vostok" station of the Institute of Marine Biology. Some results of previous studies of water chemistry, benthic community distribution and metabolism have been published earlier (Propp, 1977; Tarasoff, 1978). Sediment particle-size distribution, caloric value, chlorophyll and ATP sediment content were determined monthly at sites where metabolic rates were determined and additionally in spring all over the bay. Sediment samples (4-6 replicates) were collected by divers with 45 mm inner diameter glass tubes. The upper 2 cm layer was separated, pigments were extracted by 90% acetone and determined using a spectrophotometer (Colocoloff and Colocoloffa, 1972; Strickland and Parsons, 1972). ATP was determined by bioluminescent assay (Karl and LaRock, 1975) after benthic animals had been removed. Organic carbon in sediments was determined by dichromate oxidation and nitrogen and phosphorus photometrically after digestion with sulfuric and perchloric acids.

Four glass bell jars (15 1 capacity), covering 0.05 m$^2$ of the bottom each, were used to measure metabolic rates. The glass was treated with hydrochloric acid to diminish silicate leaching. The edges of the jars were pressed a few cm into the sediment. A similarly constructed bottle for determination of plankton metabolic rates was placed nearby, as were photocells of an integrating underwater photometer, and a plastic rod driven into the bottom with thermometer (Fig. 1). Three inert PVC tubes 4 mm inner diameter were

fastened in each jar  and on the thermometer rod with the tube ends
3-4, 12, and 20 cm above the bottom surface.  One tube came out of the
control bottle.  There was an additional tube (not shown in the drawing) of larger diameter, the end of which was 1 m above the bottom.
All the tubes were collected into a system and fastened to an anchor
and to a surface float.  The recorder in the photometer was also
placed on the float.  SCUBA was used for setting and removing the
jars, connecting and disconnecting the tubes.  On completing a series
of measurements, each jar was removed with a special scoop which retained sediment under each chamber.  In a few cases the jars were
broken while being lifted, or else some sediment was washed away
when the sea was choppy, so that not all the jars were properly incubated.  Sediment was washed through a 1 mm mesh screen, animals
were sorted according to species, counted, and each specimen weighed
separately.  Wet weights were used and molluscs were weighed with
shells and polychaetes without tubes.

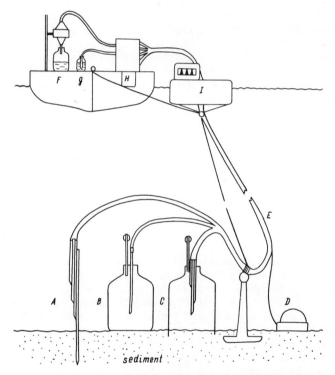

Figure 1.  A device for benthic metabolic
rate measurements:  (A) Water sampling from the
bottom layer; (B) Control bottle; (C) Experimental chamber (one of the four is shown); (D) A
photocell of an integrating photometer; (E) A
system of hoses and wires; (F) Samples for nutrient analysis; (G) samples for oxygen determination; (H) Peristaltic pump; (I) Floating
raft; see text for details.

A specially designed six-channel peristaltic pump was used for sampling water from the jars and changing it after each measurement. The pump operated on batteries and had a pumping rate of 240 ml min$^{-1}$ on each channel. Preliminary experiments showed that no bubbles were formed and no degassing of water occurred during pumping. The collected water samples (each layer of all jars, bottom water, and control bottle) were washed into glass flasks for Winkler titrations of oxygen and then through a 0.45 µm ultrafilter for frozen storage in plastic bottles and subsequent nutrient determinations. The methods of sampling, storage and analyzing followed Strickland and Parsons (1972). Nitrates were determined after reduction in a capillary column with cadmium-covered copper wire (Propp and Propp, in press). Dissolved organic carbon (DOC), dissolved organic nitrogen (DON), dissolved organic phosphorus (DOP) and amino acids were also determined in some experiments. Exchange rates obtained for these compounds were small and statistically insignificant because there was noticeable variation in the control bottle measurements, apparently caused by the death and decomposition of plankton, since unfiltered water was used for filling the vessels.

The water in the jars was changed after sampling if there was any reason to expect significant differences between the concentrations inside the jars and in the water around them. For this purpose, water was sucked at 1 m from the bottom and pumped into the middle tube of the jar. Excess water flowed out through a long, narrow compensating tube, permitting only negligible diffusion during the incubation. Underwater observations showed that no resuspension of the upper sediment layer took place. In summer it was necessary to change water after each measurement (every 9-10 h); in spring and fall water changes were required less frequently. It was usually possible to maintain dissolved oxygen and nutrient concentrations in jars similar to those in the bottom water layer, and these concentration changes could be estimated with satisfactory accuracy. Only in the first experiment (mud 1, Table 1) did oxygen concentration in one of the jars drop nearly to zero because of substantial respiration and unequal water exchange. Water sampling was generally carried out twice a day; integral light radiation was estimated simultaneously.

No formaldehyde or antibiotics were added for separate estimation of the chemical sediment oxidation or bacterial respiration, since it has been shown that the rates obtained can depend on the time interval between the addition of the poison and the measurement (Pamatmat, 1977). Moreover, these substances interfere with analytical procedures for nutrients.

Benthic metabolic rates were determined from the mean concentration of dissolved oxygen at three levels - the volume, the area, and the incubation time for each jar. Corrections for plankton metabolism were introduced if they were statistically significant. In most experiments metabolic rates determined at night were more stable than those measured during the day. They were treated separately and summed, even if the differences between night and day estimates were insignificant. This decreased the coefficient of determination.

Estimates of variation, homogeneity of the data in different jars, dependence of metabolic rates on time, temperature and other independent variables were calculated. Multiple linear regression analysis was used, since logarithmic, reciprocal and some other transformations of variables did not result in significant increase of coefficients of determination. The 0.5 significance level is used through the text, tables and figures unless otherwise stated.

Respiration was calculated for benthic molluscs after Kamluk (1974), for polychaetaes after Turpaeva *et al.* (1970), for crustacea after Suchshenja (1972), and for all other groups according to Hemmingsen (1960). Temperature corrections were calculated following Winberg (1976). Respiration was calculated for each specimen and summed for each jar.

RESULTS

The sediments at the study site were mud with a relatively high organic content (C = 1.56%, N = 0.178%, P = 0.073% dry weight) and medium-coarse sand (C = 0.198%, N = 0.058%, P = 0.026%). The anoxic layer, visible because of the black color of iron sulfide, was at 5-10 cm in the mud and 15-20 cm in sand. The upper few centimeters of sediment were bioturbated, containing numerous holes and tubes of live and dead polychaetes and phoronids. The animal composition and abundance inside the jars corresponded to bottom community distribution in the bay (Table 1; see also Tarasoff, 1978).

Sediment chlorophyll $\underline{a}$ content (corrected for phaeophytin $\underline{a}$) had an uneven distribution with the maxima along the western shore (Fig. 2). Chlorophyll $\underline{a}$ content (mg m$^{-2}$) decreased with depth (d, m), while the ratio phaeophytin $\underline{a}$ to chlorophyll $\underline{a}$ increased: (Fig. 3) (n = 31):

$$\text{Chl } \underline{a} = 151.2 - 6.49d; \ r = -.05$$

$$\frac{\text{Ph } \underline{a}}{\text{Ch } \underline{a}} = -1.91 + 0.845d; \ r = 0.65$$

Seasonal changes of chlorophyll $\underline{a}$ content were characterized by a marked maximum in the spring and a smaller peak in the fall (Fig. 4). The ATP content was also highest in spring. Chlorophyll and ATP content in sediments were correlated, although the correlation coefficients were rather low (0.28 for mud, 0.27 for sand).

Oxygen and nutrient concentrations in the bottom water, especially 3-4 cm above the sediment surface, were variable due to turbulent mixing. On the average the concentration gradients in the bottom water layer were rather small and compared well to those in the jars, but considerable variation occurred; up to 2 ml $O_2$ 1$^-$; 0.5 µM in the orthophosphate concentration, 2-3 µM in that of nitrates and ammonia and 10-15 µM in that of silicates. It was impossible to use the concentrations at different levels to calculate either significant values for oxygen and nutrient release and uptake or for eddy diffusion coefficients.

Table 1. Experiment on measuring benthic metabolic rates

| Experiment No. | Date | Water temp. (°C) | Salinity % | Depth m | No. jars | Sediment | Chlorophyll mg·m$^{-2}$ | x̄ No. Animals·m$^{-2}$ | Density x̄ biomass g wet weight·m$^{-2}$ | Dominating animal groups, % wet weight |
|---|---|---|---|---|---|---|---|---|---|---|
| M1 | 17-22 Aug 76 | 20.5 | 32 | 5 | 2 | mud | 109 | 790 | 57.2 | Polychaeta – 65<br>Bivalvia – 32 |
| M2 | 2-16 Apr 77 | 3.8 | 32.2 | 7 | 4 | mud | 256 | 1795 | 127.4 | Ophiuroidea – 45<br>Bivalvia – 27<br>Polychaeta – 24 |
| M3 | 13-20 Aug 77 | 20.1 | 32.6 | 7 | 3 | mud | 162 | 1813 | 109.1 | Bivalvia – 50<br>Polychaeta – 34 |
| M4 | 21-27 Nov 77 | 2.1 | 31.7 | 7 | 3 | mud | 130 | 4027 | 90 | Ophiuroidea – 6<br>Bivalvia – 36<br>Polychaeta – 36 |
| S1 | 26-30 Mar 77 | 0.4 | 32.5 | 3 | 4 | sand | 218 | 1795 | 127.3 | Ophiuroidea – 16<br>Polychaeta – 23<br>Bivalvia – 21<br>Phoronida – 20 |
| S2 | 22-28 Nov 77 | 2.1 | 31.8 | 3 | 4 | sand | 90 | 7420 | 70.6 | Ophiuroidea – 13<br>Phoronida – 55<br>Gastropoda – 24<br>Bivalvia – 15 |

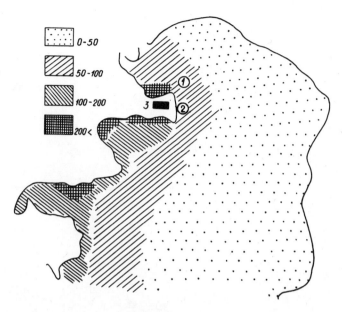

Figure 2. Distribution of chlorophyll
a (mg m$^{-2}$) in the upper layer of the sediments
of Vostok Bay. 1. site for measurements of
mud metabolism; 2. site for measurements of sand
metabolism; 3. biological sampling station.

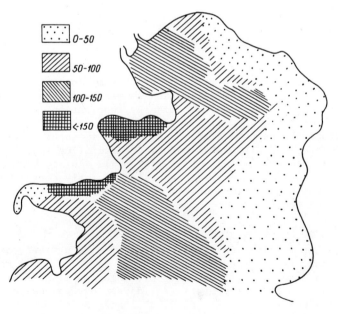

Figure 3. Distribution of benthic biomass
( g wet weight m$^{-2}$) in Vostok Bay.

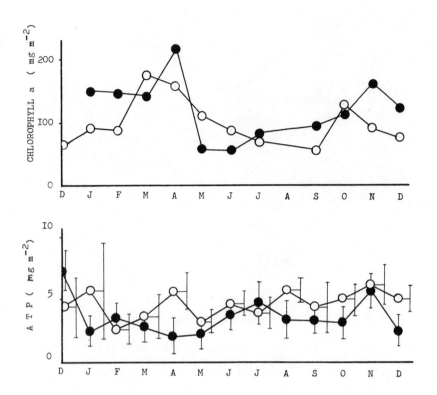

Figure 4.    Seasonal changes of the chlorophyll a and
ATP content in mud and sand.   Black circles – mud, open – sand.

Although detectable rates of benthic microalgal photosynthesis
were observed in several spring and summer measurements, photosynthesis on sand only exceeded community respiration in the spring;
at all other times community production was negative (Table 2).   On
sand, gross photosynthesis rates (y, ml $O_2$ m$^{-2}$ hour$^{-1}$) were positively correlated with total underwater radiation ($X_7$, kilojoule
cm$^{-2}$) and chlorophyll content ($x_2$, mg m$^{-2}$).   Animal biomass and duration of the measurement were not significantly correlated with
gross photosynthesis (n – number of measurements, $R_x$, $Y_i$ – total
correlation coefficient):

   $Y = -15.8 + 0.112\ x_1 + 0.096\ x_2$    n = 31   $R_y$, $x_i$ = 0.71

Benthic respiration rates changed throughout the year (Table
2).   On mud, respiration at night (y, ml $O_2$ m$^{-2}$ hour$^{-1}$) was correlated with water temperature (X, $^{\circ}$C), chlorophyll content ($X_2$, mg
m$^{-2}$), duration of measurements ($X_3$, days), oxygen concentration in
the bottom water layer ($X_4$, ml $O_2$ l$^{-1}$) and calculated animal respiration ($X_5$, ml $O_2$ m$^{-2}$ hour$^{-1}$).   In fact, the two latter variables
cannot be treated as independent:

   $Y = 22.08 + 0.504\ X_1 + 0.061\ X_2 - 2.13\ X_3 - 2.52\ X_4$ (n = 50);
$R_y$, $x_i$ = 0.85.

$$Y = 1.94 + 0.42 X_1 + 0.068 X_2 - 1.68 X_3 + 1.12 X_5 \quad (n = 50);$$
$Ry, x_i = 0.75.$

$$Y = 4.07 + 0.87 X_1 + 0.064 X_1 - 1.55 X_3 \quad (n = 83); \quad Ry, x_i = 0.75$$

Benthic animal biomass did not significantly affect the total community respiration (Table 2). The total effect of all the statistically significant factors was comparatively small, and the equations obtained cannot be used for prediction. Only chlorophyll content ($X_1$) and experiment duration ($X_2$) were significantly correlated with respiration at night. The effect of temperature could not be determined, as most measurements were made at similar temperatures:

$$Y = 33.73 - 0.05 X_1 - 1.77 X_2 \quad (n = 32) \quad Ry, x_i = 0.65$$

Table 2. Rates of respiration and photosynthesis of benthic communities (M = mud, S = sand) ml $o_2m^{-2}h^{-1}$. (NR - night respiration, NP - net photosynthesis, GP - gross photosynthesis, MD - mean daily oxygen exchange; NS - not significant at 0.05).

| Experiment number | NR | NP | GP | MD | Anim. respiration as % total com. oxygen uptake |
|---|---|---|---|---|---|
| M1 | 31.6±13.1 | -26.2±10.2 | 5.4 (NS) | -29.9±8.3 | 24 |
| M2 | 15.4±4.9 | -4.6±6.6 | 10.8±8.2 | -10±2.4 | 19 |
| M3 | 32.4±2.23 | -14.7±8.9 | 17.7±12.4 | -23.55±4.0 | 33 |
| M4 | 8.7±1.7 | -6.9±4.6 | 1.8 (NS) | -7.8±2.3 | 19 |
| S1 | 14.6±5.2 | 11.2±7.4 | -25.8±9.0 | -3.4 (NS) | 6 |
| S2 | 20.6±6.0 | -24.8±8.6 | -4.2 (NS) | -22.7±5.2 | 8 |

Nitrogen was released mainly as ammonia but, in November, a noticeable part was nitrate (Table 3) coinciding with an increase in the total amount of nitrate in the water (Propp, 1977). Nitrite release was undetectable. DON flux (including amino acids) was not measured with satisfactory accuracy, but, in any case, it would amount to only a small part of ammonia release. The average ammonia release from sand during spring was statistically insignificant with low rates observed during day and night. This is in accordance with the results of dark ammonia uptake by phytoplankton (Dugdale, 1976). Ammonia release was lower during the day. Also on mud, where measurements were reliable, rates of photosynthesis were determined, though in some cases these differences were not statistically significant. Ammonia release from mud at night (Y, $\mu mol \ m^{-2} \ day^{-1}$) was correlated with the water temperature ($X_1$, $^oC$) and experiment duration ($X_2$, days):

$$Y = 901 + 127 X_7 - 185 X_2 \quad n = 54 \quad Ry, x_i = 0.73$$

Table 3. Nutrient uptake and release ($\mu$mol m$^{-2}$ d$^{-1}$) (ML - mean for measurements in light, MD - mean for measurements in dark; MT - mean for total day; NM - not measured; NS - not significant). See Table 2 for dates and water temperatures.

| Experiment number | NH$_4$ | | | PO$_4$ | | | SiO$_4$ | | | NO$_3$ |
|---|---|---|---|---|---|---|---|---|---|---|
| | MD | ML | MT | MD | ML | MT | MD | ML | MT | MT |
| M1 | 1863 | 3526 | 2695±1797 | 355 | 1455 | 905±407 | NM | NM | NM | NM |
| M2 | 541 | 569 | 553±322 | 210 | 107 | 160±62 | 2681 | 2661 | 2670±880 | NS |
| M3 | 5442 | 2624 | 3106±590 | 633 | 433 | 540±89 | 11970 | 11505 | 11802±2450 | NS |
| M4 | 549 | -162 | 387±295 | 43 | 21 | 32 NS | 2122 | 2351 | 2237±698 | 313±205 |
| S1 | 24 | 7 | 16 NS | 11 | 17 | 14 NS | 412 | -855 | 422±224 | NS |
| S2 | 1395 | 752 | 1074±437 | 79 | -3 | 70 NS | 1398 | 1885 | 1641±375 | 718±455 |

Phosphorus was released mainly as orthophosphate, DOP flux was negligible. In most measurements release appeared higher at night than during the day, though these differences were not statistically significant. Release of phosphorus (Y, $\mu$mol $m^{-2}$ $day^{-1}$) from mud at night increased with oxygen depletion in the bottom water layer ($X_7$, ml $O_2$ $1^{-1}$) and decreased with experiment duration ($X_2$, days) and temperature ($X_3$, $^{\circ}$C); the influence of time was only significant at P = 0.07:

$$Y = 2247 - 317\ X_1 - 68\ X_2 - 28.8\ X_3 \quad (n = 57) \quad R_y, x_i = 0.66$$
$$Y = 1991 - 293\ X_1 - 26\ X_3 \quad (n = 57) \quad R_y, x_i = 0.64$$

Silicate uptake was observed only in spring on sand when high rates of photosynthesis also occurred (Table 3). In all other measurements silica release rates did not differ significantly during day and night, supporting Dugdale's (1976) suggestion that there is no direct relationship between light and silica absorption in diatoms. Release of silicate from mud at night (Y, $\mu$mol $m^{-2}$ $day^{-1}$) was only significantly correlated with temperature ($X_7$, $^{\circ}$C) and time ($X_2$, day):

$$Y = 4060 + 374\ X_1 - 819\ X_2 \quad (n = 45) \quad R_y, x_i = 0.71$$

Atomic ratios between oxygen and nutrient exchange can be calculated from these exchange rates. However, the values obtained have wide confidence intervals and are of little significance because of the necessity of calculating the differences and quotients with their errors. Therefore, linear regressions were calculated between oxygen, nitrogen (ammonia plus nitrates), orthophosphate, and silicate concentrations directly in the water samples obtained from different depths in the jars and from the bottom water layer. The independent variable of each regression equation is related to the initial concentration in the water and depends chiefly on water temperature and season. The slope, if the choice of units is appropriate, is equal to the atomic ratio in metabolism. These ratios were more stable than the metabolic rates for seaweed chambers (Table 4). They reflect processes taking place both in the water and at the sediment-water interface. Since the latter are considerably more intensive than planktonic metabolism, the necessary corrections are no more than a few percent and are within the statistically significant levels. Thus no corrections were made. Mutual and partial correlation coefficients between oxygen and nutrient concentrations were also calculated (Tables 5, 6).


Variability in Measurements

In the past, jars and vessels of various shape, size and design have been used for measuring benthic community respiration, photosynthesis and, to lesser degree, nutrient exchange. Some were equipped with mixers, others were not, some were placed *in situ* on the bottom, others were used on board ship. In some cases mixing increased oxygen consumption and nutrient release (Pamatmat, 1973; Hale, 1974; Smith, 1978). Davis (1975), however, points out that an increase in oxygen uptake and ammonia release took place only when sediment was resuspended with bottom currents exceeding 10 cm $sec^{-1}$.

Table 4.  Atomic ratios of oxygen and nutrient
concentrations (0.05) significance levels in parentheses; NM = not measured)

| Experiment number | O : N | O : P | O : Si |
|---|---|---|---|
| M1 | 6.9(5.8–8.6) | 20(24–29) | NM |
| M2 | 66(59–74) | 213(189–244) | 10(9.5–11) |
| M3 | 31(27–37) | 148(130–174) | 6.2(7.2–8.4) |
| M4 | 45(28–86) | 380(266–664) | 26(18–46) |
| S1 | 234(190–307) | 1130(900–1500) | 346(203– 1170) |
| S2 | 76(68–86) | 249(223–282) | 28(25–32) |

Table 5.  Mutual correlation coefficients between oxygen and
nutrient concentrations (NM - not measured).

| Experiment number | O : N | O : P | O : Si | N : P | N : Si | P : Si |
|---|---|---|---|---|---|---|
| M1 | -0.86 | -0.81 | NM | 0.88 | NM | NM |
| M2 | -0.81 | -0.77 | -0.89 | 0.85 | 0.82 | 0.85 |
| M3 | -0.87 | -0.80 | -0.83 | 0.92 | 0.87 | 0.87 |
| M4 | -0.53 | -0.52 | -0.85 | 0.72 | 0.40 | 0.42 |
| S1 | -0.28 | -0.54 | -0.36 | 0.35 | -0.16 | -0.16 |
| S2 | -0.89 | -0.59 | -0.73 | 0.68 | 0.38 | 0.38 |

Table 6.  Partial correlation coefficients between oxygen and nutrient concentrations (NM - not measured, NS - not significant).

| Experiment number | O : N | O : P | O : Si | N : P | N : Si | P : Si |
|---|---|---|---|---|---|---|
| M1 | -0.51 | -0.25 | NM | 0.61 | NM | NM |
| M2 | -0.36 | -0.134 NS | -0.65 | 0.54 | 0.044 NS | 0.48 |
| M3 | -0.44 | 0.05 NS | -0.31 | 0.67 | 0.18 NS | 0.18 NS |
| M4 | -0.25 | -0.107 NS | -0.81 | 0.62 | -0.11 NS | 0.03 NS |
| S1 | -0.171 | -0.59 | -0.53 | 0.17 | -0.13 NS | -0.40 |
| S2 | -0.78 | 0.04 NS | -0.57 | 0.42 | -0.20 | -0.07 NS |

Benthic metabolic processes are closely related to the structure of the sediment and adjacent water layer.  Theoretical extremes are:  (a) a pure diffusion process in a stratified sediment and an immovable adjacent water layer - a process obeying Fick's second law of diffusion; (b) an entirely bioturbated sediment with an overlying turbulent water layer.  Artificial mixing changes the metabolic rates in the first case, with little effect in the second.  In reality, the picture is more complex, because the release of nutrients occurs partly in the anoxic sediment layer which is located at different levels in different seasons.  Additionally, the thickness of the turbulent water layer contiguous to the sediment where the current is reduced only reaches a depth of about 3 cm and is extremely unstable (Wimbush and Munk, 1970; Beljakov, 1978).  The important role of bioturbation is clear from the profiles of silicate and nitrate in the pore waters in shallow and shelf sediments (Vanderborght and Billen, 1975; Grundmanis and Murray, 1977).  However, microbial activity and diffusion were sufficient to explain the nitrate profiles in the shallow sediments of the North Sea (Vanderborght and Billen, 1975).  Oxygen and nutrient gradients in the bottom water layer in our study varied considerably both in the sea and in the jars.  In the sea, periods of time were observed when concentration gradients were high, corresponding to periods when diffusional processes predominate.  These periods were of short duration in comparison to the total length of the measurements.  In only a few cases were the gradients in the jars sufficient for diffusion transfer to have occurred, while in other experiments concentrations were nearly homogeneous.  It is probable that the relative role of separate transfer processes may change over short time intervals, depending on the intensity of bioturbation and turbulent mixing.

Benthic metabolism decreased significantly with longer incubation time.  To some extent this decrease may have been due to sediment disturbance while putting the jars in place, and, in part,

it may have been caused by a decrease in average oxygen concentration during experiments despite water exchange.  Data on the effect
of disturbance and experiment duration on sediment oxygen uptake
are contradictory (Pamatmat, 1973, 1977; Smith, 1973, 1978).  If it
is assumed that actual *in situ* metabolic rates correspond to initial rates, the values obtained in longer experiments (Tables 2, 3)
(extrapolated to the initial time of measurements) would be approximately doubled.  If it is assumed that the initial increase is
caused by sediment disturbance these rates reflect the oxygen uptake under these conditions.  When the sea is rough, sediment resuspension may also occur.  Oxygen uptake is then increased and nutrient release from interstitial water may increase (Davies, 1975).
Resuspended material with phosphorus absorbed on clay may also co-
precipitate with iron oxides (Elderfield, 1976).  Periods of stagnation occur in Vostok Bay during summer when oxygen concentration
in the bottom water falls below that observed inside experimental
chambers used in this study (Propp, 1977).  Under these conditions
anaerobic processes in the sediments must assume increased importance.


DISCUSSION

The estimates of photosynthetic rates on sediments in this
study revealed pronounced seasonal fluctuations.  Oxygen produced
by photosynthesis only exceeded that used in chemical oxidation and
respiration on sand during spring when sediment chlorophyll content
and insolation were maximum.  At this time, ammonia and phosphate
release from the sand dropped to negligible rates, and silica uptake from the water was observed.  In this season, the sand community as a whole is, apparently, autotrophic and nutrients self-contained.  In experiments at other times of the year the bottom communities were heterotrophic, although, in spring and summer, benthic
diatoms were important for photosynthesis and nutrient exchange.
This conforms to measurements on sand near LaJolla (Hartwig, 1978)
and on muds in the Baltic Sea (Jansson and Wulf, 1977), where respiration also prevailed even though rates of photosynthesis were
appreciable.  If the approximate ratios of chlorophyll content,
diatom biomass, photosynthesis and respiration in Jansson and
Wulf's (1977) study are applicable to the benthic communities of
the Sea of Japan, the wet weight of diatoms may reach several grams
$m^{-2}$ and their respiration may comprise up to 30% of the total community oxygen uptake.  Making use of the dependence of photosynthesis rate on light and chlorophyll, and taking into account the decrease in these values with depth, one can calculate that photosynthesis on mud at water depths greater than 20–25 m must be negligible.  On sand, when the water transparency is highest, photosynthesis may be appreciable at a depth of 25–30 m.  In more transparent
tropical waters, noticeable photosynthesis rates of benthic microalgae have been measured as deep as 40–60 m (Plante-Cuny, 1977; 1978).
There is more chlorophyll a in sediments in Vostok Bay than in the
overlying plankton, but photosynthetic rates calculated per unit
chlorophyll are higher in the plankton.

Oxygen uptake on sand and mud in Vostock Bay was similar in magnitude to previous observations for shallow marine sediments (Davies, 1975; Smith *et al.*, 1976; Propp, 1977). In most cases, however, previously published rates (Smith, 1973; Nixon *et al.*, 1976a; Jansson and Wulf, 1977; Hartwig, 1978) were 1.5-3 times higher than we observed in Vostok Bay. Whether this is due to the experimental procedure or reflects real differences in metabolic rates is unknown. Lack of stirring inside chambers could have caused rates to be lower than would have been observed with mixing. If we assume the rates we observed are underestimates and approximate only about half of the real ones, the rates would be similar to previously published values. Those rates also approximate initial rates estimated by extrapolation. The dependence of benthic respiration on temperature and the relative role of animals are also similar to results reported earlier (Smith, 1973; Nixon *et al.*, 1976a).

Ammonia flux from shallow sediments had been determined by a few authors on the basis of the concentration gradient in the water or as a result of direct measurements (Hale, 1974; Rowe *et al.*, 1975; Davis, 1975; Nixon *et al.*, 1976a; Rowe and Smith, 1977; Propp, 1977). The rates of release and seasonal changes in our measurements are similar to those reported by Davies (1975) and, like the respiration rates, are lower than those measured by Nixon *et al.* (1976a). No significant release of nitrates was reported in these studies (Nixon *et al.*, 1976a; Jansson and Wulf, 1977), but a release was calculated on the basis of the interstitial nitrate profiles (Vanderborght and Billen, 1975; Vanderborght *et al.*, 1977b; Grundmanis and Murray, 1977). In Vostok Bay, noticeable nitrification in the sediments is limited to a few autumn and winter months when an increase in concentration of nitrate in the water also occurs (Propp, 1977).

Orthophosphate release from sand was small and statistically insignificant but it was observed constantly from mud deposits. This confirms the important role of benthic regeneration in the phosphorus cycle in coastal waters (Nixon *et al.*, this volume) High rates of phosphorus release from the sediment in anoxic environments are known from stagnant basins and some other anoxic areas (Suess, 1976). Increased phosphate release with oxygen depletion was measured on Baltic shallow mud (Jansson and Wulf, 1977), and the values obtained were similar to our results in one of the jars in the first experiment on mud, when the oxygen concentration dropped nearly to zero. A more detailed examination leads to the conclusion that there are two components of phosphorus release - aerobic and anaerobic. The first is perhaps due chiefly to the metabolism of benthic animals in the surface layer of the sediment, while the second is the result of microbial and chemical processes in the deeper layers.

Silicate flux from the sediment was determined in all experiments except the spring measurements on sand. The rate of release was high and, accounting for temperature differences, was similar to rates calculated from muds of the North Sea on the basis of silicate gradients in interstitial water (Vanderborght *et al.*, 1977a). This rate has the same order of magnitude to the overall sea average ($0.1-0.2$ g-at m$^{-2}$ yr$^{-1}$ Berger, 1976). The measured rates were, however, 2-3 times lower than the fluxes determined in the experiments carried out in Narragansett Bay (Nixon *et al.*, 1976b; Kremer and Nixon, 1978). The dependence on only temperature and time imply that silicate release is mainly a process of chemical dissolution.

These results confirm that benthic nutrient regeneration in coastal waters proceeds with considerable intensity. Most of the organic matter deposited on the sediment surface is subject to decomposition, chiefly in the thin upper aerobic sediment layer, and nutrients are returned to the water column. The quantity of the organic matter buried in the sediment is small in comparison with that subject to mineralization. This process maintains high production in coastal waters and has features similar to those of the mechanism of regeneration and maintenance of high primary production in up-welling regions (Sapojnikov, 1977).

Since Redfield (1934) first studied the ratios between oxygen uptake and the release of nitrogen and phosphorus, it has been shown that the average value 212 0 : 16 N : 1 P is applicable to aerobic regeneration processes in oceanic plankton. Of course, considerable deviation is possible in near-shore or short-term studies. Similar values of 0:N ratio have been used to calculate ammonia release from measures of oxygen uptake (Rowe *et al.*, 1975; Rowe and Smith, 1977). The results indicate that for nutrient regeneration in shallow sediments these atomic ratios vary with season and differ significantly from the theoretical ones (Table 4). Particularly high values occurred on sand where photosynthesis is important, which reaffirms the role of benthic diatoms in nutrient recycling. In most cases the 0:N ratio is considerably higher than 13-16. This may be accounted for both by a general nitrogen deficiency in the near-shore ecosystem and by denitrification in the sediments. A high rate of denitrification was inferred in the muds of the North Sea, and the flux of molecular nitrogen from the sediment approximately equaled the release of ammonia and nitrate (Vanderborght *et al.*, 1977b). However, there are few measurements of denitrification in marine sediments and any estimate of the significance of denitrification in marine ecosystems seems premature.

In Narragansett Bay the ratio between benthic oxygen uptake and ammonia release conforms to a value of 21 rather than 13-16 (Nixon *et al.*, 1976a). The authors speculated that a considerable part of nitrogen was released as DON. In our results, however, the relative role of DON in nitrogen release was small. Also, a model of nitrogen diagenesis in near-shore sediments was constructed without assuming that any forms of DON were excreted (Vanderborght *et al.*, 1977b). Anaerobic ammonia release, when organic matter oxidation is coupled with sulfate reduction, occurs after an almost complete depletion of molecular oxygen. In Vostok Bay the anaerobic layer in the sediment is usually found at a few centimeters depth, and the role of pure anaerobic processes may be significant only during short-term anoxic periods of water stagnation and in the deeper layers of the sediment.

Although in most experiments the orthophosphate-oxygen ratio differed somewhat from the theoretical value, these differences were not very great (except experiments mud I and sand I), considering the significance levels of the estimates. However, the N:P ratios differ considerably from 16. Such relative surplus of phosphorus in comparison with nitrogen occurs in stagnant basins (Suess, 1976) and has been also observed in Narragansett Bay, where the average ratio was approximately 6 (Nixon *et al.*, 1976a), a value similar to that in our measurements. They also found high nitrogen and phosphorus fluxes in relation to oxygen uptake in summer on mud, indicating that the significance of anaerobic processes increases with high temperature, i.e., when oxygen concentration decreases and rate of chemical and biological oxidation rises.

Mutual correlation coefficients between oxygen, nitrogen (ammonia plus nitrate), orthophosphate and silicate concentrations are comparatively high, except sand I experiment (Table 5). Partial correlation coefficients (Table 6) show, however, that only some of these concentrations are significantly related to each other. The most significant correlations are between oxygen and silicate and nitrogen and phosphorus. Although the partial correlation coefficients may be interpreted in various ways, the connection between oxygen consumption and silicate release shows the important role which diatoms may play with respect to this nutrient. It is known that the dissolution of planktonic diatom frustules is rapid, but only after the destruction of the cell wall (Lewin, 1961; Conway *et al.*, 1977). The cell wall contains little nitrogen or phosphorus and the release of ammonia and orthophosphate occurs as a result of microbial and chemical decomposition of organic matter and, to a smaller extent, benthic animal excretion. Only a small part of oxygen is used directly for the oxidation of nitrogen-containing substances and its utilization connected with phosphorus release is negligible.

In summary, oxygen consumption and the regeneration of nutrients cannot be viewed as a single diffusion limited process. The direct role of macrobenthic animals in their exchange is relatively small, and flux is related mainly to the microbial and chemical processes in the sediment.

REFERENCES

Ankar, S. and R. Elmgren. 1977. The soft bottom subsystem. *In* Ecosystem Analysis of a Shallow Sound in the Northern Baltic – A Joint Study by the Åsko Group. Stockholm. 18: 72–80.

Beljakov, L. N. 1978. Experimental investigation of dynamic structure of benthic boundary layer. Problemi Arctici i Antarctici. 54: 57–61. (*In* Russian).

Berger, W. H. 1976. Biogenous Deep Sea Sediments: Production, Preservation and Interpretation. pp. 266–388. *In* Chemical Oceanography. J. P. Riley and R. Chester (eds.), 2nd edition, v. 5. Academic Press.

Colocoloff, M. and Ch. Colocoloffa. 1972. Recherches sur la production primaire d'un fond sableux 2-methodes. Tethys. 4: 779–790.

Conway, H. L. and J. I. Parker, E. M. Yaguchi and D. L. Mellinger. 1977. Biological utilization and regeneration of silicon in Lake Michigan. J. Fish. Res. Board Can. 34: 537–544.

Davies, J. M. 1975. Energy flow through the benthos in a Scottish

Sea Loch.  Mar. Biol. 31:  353–362.

Dugdale, R. C.  1976.  Nutrient cycles. p. 141–172.  *In* D. H. Cushing and J. J. Welch (eds.), The Ecology of the Seas.  Blackwell Sci. Publications.

Elderfield, H.  1976.  Hydrogenous material in marine sediments; excluding manganese nodules. p. 137–216.  *In* J. P. Riley and R. Chester (eds.), Chemical Oceanography 2nd edition. v. 5.  Academic Press.

Hale, S. S.  1974.  The role of benthic communities in the nutrient cycles of Narragansett Bay. 123 pp.  Master's thesis. University of Rhode Island. (not seen).

Hargrave, B. T.  1973.  Coupling carbon flow through some pelagic and benthic communities.  J. Fish. Res. Board Can. 30:  1317–1326.

Hartwig, E. O.  1974.  Physical, chemical and biological aspects of nutrient exchange between the marine benthos and the overlying water.  174 pp.  Ph.D. dissertation, University of California, San Diego.  Scripps Institution of Oceanography, LaJolla, California. (not seen).

Hartwig, E. O.  1976a.  The impact of nitrogen and phosphorus release from a siliceous sediment on the overlying water.  *In* M. L. Wiley (ed.), Estuarine Processes. Vol. I.  Uses, stresses, and adaptation to the estuary.  Academic Press.

Hartwig, E. O.  1976b.  Nutrient cycling between the water column and a marine sediment.  I. Organic carbon.  Mar. Biol. 34:  285–295.

Hartwig, E. O.  1978.  Factors affecting respiration and photosynthesis by the benthic community of a subtidal siliceous sediment.  Mar. Biol. 46:  283–293.

Hemmingsen, A. M.  1960.  Energy metabolism as related to body size and respiratory surface and its evolution.  Rep. Steno Mem. Hospital and Nordisk Insulinlab. (Copenhagen). 9:  3–110.

Jansson, B. O. and F. Wulf (Principal investigators).  1977.  Ecosystem Analysis of a Shallow Sound in the Northern Baltic – A Joint Study by the Åsko Group.  Stockholm. 18:  160 pp.

Kamluk, L. V.  1974.  Metabolic rate of free-living flat and ringed vermes and factors determining it.  J. Obschej Biologii. 35:  874–875 (*In* Russian).

Karl, D. M. and P. LaRock.  1975.  Adenosine triphosphate measurements in soil and marine sediments.  J. Fish. Res. Board Can. 32:  599–607.

Kremer, J. N. and S. W. Nixon.  1978.  A Coastal Marine Ecosystem. Simulation and Analysis.  *In* W. D. Billings, F. Golley, O. L. Lange and J. S. Olson (eds.), Ecological Studies. 24:  1–217. Springer-Verlag.

Lewin, J. C.  1961.  The dissolution of silica from diatom walls. Geochim. Cosmochim. Acta. 21:  182–195.

Nixon, S. W., C. A. Oviatt and S. S. Hale.  1976a.  Nitrogen regeneration and the metabolism of coastal marine bottom communities. p. 269–283.  *In* J. M. Anderson and A. Macfadyen (eds.), The Role of Terrestial and Aquatic Organisms in Decomposition Processes. Blackwell Sci. Publications.

Nixon, S. W., C. A. Oviatt, J. Garber and V. Lee.  1976b.  Diel metabolism and nutrient dynamics in a salt marsh embayment. Ecology. 57: 740–750.

Pamatmat, M. M. 1973. Benthic community metabolism on the continental terrace and in the deep sea in the North Pacific. Int. Revue ges. Hydrobiol. 58: 345–368.

Pamatmat, M. M. 1977. Benthic community metabolism: a review and assessment of present status and outlook. p. 89–111. *In* B. C. Coull (ed.), Ecology of Marine Benthos. Belle W. Baruch Library in Marine Science, No. 6. University of South Carolina Press, Columbia.

Plante-Cuny, M. R. 1977. Pepartition a la surface et au sein du sediment de la chlorophylle "a" et des pheopigments de quelque substrats meubles tropicaux immerges. J. rech. oceanogr., 2, n.2: I–II.

Plante-Cuny, M. R. 1978. Pigmentes photosynthetiques et production primair des fonds meubles neretiques d'une region tropicale (Nosy-Be, Madagaskar). J. rech. oceanogr., 3, n. I: 1–14.

Propp, M. V. 1977. Exchange of energy, nitrogen and phosphorus between water, bottom and ice in a near-shore ecosystem of the Sea of Japan. Helgolander wiss. Meeresunters. 30: 598–610.

Propp, M. V. and L. N. Propp. In press. Improved capillary column for nitrate – nitrite reduction in fresh and sea waters. Okeanologia (Russian).

Redfield, A. C. 1934. On the proportions of organic derivates in sea water and their relation to the composition of plankton. pp. 176–192. *In* James Memorial Volume, University Press, Liverpool.

Rhoads, D. C., R. C. Aller and M. B. Goldhaber. 1977. The influence of colonizing benthos on physical and chemical diagenesis of the estuarine seafloor. pp. 113–138. *In* B. C. Coull (ed.), Ecology of Marine Benthos. Belle W. Baruch Library in Marine Science, No. 6, University of South Carolina Press, Columbia.

Rittenberg, S. C., K. O. Emery and W. L. Orr. 1955. Regeneration of nutrients in sediments of marine basins. Deep-Sea Res. 3: 23–45.

Rowe, G. T., C. H. Clifford, K. L. Smith, Jr. and P. L. Hamilton. 1975. Benthic nutrient regeneration and its coupling to primary productivity in coastal waters. Nature. 225: 215–217.

Rowe, G. T. and K. L. Smith, Jr. 1977. Benthic-pelagic coupling in the Mid-Atlantic Bight. pp. 55–66. *In* B. C. Coull (ed.), Ecology of Marine Benthos. Belle W. Baruch Library in Marine Science, No. 6, University of South Carolina Press, Columbia.

Sapojnikov, V. V. 1977. Concentrations of phosphates as a result of "biofilter" process in the high productivity regions of the ocean. pp. 57–59. *In* B. A. Skopinzev and V. N. Ivanenkov (eds.), Chimiko-Okeanologicheskie Issledovania. Moskow. Nayka. (*In* Russian).

Smith, K. L., Jr. 1973. Respiration of a sublittoral community. Ecology. 54: 1065–1075.

Smith, K. L., Jr. 1978. Benthic community respiration in the N. W. Atlantic Ocean: *in situ* measurements from 40 to 5200 m. Mar. Biol. 47: 337–347.

Smith, K. L., Jr., C. H. Clifford, A. H. Eliason, B. Walden, G. T.

Rowe and J. M. Teal. 1976. A free vehicle for measuring benthic community metabolism. Limnol. Oceanogr. 21: 164-170.

Strickland, J. D. H. and T. R. Parsons. 1972. A practical handbook of sea-water analysis. 2nd edition. Bull. Fish. Res.
Board Can. 167: 1-310.

Suchshenja, L. M. 1972. Respiration intensity of Crustacea. 195
pp. Kiev. Naykova dumka. (*In* Russian).

Suess, E. 1976. Nutrients near the depositional interface. p. 57-
80. *In* I. N. McCave (ed.), The Benthic Boundary Layer. Plenum
Press.

Tarasoff, F. G. 1978. Distribution and trophic zonation of the
soft bottom communities in the Vostok Bay (Sea of Japan). Bio-
logia Morja. 6: 16-23. (*In* Russian).

Turpaeva, E. P., I. N. Soldatova, and R. G. Simkina. 1970. Black
Sea bivalvia metabolic intensity, Zoologicheskij Jurnal. 49:
1571 - 1572. (*In* Russian).

Vanderborght, J. P. and G. Billen. 1975. Vertical distribution of
nitrate concentration in interstitial water of marine sediment
with nitrification and denitrification. Limnol. Oceanogr. 20:
953-961.

Vanderborght, J. P., R. Wollast and G. Billen. 1977a. Kinetic
models of diagenesis in disturbed sediments. Prt. I. Mass
transfer properties and silica diagenesis. Limnol. Oceanogr.
22: 787-793.

Vanderborght, J. P., R. Wollast and G. Billen. 1977b. Kinetic
models of diagenesis in disturbed sediments. Prt. 2. Nitrogen diagenesis. Limnol. Oceanogr. 22: 794-803.

Wimbush, M. and W. Munk. 1970. The benthic boundary layer. p.
731-758. *In* A. E. Maxwell (ed.), The Sea. 4. Wiley-Inter-
science.

Winberg, G. G. 1976. The dependance of aquatic poikilotherm animals' metabolic rate from body weight. Obschej Biologii. 37:
56-69. (*In* Russian).

# Relationships of Tube-Dwelling Benthos with Sediment and Overlying Water Chemistry

Robert C. Aller

ABSTRACT: The construction and irrigation of
tubes by macrobenthos result in three-dimension-
al concentration gradients of pore water solutes
in the inhabited zone of marine sediments. By
idealizing the geometry of diffusion in such
cases it is possible to derive a two-dimensional
cylindrical coordinate model that adequately, and
more or less realistically, describes transport
of solutes such as microbial metabolites, near
the sediment-water interface.

The model fits extraordinarily well field
data on the distribution of pore water constituents. The presence of even a few tens of bur-
rows/$m^2$ has radical effects on the build-up of
interstitial reaction products. Large burrows or
relatively densely packed burrows have the greatest influence on sediment chemistry. The influence of burrows on sediment-water exchange rates
depends on the particular element involved. Some
elements, such as nitrogen, show relatively
little increase in flux due to macrofaunal bioturbation while others, for example silicon, are
dramatically increased. Two end-member descriptions of such elements are proposed: reaction

controlled and transport controlled. The benthic
community may, therefore, differentially couple
to various components of the pelagic community
depending on the limiting elements involved.

Animals may best control the chemical environment of their burrows and reduce the amount of irrigation work required to do this either by: 1)
building a large burrow or 2) crowding close together. This latter strategy is apparently used
by opportunistic colonizers of barren sea floor
or polluted substrata.


## INTRODUCTION

The composition of a sedimentary deposit represents a balance
between transport and reaction processes taking place within and
around it. In this paper I examine how the construction of tubes
by infaunal macrobenthos influences the transport of interstitial
solutes in sediments and thereby the chemical properties of those
deposits, their overlying waters, and the adaptations of the infauna themselves. The distribution and build-up of one common pore-
water constituent, ammonium ion, in the presence of a relatively
simple bottom community, will be used to demonstrate the principles
involved. These principles will then be shown to have general ap-
plicability.

Ammonia is produced in sediments and released into pore waters
by metabolic reactions such as (Richards, 1965):

$$6(CH_2O)_x(NH_3)_y(H_3PO_4)_z + 3xSO_4^{=} \rightarrow 6xHCO_3^{-} + 3xHS^{-} + 6yNH_4^{+} + \quad (1)$$

$$6zHPO_4^{=} + (3x + 12z - 6y)\ H^{+}$$

The C:N:P stoichiometry of this reaction, indicated by the variables
$x$, $y$, and $z$, is known to change both with depth in sediment and the
depositional environment (Sholkovitz, 1973; Hartmann *et al.*, 1973;
Martens *et al.*, 1978). Following release, $NH_4^{+}$ can undergo further
biogenic reactions, be adsorbed onto particles (Mortland and
Wolcott, 1965; Nommik, 1965; Müller, 1977; Rosenfeld, 1979), diffuse
into overlying waters where reincorporation into plankton material
takes place (Rowe *et al.*, 1975; Hale, 1975; Hartwig, 1976; Nixon *et
al.*, 1976), or in some cases, diffuse into other regions of the
sediment body.


## STUDY SITE

Pore water ammonium profiles, sediment-water fluxes, and ammonium production rates to be used for illustration here were obtained
at Station 5 in Mud Bay, South Carolina (Fig. 1). Mud Bay is a
shallow, approximately 2 m depth at high tide, muddy embayment on
the north side of the larger, deeper Winyah Bay (~10-12 m maximum
depth; Johnson, 1970). This latter estuarine bay receives its fresh

water run-off from the Pee Dee, Sampit, Black, and Waccamaw Rivers, whose confluence is approximately ~10-15 km up river of Mud Bay. The salinity of Mud Bay is highly variable; no long-term water monitoring data are available but the pore water chlorinities measured in 1 m length cores from this area indicate an average long-term chlorinity of ~0.25 M (M = moles/liter). This is the asymptotic Cl⁻ value at depth (see estuarine sediment Cl⁻ models of Holdren *et al.*, 1975). Surface water values at the time of sampling reported here were 0.2 to 0.3 M Cl⁻. The yearly temperature range is presumably similar to that of the adjacent Winyah Bay: 6-30°C (Johnson, 1970). At the time of sampling T = 29°C.

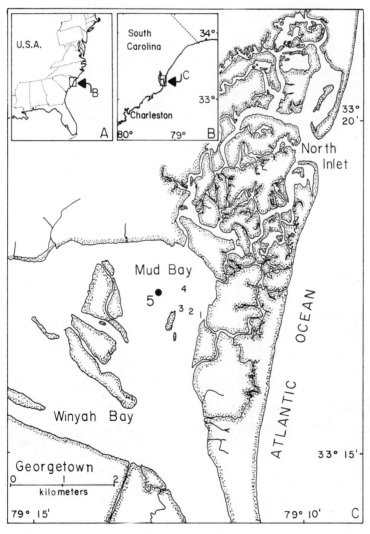

Figure 1. Map of Mud Bay, South Carolina showing location of sampled stations.

On the basis of six box cores (area sampled = 0.022–0.0297 $m^2$) taken at Station 5 from June 24–July 6, 1978, the macrofauna ($\gtrsim$1 mm) are composed predominately of the bivalves *Tellina texana* (43 $\pm$ 13/$m^2$), *Macoma balthica* (24 $\pm$ 24/$m^2$), and the capitellid *Heteromastus filiformis* (108 $\pm$ 101/$m^2$; range = 45–269/$m^2$). These same fauna characterize four additional stations (Fig. 1) sampled by grab (0.052 $m^2$). At these four stations *Heteromastus* abundances are similar to Station 5 and range from ~211 to 842/$m^2$. Other faunal elements include two extremely small oligochaete species (67–268/$m^2$; 0–555/$m^2$), at least one, possibly two, small nemertean species (19–180/$m^2$), and at Stations 1–4, additional nereid polychaetes *Nereis succinia* and *Laeoneris culveri* (0–38/$m^2$; 0–111/$m^2$) and the spionid polychaete *Scolecolepides virens* (0–19/$m^2$).

The fine-grained bottom sediment is almost entirely pelletized into fecal material $\geq$0.5 mm in diameter. These appear to be predominately feces of *Heteromastus*. The upper few centimeters of sediment are generally flocculent and yellow-orange in color and are underlain by brown-black material. *Spartina* and other marsh debris are abundant.

Because of the shallow depth of the Bay, wind-generated waves can disturb the upper few centimeters of sediment. This is reflected in scour surfaces and depositional laminations seen in the upper portions of x-radiographs. Otherwise the sediment is generally devoid of physically-formed sedimentary structures because of a relatively high degree of bioturbation (Moore and Scruton, 1957). Burrow structures formed by *Heteromastus* and nereids are commonly observed.

METHODS

Handheld Plexiglas box corers were used to obtain undisturbed samples of bottom sediment (Aller, 1980). Cores were returned to the lab (~ 2 h) and placed in a $N_2$-purged glove bag for processing. Sediment was removed in 1, 2, or 3 cm thick depth intervals, packed into centrifuge bottles or tubes depending on sample size, and pore water separated out by use of a Sorvall centrifuge (20 min., 5000 rpm). Pore water was removed and filtered under $N_2$ through 0.4 μm pore size Nucleopore filters. Filtered samples were placed immediately in a refrigerator and a subsample fixed with phenol within 24 hours (Degobbis, 1973). $NH_4^+$ analyses were made by use of the phenol-hypochlorite method (Solorzano, 1969).

Ammonium production rates as a function of depth in sediment at Station 5 were estimated by anoxically incubating sediment samples from selected depth intervals and opening them at successive times thereafter (see Martens and Berner, 1974; Goldhaber *et al.*, 1977; Billen, 1978; Aller and Yingst, 1979). Pore water separation and analyses were made as previously described.

Additional analyses, experiments, and a more detailed description of techniques will be given in later publications (Ullman and Aller; Aller and Ullman in prep.).

RESULTS AND DISCUSSION

Distribution of pore water solutes can be quantitatively described by transport-reaction models that take into account diffusion, advection (~ sedimentation rate) and reaction (Berner, 1974, 1980; Lerman, 1976). These models have been successfully used to provide insight into the general processes controlling the composition of sediments and also allow quantitative checks on interpretations of individual pore water constituent profiles. In this paper, several transport-reaction models will be presented in order to illustrate interactions of macrobenthos with sediment and overlying water chemistry.

Most quantitative or qualitative interpretations of sediment composition assume, explicitly or implicitly, that the sediment body is analogous to a one-dimensional porous slab accreting upward at a rate determined by net sedimentation. The slab is bounded above by sea water and below by a relatively impermeable basement. In such descriptions, net transport occurs only in the vertical dimension. The transport-reaction equation for the vertical distribution of a constituent like $NH_4^+$ would be given in this case by (compaction ignored):

$$\frac{\partial C}{\partial t} = \frac{D}{1+K} \frac{\partial^2 C}{\partial x^2} - \omega \frac{\partial C}{\partial x} + \frac{R}{1+K}(x,t) \tag{2}$$

where $C$ = pore water concentration; $x$ = space coordinate, origin fixed at sediment-water; interface, positive axis into sediment; $t$ = time; $\omega$ = sedimentation rate; $D$ = molecular diffusion constant modified for tortuosity, etc.; $K$ = linear adsorption coefficient; and $R(x,t)$ = reaction term.

Generally, modified forms of equation 2 are solved at steady state ($\partial C/\partial t = 0$) and the solution used in conjunction with available data to determine relationships between known and unknown variables in the equation. For example, $C(x)$, $D$, and $\omega$ may all be known, thereby allowing calculation of $R(x)$, *i.e.*, metabolic activity as a function of depth. In some instances, such as highly seasonal environments, the nonsteady state case of equation 1 must be considered (Lasaga and Holland, 1976; Holdren *et al.*, 1975; Aller, 1980). For reasons that will become clear later only the steady state will be considered here.

The maximum value of the sedimentation rate $\omega$ in Mud Bay can be estimated from pore water $SO_4^=$ gradients below the zone of bioturbation (~ 0-15 cm) (Berner, 1978). These estimates, based on data to be presented in later publications, give $\omega \lesssim 0.5$ cm/yr. Because the distribution of pore water $NH_4^+$ in the upper 0 to 20 cm are only of interest for the present, $\omega$ is sufficiently small (see Lerman, 1975) relative to rates of diffusion. Thus, equation 2 can be simplified at steady state to:

$$0 = D \frac{\partial^2 C}{\partial x^2} + R(x) \tag{3}$$

The values of $D$ and $R$ to be used in equation 3 for predicting $NH_4^+$ distributions in Mud Bay sediments are determined as follows.

The incubation experiments allow estimates of $R(x)$ [T ~ 22°C] over the upper 0–15 cm.  Rates obtained in sediment incubated from different depth intervals (Station 5) are plotted in Figure 2.  These can be fit by a function of the form $R = R_o \exp(-\alpha x) + R_1$, where $R_o$, $R_1$, and $\alpha$ are constants so that $R/(1 + K)$ ~ $23/(1 + K) \exp(-0.61 x) + 0.69/(1 + K)$ mM/yr with x expressed in cm.  Taking the adsorption coefficient $K$ ~ 1 (Rosenfeld, 1979) then $R - R_o \exp(-\alpha x) + R_1$ is $46 \exp(-0.61 x) + 1.38$ mM/yr.

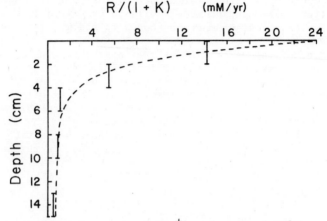

$$R/(1 + K) \qquad (mM/yr)$$

Figure 2.   Rate of $NH_4^+$ production (~22°C) as a function of depth in incubation experiments at Station 5.  The rate as measured includes the effect of adsorption here indicated by the factor $1 + K$.  The curve plotted is $R = 23 \exp(-0.61 x) + 0.69$ mM/yr, where x is measured in cm.

A value for D can be estimated from the infinite dilution data of Li and Gregory (1974) with a correction for sediment formation factor (e.g., effects of tortuosity, porosity) estimated from the water content in the upper 25 cm of sediment and the use of Archie's law (Manheim, 1970; Manheim and Waterman, 1974; Krom and Berner, 1980).  Water contents (weight loss at 80°C) average about 70% in the top 25 cm of Mud Bay sediments.  Assuming a sediment particle density of ~ 2.5 g/cc, then a formation factor of ~ 1.4 is calculated from $1/\phi^2$ (empirical Archie's Law relation) where $\phi$ = porosity.  This gives $D$ ~ $1.3 \times 10^{-5}$ cm$^2$/sec at 22°C.

With the boundary conditions $x = 0$; $C = C_o$, and $x = L$; $\alpha C/\alpha x = B$, then the concentration profile predicted by equation 3 with $R = R_o \exp(-\alpha x) + R_1$ is:

$$C = C_o + \frac{R_o}{D\alpha^2} [1 - \exp(-\alpha x)] + (B + \frac{R_1 L}{D} - \frac{R_o \exp(-\alpha L)}{D\alpha})x - \frac{R_1 x^2}{2D}$$

The boundary conditions used assume 1) a constant overlying water solute concentration $C_o$ and 2) that at some depth L the concentration gradient is known and fixed at a value B.  This latter condition requires a continuity of solute flux between the modeled zone and underlying sediment.

The pore water gradient in the vertical dimension predicted by this model (D, R values as above, L = 15 cm, the approximate depth of burrowed zone at Station 5, and B = 0.011 µmoles/cm$^4$, measured from gradient between 13-20 cm in Figure 3) is shown along with two measured pore water profiles from Station 5 in Figure 3. The poor fit implies that the assumptions of this simple model do not represent a good characterization of processes influencing NH$_4^+$ concentrations in surface sediments of Mud Bay.

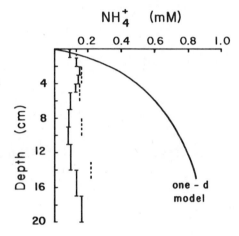

Figure 3. NH$_4^+$ pore water profiles obtained at two separate times from Station 5. The continuous solid curve is a predicted fit to the data of a one-dimensional transport-reaction rate of Figure 2 and assuming steady-state.

Disagreements of this general type between modeled and observed distributions of pore water constituents near the sediment water interface have been found in many environments. Many investigators have recognized that at least part of the disagreement comes about because of the mixing of sediments by benthos and a resulting transport of solutes by processes other than molecular diffusion. One approach to alleviating the discrepancies has therefore been to simply adjust the diffusion coefficient D to a new apparent value. The apparent value, which is generally larger than the molecular diffusion coefficient, reflects increased effective transport by physical or biogenic reworking of sediment. The use of a larger apparent D results in a lowering of model concentrations in the upper zone to values closer to those actually observed (Hammond *et al.*, 1975; Vanderborght *et al.*, 1977; Goldhaber *et al.*, 1977; Aller, 1978a).

Such an empirical approach works well when disturbance of sediments is limited to the upper few centimeters, as for example in the presence of highly mobile deposit-feeders. Unfortunately, as the depth interval to which the apparent D is applied increases in thickness, discrepancies between model and observed concentration distributions again arise (Aller, 1980). Simple changes in D are also not capable of explaining maxima in pore water profiles as observed in Figures 3 and 6 (expanded scale in the latter figure for clarity).

In some cases various investigators assume that deep-living, tube-dwelling animals exchange volumes of pore water with overlying water during irrigation of burrows (e.g., Grundmanis and Murray, 1977; Hammond and Fuller, 1979; McCaffrey *et al.*, 1980). Pore water-overlying water transport is modeled as a rapid advective process rather than a diffusive one. Except for use in sandy sediments, I think these latter models generally ignore Darcy's Law and the problem of permeability. They are useful parameterizations and certainly reflect a component of transport, but I think they can potentially lead to misconceptions about solute movement in the upper regions of sedimentary deposits.

An alternative approach is to postulate that, in the presence of relatively sedentary, tube-building macrobenthos, sediments behave not as slabs dominated by one-dimensional transport but rather as a body permeated by cylinders (Aller, 1977, 1978a and b). The fluid in these cylinders is held by animal irrigation activity at close to, but not identical to, the composition of overlying sea water. The implications of this simple change in transport geometry to the chemistry of sediments, sediment-water exchange, and animal adaptations will now be briefly developed.

Effect of burrows on sediment chemistry

Mud Bay is used as an example in this paper because *Heteromastus filiformis* is the main faunal element inhabiting the region below 2-3 cm. This species forms a relatively simple, largely vertical burrow (Schäfer, 1972). Considering the general form of these burrows, the inhabited sediment zone (~ 0-15 cm) in Mud Bay may be initially idealized as a body composed of packed hollow cylinders (Figure 4). The inner radius of each cylinder $(r_1)$ is determined by the radius of the average burrow size, while the outer radius $(r_2)$ is simply half the distance between any two burrow axes (Aller, 1978a). A small portion of sediment at the outer intersection of any three cylinders is unaccounted for and will be ignored. Looking down on the sediment-water interface, the surface would ideally appear like the top of a case of cans (e.g., soda cans); at the center of each can is a hollow straw. Animals do not, of course, space themselves evenly in sediment (Jumars *et al.*, 1977) and the results of this, as well as possible biogeochemical reasons for different animal spacing, will become clear as the model is developed.

The transport problem for the upper zone of the sediment now becomes identical to the problem of diffusion in a single finite, hollow cylinder whose outer boundary is effectively impermeable, that is, concentrations go through a maximum or minimum when sediment is located halfway between any two burrows. The transport-reaction equation (3) with the appropriate reaction term for the vertical zone $0 \leq x \leq L$ then becomes, at steady state:

$$\frac{\partial C}{\partial t} = 0 = D\frac{\partial^2 C}{\partial r^2} + \frac{D}{r}\frac{\partial C}{\partial r} + D\frac{\partial^2 C}{\partial x^2} + R_0 e^{-\alpha x} + R_1 \qquad (5)$$

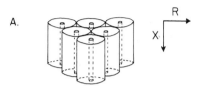

Figure 4. (A) Sketch of upper region of sediment idealized as packed hollow cylinders; (B) Vertical cross-section of sediment with idealized diffusion geometry of (A); (C) The single hollow cylinder visualized as the average microenvironment within the upper burrowed region of sediment. The dimensions shown correspond to those used in the transport-reaction model described in the text.

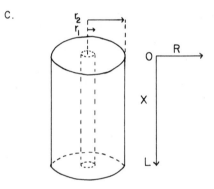

The boundary conditions are taken as:

$$r = r_1 , \quad C = C_o ; \quad x = 0 , \quad C = C_o \tag{a}$$

$$r = r_2 , \quad \partial C/\partial r = 0 \tag{b}$$

$$x = L, \quad \partial C/\partial x = B \tag{c}$$

These conditions specify that : a) the concentration of the solute is constant along the sediment-water interface and within a burrow; b) concentrations go through a maximum or minimum halfway between any two burrows; and c) there is continuity of flux between the burrowed zone and underlying sediment. The concentration of solutes in the water within the burrow may not be exactly that of overlying sea water, as assumed here, but is not expected to be radically different from overlying water as long as the burrow is irrigated. In addition the burrow wall is assumed permeable and no radial component to R, the reaction rate, is assumed to exist (see

Aller and Yingst, 1978, for contrary example). The solution to equation 5 is, by separation of variables:

$$C = C_o + Bx - \frac{2B}{L} \sum_{n=o}^{\infty} \frac{(-1)^n U_o(\lambda_n r)}{\lambda_n^2 U_o(\lambda_n r_1)} \sin (\lambda_n x) \tag{6}$$

$$+ \frac{2}{LD} \sum_{n=o}^{\infty} \left[ \frac{G_n}{\lambda_n^2} \frac{U_o(\lambda_n r)}{U_o(\lambda_n r_1)} - 1 \right] \sin (\lambda_n x)$$

Where:

$$\lambda_n = (n + 1/2) \frac{\pi}{L}$$

$$G_n = \left[ \left( \frac{R_o e^{-\alpha L}(-1)^n - \lambda_n R_o}{\alpha^2 + \lambda_n^2} - \frac{R_1}{\lambda_n} \right) \right]$$

$$U_o(\lambda_n r) = K_1(\lambda_n r_2) I_o(\lambda_n r) + I_1(\lambda_n r_2) K_o(\lambda_n r)$$

The functions $I_v(z)$ and $K_v(z)$ are the modified Bessel functions of the first and second kind respectively of order v (see e.g., Abramowitz and Stegun, 1964, for values).

The solution (6) gives the radial and vertical pore water distribution around a single burrow. The measured pore water profiles on the other hand, are simple average vertical distributions. If a box core integrates over a sufficient number of burrows (average sediment microenvironment), then the average vertical pore water concentration is that of a single model cylinder, so that over any vertical interval $x_1$-$x_2$:

$$\overline{C} = \frac{\displaystyle\int_{x_1}^{x_2} \int_{r_1}^{r_2} C \; rdrdx}{\displaystyle\int_{x_1}^{x_2} \int_{r_1}^{r_2} rdrdx} \tag{7}$$

where C is given by equation 6.

Based on measured diameters of *Heteromastus* taken from Station 5, the average inner burrow radius at that station must be $r_1 \sim 0.05$ cm. The outer radius $r_2$ may be estimated in two ways. If N = number of individuals per $m^2$, then assigning equal circular areas to each individual gives $r_2 = 1/(\pi N)^{\frac{1}{2}}$ in meters. An alternative method is to measure the total length $\ell$ of burrows/$cm^3$ giving $r_2 = 1/(\pi \ell)^{\frac{1}{2}}$.

This latter method is employed by soil scientists in the estimation
of outer root radii (Nye and Tinker, 1977; Gardner, 1960; Cowan,
1965) and has been used for the multiple species case in marine
sediments (Aller, 1977).  In this study $r_2$ is estimated from N.
Using observed abundances of *Heteromastus* in Mud Bay, reasonable
initial estimates of $r_2$ are 2-5 cm over the Bay as a whole with 4-5
cm being initial reasonable estimates for populations at Station 5
where pore water profiles are available.

To illustrate the model, radial $NH_4^+$ concentration gradients
around a burrow of length L = 15 cm, inner radius $r_1$ = 0.05 cm,
half distance between burrows $r_2$ = 3 cm, vertical gradient at L, B
= 0.011 $\mu$moles/$cm^4$ (as before), reaction rate R = 46 exp (-0.61x) +
1.38 mM/yr. and D = 1.3 x $10^{-5}$ $cm^2$/sec (molecular value) have been
plotted for particular depth intervals in Figure 5A.  The average
vertical concentration gradient measured in one centimeter intervals in such a cylinder is shown in Figure 5B.  Note that this gradient is radically different from the no burrow case that is plotted
in Figure 3.  Without altering the primary data such as molecular
diffusion coefficients, the cylinder model reproduces many of the
features of the observed profiles.  In particular the appropriate
range of concentration in the top 0-15 cm is attained, and the general form of the profile is generated.  By allowing for, as benthic
populations go, a small degree of spatial variability in population
abundances or depth of burrowing, the observed profiles can in fact
be fit almost exactly (Fig. 6).  This is remarkable agreement with
very minor and reasonable variation in parameters.  The number of
tubes required by the model is slightly higher than the measured
abundance of *Heteromastus* at Station 5, presumably due to the
presence of other burrowers such as nereids and the building by
*Heteromastus* of multiple or branched tubes.

To illustrate the effect of different abundances or sizes of
burrows on the build-up of pore water concentrations, equation 7
is used in Figure 7 to predict the average $NH_4^+$ pore water concentrations expected in the upper 0-15 cm of typical Mud Bay sediment
for various values of $r_1$ and $r_2$.  This average concentration would
be the value obtained if the upper 15 cm were collected and squeezed
as a single sample (ignoring minor porosity effects).  All terms in
the model except either $r_1$ or $r_2$ are kept constant and the ammonium
production rate shown in Figure 2 is assumed.  Model values are
bounded above by the one dimensional vertical case (equation 4 averaged over 0-15 cm) and below by the assumed overlying water value
0.2 $\mu$M.

Only two or three curves of the possible families have been
plotted to illustrate model behavior (Fig. 7).  In one set of
curves, $r_2$ was held constant at either 2, 3, or 4 cm and $r_1$ allowed
to vary from 0.01 cm to 1.0 cm.  This range of burrow size is about
that of an oligochaete burrow to that of a large crustacean burrow.
In the other set of curves, burrow radius was held at either 0.05
cm (*Heteromastus*) or at 0.5 cm (*Uca* size) and population abundance
or, equivalently, average interneighbor distances allowed to vary.
At zero burrow radius or at zero population abundances, curves
asymptote to the one-dimensional case as expected.

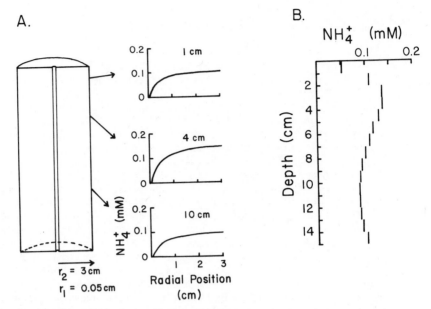

Figure 5. (A) Illustrative radial concentrations at distinct depths for hollow cylinder of typical size for Mud Bay sediments. The inner radius $r_1$ = 0.05 cm, similar to *Heteromastus*; the outer radius $r_2$ = 3 cm is a reasonable value for the mean half distance between burrows in this environment; (B) The average concentrations in the cylinder of (A) shown only as a function of vertical position as would be the case in normal sampling of sediments.

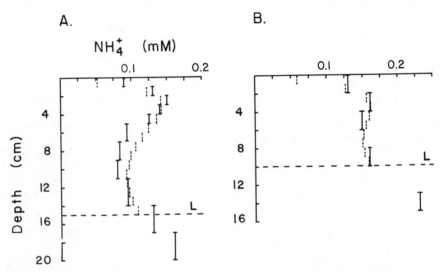

Figure 6. Hollow cylinder model fits (dotted bars) to pore water $NH_4^+$ profiles (solid lines) at Station 5. The depth of the burrowed zone is given as L, $r_1$ = 0.05 cm in (A) and (B), $r_2$ = 3.1 cm, (N = 331/$m^2$) in (A) and 3.3 (N = 292/$m^2$) in (B); the production rate is that of Figure 2 in both cases. The fit is at steady state. (A) = core collected July 6, 1978. (B) = core collected for incubation experiments July 9, 1978.

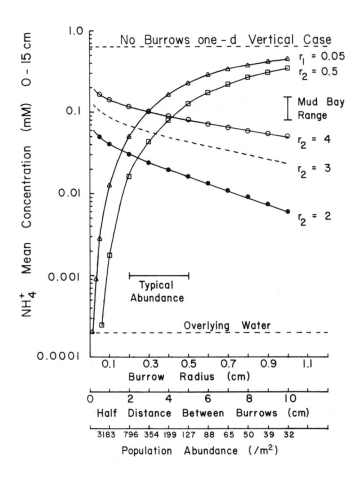

Figure 7. Predicted average $NH_4^+$ concentration in upper 0-15 cm (burrowed zone) of sediments as a function of either variable burrow size ($r_1$) with fixed abundance ($r_2$, N) or variable abundance ($r_2$, N) with fixed burrow size ($r_1$). The $NH_4^+$ production rate is that of Figure 2. The typical range of abundance or interburrow half distance for Mud Bay is indicated as the horizontal bar. The observed range of concentrations in the upper 15 cm is indicated as the vertical bar. Limits on the model concentrations are the overlying water concentration (lower limit) and the value of the one-dimensional model (upper limit, see Figure 3). The curve believed to best describe Mud Bay is $r_1 = 0.05$ cm.

The build-up of concentrations is extremely sensitive to burrow abundance (packing) and somewhat less sensitive to, but still an exponential function of, burrow size. The presence of even a few tens of burrows/$m^2$ of radius 0.05 cm can reduce the mean pore water solute concentrations by factors of 3 from those expected on the basis of a one-dimensional model only. The observed ranges of *Heteromastus* abundances and the ranges of $NH_4^+$ concentrations found in the upper 15 cm of Mud Bay pore waters are plotted as range bars on the graph. These demonstrate that, using the 0.05 cm burrow radius curve, the simple model outlined here is unusually good at predicting expected concentrations. Aside from agreeing with the observed data, one of the main strengths of the model is that it represents a relatively realistic, albeit idealized, description of solute transport in the bioturbated zone.

The sensitivity of solute concentrations to interburrow distances suggests that spatial heterogeneity in sediment pore water chemistry can be expected to track rather minor alterations in benthic population abundances, i.e., patches of animals. Further comments on this will be made later.

### Effect of Burrows on Sediment-Overlying Water Exchange

The above considerations demonstrate that the presence of irrigated burrows near the sediment-water interface can have significant influences on the distribution and build-up of pore water constituents in sedimentary deposits. Because the diffusion of nutrients and dissolved constituents from sediments into overlying waters is known to have important effects on biological and chemical properties of overlying waters (Rowe *et al*., 1975; Hartwig, 1976; Hale, 1975; Nixon *et al*., 1976; Aller, 1980), it is reasonable to assess how sediment-water fluxes of pore water derived solutes might be affected by the presence of biogenic tube structures. It can be shown that for some elements the flux will be slightly altered while for others very large changes can be brought about.

Consider equation 3 and its solution, equation 4 which describes the $NH_4^+$ concentration distribution at steady state in one dimension. The flux across the sediment water interface is given by Fick's first law evaluated at x = 0 and is (ignoring advection and assuming that porosity at the interface is $\Phi = 1$):

$$J = - D \left(\frac{\partial C}{\partial x}\right)_{x = 0}$$

Evaluating J from equation 4 shows that as $\alpha L$ becomes >1 and if B is small, then $|J| \sim R_0/\alpha + LR_1$ at steady state. This means that any constituent that has a fixed reaction term of the form $R = R_0 \exp(-\alpha x) + R_1$ will have a steady state flux determined by reaction rate independent of the transport regime. $NH_4^+$ appears to be such an element (e.g. Billen, 1978; Aller and Yingst, 1979; Aller, 1980). For example, in Mud Bay $R_0/\alpha + LR_1$ (from Fig. 2) gives $J \sim 2.8$

$\mu$moles/m$^2$/day, while direct flux measurements of $NH_4^+$ released in incubated box cores (4-6 h) give J ~ 2.5 $\mu$moles/m$^2$/day. A similar correspondence is found for sediments of Long Island Sound (Aller and Yingst, 1979; Aller, 1980).

Because the microorganisms which catalyze the production of $NH_4^+$ do not exist independent of the buildup of associated metabolites in the sediment, the production rate of $NH_4^+$ cannot be entirely independent of the transport regime. However, based on the kinds of measurements noted above, the effect on production rate seems to be relatively small, perhaps a factor of two or so. An effect of this magnitude becomes hard to detect in actual field data due to problems of spatial heterogeneity and cumulative error of measurement. The stimulation, however small, of microbial activity by macrofaunal influence on the transport of interstitial solutes would represent one component of the "microbial gardening" effect proposed by some investigators (Yingst and Rhoads, 1980). This stimulation can be represented by the dependence of the reaction rate R on concentrations of the solutes $C_i$ in pore water and remains to be determined in detail.

In contrast to the previous example there are a number of interstitial solutes whose production rate is highly sensitive to the transport regime. Dissolved Si, an important and possibly limiting nutrient in overlying waters, is one such case. If a solid phase silica source in the sediment is not limiting, dissolved Si is produced at a rate described by equations of the form: $R = k \ (C_{eq} - C)$ where k is an apparent first order rate constant and $C_{eq}$ is an equilibrium limiting solubility value (see Hurd, 1973; Wollast, 1974; Schink *et al.*, 1975). A transport-reaction equation analogous to equation 4 with the appropriate reaction term can be written to describe the Si concentration distribution. When this is done and assuming B = 0; $\frac{k}{D} L \rightarrow$ large, then with $C_o$ = overlying water Si concentration:

$$J \sim \sqrt{kD} \ (C_{eq} - C_o) \tag{10}$$

Unlike the case of $NH_4^+$, the diffusion coefficient D remains in the expression for J demonstrating simply, that, in the absence of a low limit on a solid phase source, the dissolved Si flux from sediment to overlying water is subject to transport control. This can again be substantiated or at least supported by data from Mud Bay, where, from the same incubation experiments used to estimate the $NH_4^+$ production rates, it was possible to obtain rough values for k and $C_{eq}$. These are k ~ 0.02 ± .01/day and $C_{eq}$ = 687 $\mu$M, which are in general agreement with reported values from other environments (Schink *et al.*, 1975). Taking $D_{Si}$ ~ 4 x $10^{-6}$cm$^2$/sec (Fanning and Pilson, 1974) and $C_o$ = 74 $\mu$M; then equation 10 predicts J ~ 0.5 $\mu$moles/m$^2$/d. Measured Si fluxes are ~ 5 $\mu$moles/m$^2$/d. This 10-fold discrepancy from the one-dimensional molecular transport model supports the contention that Si fluxes are transport-controlled and that macrofaunal activities such as burrow formation may greatly influence the sediment-water exchange rates in such cases. The actual Si data, flux data, and modeling will be published in detail in the future.

Other elements, such as Mn, are influenced dramatically by particle reworking in addition to pore water transport. Some of these, Fe, Mn, and Ca, are discussed in Aller (1978a).

It is therefore possible to delineate ideal end-member elements such that sediment-water fluxes will either be reaction rate controlled or transport controlled. No pore water constituent will fit either end-member case exactly for reasons mentioned previously, but clearly some solutes, such as $NH_4^+$, tend more toward the reaction control end-member than others such as Si. This means that the burrow formation and sediment reworking activities of a given benthic community may have differential stimulation effects on the overlying planktonic community; diatoms, for example, may experience a greater degree of coupling to benthic communities than do other components of the plankton.

### Effect of Sediment Chemistry on the Adaptations of Tube-Dwellers

One of the least understood aspects of animal-sediment interactions is the relationship of animal behavior or adaptations to sediment chemistry. The theoretical framework derived above can be used to illustrate or at least suggest certain fundamental interactions.

Previous work demonstrated that for a field population of *Amphitrite ornata* the flux of metabolites from surrounding sediment (or excretion by burrow inhabitants) into the burrows of individual *Amphitrite* was balanced by irrigation rates such that (Aller and Yingst, 1978):

$$F = \frac{JA}{V_o} = \frac{v}{V_o} (C_B - C_o) \tag{11}$$

where: $J$ = flux of solute (from burrow walls (mass/area/time)); $A$ = area of inner burrow wall; $V_o$ = volume of burrow; $F$ = rate of change of burrow water concentration (mass/volume/time); $v$ = irrigation rate (volume/time); $C_B$ = steady state concentration of particular solute in burrow; and $C_o$ = concentration of solute in overlying water reservoir. When irrigation ceases, as occurs during intertidal periods, burrow water begins to change at a rate at least equal to $F$ (Aller and Yingst, 1978). If excretion by inhabitants is also important for a particular case, then burrow water concentrations will change even faster than that predicted by the diffusion rate of material from surrounding sediment. Assuming that an animal has a particular range of physiological tolerance of $C_B$ values for a given solute, then the magnitude of $F$ is important in determining where and under what conditions that animal can exist. The separate cases of intertidal and subtidal animals will now be examined again using $NH_4^+$ as an example.

The value of $F_{NH_4^+}$ for burrows of a given size and distance apart in Mud Bay can be calculated using equation 4 assuming the values for D, L, R, $C_o$, and B used previously. Fick's first law gives:

$$J(r_1,x) = -D \left(\frac{\partial C}{\partial r}\right)_{r = r_1} \tag{12}$$

This can be integrated over the burrow length to obtain the mean flux into (or out of) a burrow from surrounding sediment.

The resulting values of $F_{NH_4}{}^+$ are plotted in Figure 8 for a constant burrow radius of 0.05 cm and variable distance between burrows (variable $r_2$). The case of a fixed distance between burrows and variable burrow size (variable $r_1$) is also plotted. The normal range of animal abundance in Mud Bay is indicated as the horizontal bar at the base of the graph. These are approximate values of F because as burrow water approaches pore water solute concentrations, the flux from surrounding sediment will decrease as required by equation 12. However, they serve to illustrate the principles involved.

The plot has the most implications for intertidal animals that are periodically exposed making irrigation impossible. Because overlying water concentrations of $NH_4{}^+$ are ~ 0.2 µM it is obvious that burrow inhabitants can experience major excursions in concentrations of burrow water $NH_4{}^+$ above that of sea water during even a short intertidal period (see Aller and Yingst, 1978). In the case of the common burrow sizes and abundances of Mud Bay that might become exposed during low tide, the burrow water would quickly attain the composition of surrounding interstitial water (~ 1 h) after cessation of irrigation (Figure 8).

The plot reveals that there are two strategies available for lowering the rate of change of burrow water concentration. These are essentially strategies to lower the surrounding pore water concentrations and reduce the concentration gradient using the same principles as outlined in Figure 7. One is simply to increase the size of the burrow and the other is to increase population densities, that is, to crowd together. A third strategy is, of course, for the animal to build an impermeable boundary between itself and the surrounding sediment. The latter course of action is represented in such features as thick burrow linings, tube cements, or in the case of bivalves – mantle fusion. These adaptations may have multiple functions and I do not mean to imply they are solely a response to sediment chemistry.

For animals that remain submerged at all times, the value of F per se may not be the most important factor to consider, but instead the variable $F \cdot V_0$ becomes appropriate. If $C_B$ and $C_0$ are constant or remain within small ranges, then the value $F \cdot V_0$ is essentially irrigation rate. A plot of this variable can be made in the same way as before to illustrate how irrigation rate must vary with F in order to maintain a particular $C_B$, that is, burrow environment. This is a measure of the work an animal must spend in order to 'stat' its chemical environment with regard to fluxes from surrounding sediment. Burrow length is held constant at 15 cm as before. Figure 9 illustrates that at small burrow sizes a crowding strategy can greatly reduce the irrigation work required to maintain a particular $C_B$. At low population densities, however, increasing the size of the burrows lowers irrigation requirements below those of smaller burrows. In the particular case of *Heteromastus* ($r_1 = 0.05$ cm), when individuals are greater than ~3.5 cm apart *Heteromas-*

*tus* could best control its chemical habitat by either enlarging its burrow to $r_1 > 0.35$ cm or by reproducing and thereby adding more burrows into its general environment.

Varying burrow length produces the expected changes in F and $F \cdot V_O$ (Figure 10). Irrigation requirements increase for longer burrows because total burrow volume increases. The value of F also begins to decrease for the same reason and, of course, because the burrow is extending into sediment with a lower and lower surrounding production rate of metabolites ($R = R_O \exp(-\alpha x) + R_1$ with x large).

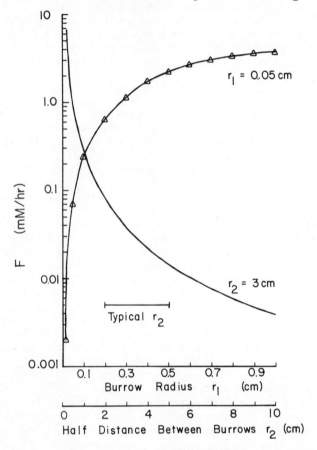

Figure 8. Corresponding graphs to Figure 7 but illustrating the rate of change of burrow water concentrations that would occur in the absence of irrigation at the production rate of Figure 2 and for either fixed $r_1$ (0.05 cm) or fixed $r_2$ (3 cm).

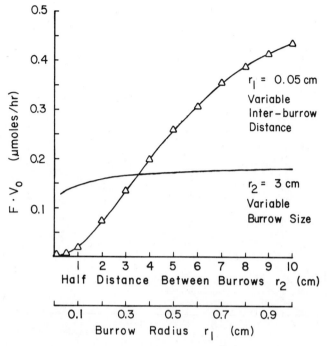

Figure 9. The rate of irrigation $F \cdot V_o$
required to hold a burrow at constant concentration of solutes in the presence of a flux
from surrounding sediment. The irrigation
rate required of individuals can be greatly
reduced by having large numbers of burrows of
a given size (e.g., $r_1$ = 0.05 cm). At low
abundances, increasing the size of the burrow
results in relatively lower irrigation requirements. Note that abundance is the more dominant effect.

These considerations may give some insight into commonly observed animal distributions and adaptations. Early colonizers of
defaunated bottoms (often dump sites or sediment that incubated
without macrofauna, and so, is high in interstitial metabolites, or
polluted substrates) are characteristically small tube dwellers and
are densely crowded (Grassle and Grassle, 1974; McCall, 1977; Rhoads
*et al.*, 1977; Pearson and Rosenberg, 1978). Figures 7-10 demonstrate that this is an extremely effective strategy for a population
to alter the chemistry of the surrounding sediment and lower the
work required of individuals in the population to maintain burrow
waters within physiological tolerances. Food requirements can be
expected to place a lower limit on this otherwise beneficial crowding. This kind of observation suggests the possibility of group
selection mediated through sediment chemistry and should be investigated further.

With regard to controlling the chemical habitat, the alternative to small size and crowding is to build a large burrow. The
absence of such animals in polluted or especially organic-rich sediments presumably reflects the low $O_2$ tensions in such areas and

the relatively unfavorable low surface to volume ratio of the larger
infauna (e.g. Pearson and Rosenberg, 1978). The absence of larger
infauna in recently recolonized natural bottom regions is usually
attributed to more conservative life history patterns (McCall,
1977).

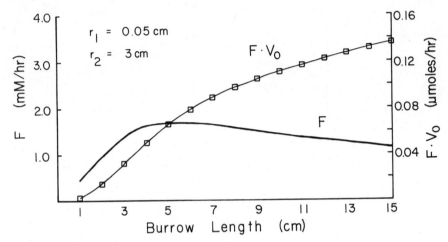

Figure 10. The effect of changing burrow length on the
potential rate of change of burrow water $NH_4^+$ concentrations
(F) or irrigation rate ($F \cdot V_0$). Mud Bay production rate (Figure 2) with $r_1$ = 0.05 cm and $r_2$ = 3 cm.

Therefore 'opportunistic' species are adapted at the population
level to colonize chemically hostile sedimentary environments. By
adding neighbors, i.e., reproducing, these species are capable of
utilizing some of the simple consequences of diffusion geometry to
radically influence their chemical micro-habitat and lower the
stress, both of physical work and chemical quality, on each individual. Larger animals having more conservative life histories (better
competitors at the individual level) and occurring in lower densities are, from these same standpoints, adapted to a more benign
chemical environment in the sedimentary habitat. These apparent
differences in adaptation to environment lend support to the concept of succession in benthic communities.

ACKNOWLEDGMENTS

This study was made possible by an Alfred P. Sloan Research
Fellowship (Sloan Foundation), the Louis B. Block Fund (Chicago),
and the Gurley Paleontology Fund (Chicago). I thank W. J. Ullman
for aid in the mud, water, and laboratory. Thanks go to V. Barcilon
for advice on math and to R. A. Berner for critical comments on the
manuscript. A congenial working atmosphere was provided by colleagues at the Baruch Marine Field Laboratory.

REFERENCES

Abramowitz, M., and I. A. Stegun (eds.). 1964. Handbook of Mathematical Functions with Formulas, Graphs, and Mathematical Tables. Dover Publ., Inc., New York. 1,045 pp.

Aller, R. C. 1977. The influence of macrobenthos on chemical diagenesis of marine sediments. Ph.D. dissertation Yale University, New Haven, Conn. 600 pp.

Aller, R. C. 1978a. Experimental studies of changes produced by deposit feeders on pore water, sediment, and overlying water chemistry. Am. Jour. Sci. 278: 1185-1234.

Aller, R. C. 1978b. The effects of animal-sediment interactions on geochemical processes near the sediment-water interface. *In* M. L. Wiley (ed.), Estuarine Interactions. Academic Press. 157-172.

Aller, R. C. 1980. Diagenetic processes near the sediment-water interface of Long Island Sound. I: Decomposition and nutrient element geochemistry. *In* B. Salzman (ed.), Physics and Chemistry of Estuaries: Studies in Long Island Sound. Advances in Geophysics. Vol. 22. Academic Press, New York. (in press).

Aller, R. C., and J. Y. Yingst. 1978. Biogeochemistry of tube-dwellings: a study of the sedentary polychaete *Amphitrite ornata* (Leidy). J. Mar. Res. 36: 201-254.

Aller, R. C. and J. Y. Yingst. 1979. Relationships between microbial distributions and the anaerobic decomposition of organic matter in surface sediments of Long Island Sound, U.S.A. Mar. Biol. (in press).

Berner, R. A. 1974. Kinetic models for the early diagenesis of nitrogen, sulfur, phosphorus and silicon in anoxic marine sediments. *In* E. D. Goldberg (ed.), The Sea, Vol. 5, Marine Chemistry. John Wiley and Sons, New York. 427-450.

Berner, R. A. 1978. Sulfate reduction and the rate of deposition of marine sediments. Earth Planet. Sci. Letters 37: 492-498.

Berner, R. A. 1980. Early Diagenesis: A Theoretical Approach. Princeton University Press. (in press).

Billen, G. 1978. A budget of nitrogen recycling in North Sea sediments off the Belgian Coast. Estuarine Coastal Mar. Sci. 7: 127-146.

Cowan, I. R. 1965. Transport of water in the soil-plant atmosphere system. J. Appl. Ecol. 2: 221-239.

Degobbis, D. 1973. On the storage of sea water samples for ammonia determination. Limnol. Oceanogr. 18: 146-150.

Fanning, K. A., and M. E. Pilson. 1974. The diffusion of dissolved silica out of deep-sea sediments. J. Geophys. Res. 79: 1293 1297.

Gardner, W. R. 1960. Dynamic aspects of water availability to plants. Soil Sci. 89: 63-73.

Goldhaber, M. B., R. C. Aller, J. K. Cochran, J. K. Rosenfeld, C. S. Martens, and R. A. Berner. 1977. Sulfate reduction diffusion and bioturbation in Long Island Sound sediments: report of the FOAM group. Am. Jour. Sci. 277: 193-237.

Grassle, J. F., and J. P. Grassle. 1974. Opportunistic life histories and genetic systems in marine benthic polychaetes. J. Mar. Res. 32: 253-284.

Grundmanis, V., and J. W. Murray. 1977. Nitrification and denitrification in marine sediments from Puget Sound. Limnol. Oceanogr. 22: 354-368.

Hale, S. S. 1975. The role of benthic communities in the nitrogen and phosphorus cycles of an estuary. *In* Mineral Cycling in Southeastern Ecosystems. Proc. Symp. at Augusta, Georgia. May 1-3, 1974, ERDA Symp. Series. 291-308.

Hammond, D. E., H. J. Simpson, and G. Mathieu. 1975. Methane and Radon-222 as tracers for mechanisms of exchange across the sediment-water interface in the Hudson River Estuary. *In* T. M. Church (ed.), Marine Chemistry in the Coastal Environment. ACS Symp. Series. 18: 119-132.

Hammond, D. E., and C. Fuller. 1979. The use of Radon-222 as a tracer in San Francisco Bay. AAAS Symposium volume on San Francisco Bay. (in press).

Hartmann, M., P. Müller, E. Suess, and C. H. Van der Weijden. 1973. Oxidation of organic matter in recent marine sediments. "Meteor" Forschungsergeb. Reihe C 12: 74-86.

Hartwig, E. O. 1976. The impact of nitrogen and phosphorus release from a siliceous sediment on the overlying water. *In* M. L. Wiley (ed.), Estuarine Processes, Vol. 1. Academic Press. 103-117.

Holdren, G. R., Jr., O. P. Bricker, and G. Matisoff. 1975. A model for the control of dissolved manganese in the interstitial waters of Chesapeake Bay. *In* T. M. Church (ed.), Marine Chemistry in the Coastal Environment. ACS Symposium Series. 18: 364-381.

Hurd, D. C. 1973. Interaction of biogenic opal sediments and sea water in the central equatorial Pacific. Geochim. Cosmochim. Acta. 37: 2257-2282.

Johnson, F. A. 1970. A reconnaissance of the Winyah Bay estuarine zone, South Carolina. South Carolina Water Resources Commission Report No. 4, 36 pp.

Jumars, P. A., D. Thistle, and M. L. Jones. 1977. Detecting two-dimensional spatial structure in biological data. Oecologia 28: 109-123.

Krom, M. O., and R. A. Berner. 1980. The diffusion coefficients of sulfate, ammonium, and phosphate ions in anoxic marine sediments. Limnol. Oceanogr. (submitted).

Lasaga, A. C., and H. D. Holland. 1976. Mathematical aspects of non-steady state diagenesis. Geochim. Cosmochim. Acta. 40: 257-266.

Lerman, A. 1975. Maintenance of steady state in oceanic sediments. Am. Jour. Sci. 275: 609-635.

Lerman, A. 1976. Migrational processes and chemical reactions in interstitial waters. *In* E. D. Goldberg, I. N. McCave, J. J. O'Brien, J. H. Steele (eds.), The Sea, vol. 6, Marine Modeling. John Wiley & Sons, New York. 695-738.

Li, Y-H, and S. Gregory. 1974. Diffusion of ions in sea water and in deep-sea sediments. Goechim. Cosmochim. Acta. 38: 703-714.

Manheim, F. T. 1970. The diffusion of ions in unconsolidated sediments. Earth Planet. Sci. Lett. 9: 307-309.

Manheim, F. T., and L. S. Waterman. 1974. Diffusimetry (diffusion constant estimation) on sediment cores by resistivity probe. *In* Initial Reports of the Deep Sea Drilling Project. 22: 663-670.

Martens, C. S., and R. A. Berner. 1974. Methane production in the interstitial waters of sulfate-depleted marine sediments. Science. 185: 1167-1169.

Martens, C. S., R. A. Berner, and J. K. Rosenfeld. 1978. Interstitial water chemistry of anoxic Long Island Sound sediments. 2. Nutrient regeneration and phosphate removal. Limnol. Oceanogr. 23: 605-617.

McCaffrey, R. J., A. C. Myers, E. Davey, G. Morrison, M. Bender, N. Luedtke, D. Cullen, P. Froelich, and G. Klinkhammer. 1980. The relation between pore water chemistry and benthic fluxes of nutrients and manganese in Narragansett Bay, Rhode Island. (in press).

McCall, P. L. 1977. Community patterns and adaptive strategies of the infaunal benthos of Long Island Sound. J. Mar. Res. 35: 221-266.

Moore, D. G., and P. C. Scruton. 1957. Minor internal structures of some recent unconsolidated sediments. Amer. Assoc. Petroleum Geol. Bull. 41: 2723-2751.

Mortland, M. M., and A. R. Wolcott. 1965. Sorption of inorganic nitrogen compounds by soil materials. *In* W. V. Bartholomew and F. E. Clark (eds.), Soil Nitrogen Agronomy. 10: 150-197.

Müller, P. J. 1977. C/N ratios in Pacific deep-sea sediments: Effect of inorganic ammonium and organic nitrogen compounds sorbed by clays. Geochim. Cosmochim. Acta. 41: 765-776.

Nixon, S. W., C. A. Oviatt, and S. S. Hale. 1976. Nitrogen regeneration and the metabolism of coastal marine bottom communities. *In* J. M. Anderson and A. Macfadyen (eds.), The Role of Terrestrial and Aquatic Organisms in Decomposition Processes. 17th Symp. Brit. Ecol. Soc. 269-283.

Nye, P. H., and P. B. Tinker. 1977. Solute Movement in the Soil-Root System. Studies in Ecology V. 4. University of Calif. Press, Berkeley and Los Angeles. 342 pp.

Nõmmik, H. 1965. Ammonium fixation and other reactions involving a nonenzymatic immobilization of mineral nitrogen in soil. *In* W. V. Bartholomew and F. E. Clark (eds.), Soil Nitrogen, Agronomy. 10: 198-258.

Pearson, T. H., and R. Rosenberg. 1978. Macrobenthic succession in relation to organic enrichment and pollution of the marine environment. Oceanogr. Mar. Biol. Ann. Rev. 16: 229-311.

Richards, F. A. 1965. Anoxic basins and fjords. *In* J. P. Riley and G. Shirrow (eds.), Chemical Oceanography, Vol. 1. Academic Press, New York. 611-645.

Rhoads, D. C., R. C. Aller, and M. B. Goldhaber. 1977. The influence of colonizing benthos on physical properties and chemical diagenesis of the estuarine seafloor. *In* B. C. Coull (ed.), Ecology of Marine Benthos, Univ. of South Carolina Press, Columbia.

Rosenfeld, J. K. 1979. Ammonium adsorption in nearshore anoxic sediments. Limnol. Oceanogr. 24: 356-364.

Rowe, G. T., C. H. Clifford, K. L. Smith, Jr., and P. L. Hamilton. 1975. Benthic nutrient regeneration and its coupling to primary productivity in coastal waters. Nature. 255: 215–217.

Schäfer, W. 1972. Ecology and paleoecology of marine environments, G. Y. Craig (ed.), I. Oertel (translator), Univ. of Chicago Press. 568 pp.

Schink, D. R., N. L. Guinasso, Jr., and K. A. Fanning. 1975. Processes affecting the concentration of silica at the sediment-water interface of the Atlantic Ocean. J. Geophy. Res. 80: 3013–3031.

Sholkovitz, E. 1973. Interstitial water chemistry of the Santa Barbara Basin sediments. Geochim. Cosmochim. Acta. 37: 2043–2073.

Solórzano, L. 1969. Determination of ammonia in natural waters by the phenolhypochlorite method. Limnol. Oceanogr. 14: 799–801.

Vanderborght, J. P., R. Wollast, and G. Billen. 1977. Kinetic models of diagenesis in disturbed sediments: Part I. Mass transfer properties and silica diagenesis. Limnol. Oceanogr. 22: 787–793.

Wollast, R. 1974. The silica problem. *In* E. D. Goldberg (ed.), The Sea, V. 5, Marine Chemistry. J. Wiley & Sons, New York. 359–392.

Yingst, J. Y., and D. C. Rhoads. 1980. The role of bioturbation in enhancing bacterial growth rates in sediments. *In* K. R. Tenore and B. C. Coull (eds.), Marine Benthic Dynamics. Univ. of South Carolina Press, Columbia.

DETRITUS

# The Role of Detritus and the Nature of Estuarine Ecosystems

John R. Sibert
Robert J. Naiman

ABSTRACT: The detritus concept in ecology has
gone through changes attributable to theoretical
and methodological advances. Prior to 1940, the
importance of detritus in the nutrition of benthic
communities was accepted by the scientific community. There was some understanding of the role
of bacteria in nourishing invertebrates but this
notion did not receive wide acceptance. In the
1940's and 1950's, attention was directed away
from the study of detritus and bacteria toward
photosynthesis and factors regulating primary
productivity. In the 1960's, more attention was
given to detritus and associated bacteria. The
modern concept differs from the original only by
increased appreciation of the role of bacteria in
the detritus food chain.

Estuaries and detritus both appeared early
in the history of the Earth prior to the origin
of life. The first food chains, in the early Precambrian era, were detritus-based. Photosynthesis developed later, possibly as a mechanism to
augment the limited supplies of detritus. As the
Earth's atmosphere became increasingly oxidized,

the world ecosystem became increasingly dependent
on photosynthesis to supply organic matter.
Modern detritus-based food chains are direct de-
scendants of the primordial ecosystem.  Estuaries
receive large subsidies of imported organic matter
and have well-developed detritus food chains.
The estuarine ecosystem should be considered to
have two productive bases:  detritus processing
by microbes and photosynthesis.


INTRODUCTION

In recent years, understanding of the importance and structure
of detritus-based food chains has increased (see, for example,
Fenchel and Jorgensen, 1977; Tenore, 1977).  Nevertheless, detritus-
based food chains are often treated in practice as unwanted step-
children of more legitimate photosynthetic food chains.  Both Mann
(1972a) and Pomeroy (1974), realizing this neglect, have pleaded
for a more realistic view of the role of detritus and microbial processes in aquatic ecosystems.  In this article we attempt to put
the central role of detritus into perspective by briefly examining
the historical development of modern thought about the role of detritus in ecosystem structure and function as reflected by a sampling of major ecology textbooks; by considering the biogeological
history of detrital food chains; and by suggesting an interpretation
of food chain dynamics in estuarine ecosystems which we hope will
stimulate fruitful research.  Our discussion will be most relevant
to estuarine benthic systems but some of our remarks certainly apply
to aquatic systems in general, and possibly even to the terrestrial
environment.  The term "detritus" is used here in the broad sense
of Wetzel (1975) to include "all forms of organic carbon lost by
nonpredatory means from any trophic level (includes egestion, excretion, secretion, and so forth) or inputs from sources external
to the ecosystem that enter and cycle in the system (allochthonous
organic carbon)."  This definition removes the highly arbitrary
"particulate" restriction from existing definitions of detritus.
The terms "heterotroph", "decomposer", "bacteria", and "microbe"
are used interchangeably to mean "microorganisms" which require organic carbon compounds for energy metabolism, for synthesis of pro-
toplasm, or for both (i.e., osmotrophic, or microphagotrophic, chemo-
organotrophic prokaryotes, although some enkaryotes should perhaps
be included also).  The detritus food chain is any route by which
chemical energy contained within detrital organic carbon becomes
available to the biota (Wetzel, 1975).


HISTORY OF THE DETRITUS CONCEPT

Kuhn (1970) presents some attributes of what he calls scientific communities.  Members of a scientific community have undergone

similar educations and professional initiations and have absorbed the same technical literature and drawn the same lessons from it. These shared experiences include a common set of "exemplars" or practical examples of didactic value used to illustrate a concept or an approach. One might expect that an examination of textbooks, and the exemplars they contain, would reflect the attitudes of a scientific community. What follows is a brief examination of some textbooks of ecology with the aim of deducing the attitude of the community of ecologists towards the role of detritus and microbes in ecosystems.

Three periods may be recognized in the historical development of the detritus concept in ecology: prior to 1940, from 1940 to 1960, and since 1960. Shelford's (1929) textbook is representative of the early period. The fact that detritus is considered to be the principal source of nutrition for benthic animals is based on early Danish work on the distribution and feeding of larger benthic fauna. Shelford uses the paper by Blegvad (1914) as an exemplar and quotes extensively from it. There are few allusions to bacteria in the entire book and no mention of their role in animal nutrition.

Some progress occurred in understanding the role of bacteria in the 1930's. G. E. MacGinitie (1932, 1935), Clarke and Gellis (1935), and others were able to show that bacteria can transform detritus into food capable of nourishing consumers. This concept was not widely embraced by the ecology community at large, however, primarily because of technical problems associated with working with natural bacterial systems.

In the 1940's, appreciation of the central role of detritus regressed. Lindeman (1942) published his influential paper on energy flow which crystalized the concept of trophic level in ecological thinking. From the Lindemanian viewpoint, heterotrophs are seen as regenerators of inorganic nutrients, and organic matter is "mineralized" so that primary production by photosynthesis can proceed. In the ecology text of Allee *et al.* (1949), Lindeman's paper is presented as an exemplar and discussed in great detail. Exactly five discrete trophic levels are defined with heterotrophs occupying the level above carnivores as decomposers. There is no indication that detritus or bacteria are important as food, although there is some uncertainty expressed about the ability of invertebrates to utilize dissolved organic matter.

The three editions of E. P. Odum's widely used textbook provide an interesting example of the evolution of the detritus concept. Odum's first edition (1953), reflects the same Lindemanian approach as Allee *et al.* (1949). There is no discussion of bacteria or detritus in the section on energy flow and food chains. The only mention of bacteria is in relation to their roles as "reducer (decomposer) organisms" in the "mineralization" of organic matter. In the second edition (Odum, 1959), the Lindemanian concept is reinforced with a new exemplar taken from H. T. Odum (1956). This exemplar is, of course, the widely reproduced diagram of energy flow in "steady-state" communities, which further crystalizes the role of bacteria and detritus in mineralization. There are no flows in the diagram leading from the "decomposer" component to any secondary producer

box.  Nevertheless, three textbook roles are found for bacteria:
1) mineralization of organic matter; 2) production of ectocrine substances such as vitamins and antibiotics; and 3) production of food.
The third role is amplified in a discussion of "saprophytic food
chains" and a qualitative example is given in the form of fiddler
crabs exploiting marsh plant detritus.  Trophodynamic aspects are
developed briefly and there is explicit recognition that allochthonous detritus represents an energy subsidy, which compensates for
decreased efficiency of a longer pathway.  Generally the treatment
of detritus is up-to-date but cursory.

In the 1960's considerable progress in understanding the role
of bacteria and detritus in food chains was achieved, and this understanding is well reflected in Odum's third edition (1971).  Bacteria are given the same basic function as in the second edition
but the "food production" role is developed in much greater detail.
The role of bacteria in transforming detritus into nutritional particles is discussed, and overall ecosystem productivity is clearly
related to detrital food chains.

Several factors contributed to these shifts in emphasis.  The
appealing simplicity of Lindeman's concept of trophic levels tended
to focus attention on photosynthesis and primary productivity as
the base of food chains.  In addition, the introduction of two powerful techniques in the 1950's:  the use of chlorophyll to estimate
photosynthetic biomass (Richards and Thompson, 1952), and the use
of $^{14}C$ to measure primary productivity (Steeman-Nielsen, 1952)
tended to reinforce the assumed overwhelming importance of autotrophic food chains.  This effective combination of theory and technique catalyzed a rapid expansion of fruitful work on primary productivity and factors regulating its rate.  Neglect of bacteria and
detritus was concomitant with emphasis on primary production.  This
trend was not reversed until the 1960's when rapid analytical techniques were introduced for study of microbial processes.  The advent
of the $^{14}C$-glucose bioassay for microbial activity (Parsons and
Strickland, 1962) and the ATP assay (Holm-Hansen and Booth, 1966)
provided basic techniques for study of microbes nearly equivalent
to those previously developed for autotrophs, and stimulated an
immense research effort on microbes and detritus.

Uncertainty about the proper role of detritus and heterotrophic processes in ecosystems can be attributed to excessively rigid
interpretation of the trophic level concept and lack of precision
in definition of ecosystem boundaries.  If the role of microbes is
viewed principally as one of "mineralization," then the proper place
for bacteria and detritus would be off to one side of the classical
diagram.  Inputs would be organic matter from all trophic levels and
outputs would be respiration and regenerated nutrients.  In this
view, microbes are essentially energy sinks with no useful outputs
other than nutrients.  On the other hand, if the microbial role is
also one of providing food for consumer species, then their proper
level is equivalent to "primary producers."  Input is organic matter
from all trophic levels and output is particulate food for consumers
in addition to respiration and nutrients.

The degree to which microbes are energy sinks depends on the
efficiency with which systems retain detritus, bacterial efficiency

in utilizing the detritus, and the efficiency at which bacteria are grazed. These dependencies are illustrated in Figure 1 for two levels of efficiency. Initially, it is conservatively assumed that 80% of the primary productivity (P) goes to detritus and that only 20% is grazed directly (Mann, 1972b); that is, the total amount of grazed material is only 0.2(P) in the absence of a detritus food chain. If consumer species graze microbes, then for the low efficiency case, there is an added increment of 0.016(P) in the amount of material eventually eaten by consumers. If the system receives an imported subsidy of organic carbon (I), then there will be a further increment of 0.016(I) in the low efficiency case to the amount of material eventually grazed. In this low efficiency case, total grazed material would be doubled if imported carbon were to exceed *in situ* primary production by a factor of 10, as it does for the Nanaimo Estuary, British Columbia (Naiman and Sibert, 1979). For the high efficiency case, microbially processed carbon increases the total amount of grazed material by 0.22(P), which is nearly equal to the directly grazed plant material. Imported carbon produces a further increment of 0.28(I) in the high efficiency case, and has the effect of greatly stimulating total grazed material in systems where any appreciable loadings of allochthonous detritus occur.

Figure 1. The total amount of material directly grazed by consumers depends on the net efficiency of both the microbial and autotrophic food webs. In this example 20% of autochthonous primary production (P) is grazed directly by consumers (0.2 · P) and 80% enters the detrital food chain (0.8 · P). Allochthonous material (I) enters the detrital food chain directly. Grazed material is the total amount of carbon actually eaten by higher consumers. Carbon transfer through the detrital food chain is shown for two sets (low and high) of assumed, but realistic efficiencies; the high efficiency case is in parentheses. The results indicate that even for the low efficiency case the detrital food chain has the potential to significantly increase the total amount of grazed material. For the high efficiency case the total carbon consumed by higher animals is greatly enhanced.

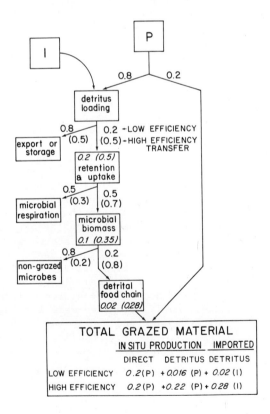

| TOTAL GRAZED MATERIAL | | |
|---|---|---|
| IN SITU PRODUCTION | | IMPORTED |
| DIRECT | DETRITUS | DETRITUS |
| LOW EFFICIENCY 0.2(P) | +0.016 (P) | + 0.02 (I) |
| HIGH EFFICIENCY 0.2(P) | +0.22 (P) | + 0.28 (I) |

Whether microbes can be considered to be "primary producers" also depends on the spatial scale of the ecosystem and the nature of its boundaries.  On a planetary scale, or in a system closed to all inputs except sunlight, microbes cannot be considered "primary producers" because they do not transform $CO_2$ into organic carbon; they merely recycle organic carbon previously fixed by photosynthesis.  However, on an ecosystem level, where the system receives substantial subsidies of organic matter from outside, microbes are effectively functioning as "primary producers"; potential food particles are conditioned and their food quality enhanced.  Whether this view is acceptable depends on whether a system, as perceived by the investigator, can receive allochthonous subsidies of organic matter, or whether areas in which the organic matter was originally produced should be considered to be part of the system.

Estuaries receive substantial imports of nutrients and organic matter from both marine and terrestrial sources.  It seems simpler to exclude the adjacent ocean and watershed from consideration of processes operating in the estuarine ecosystem, than to consider the much larger and more complex system of sea and forest.  From this point of view, material imported by the estuarine ecosystem may be considered a driving force, and it is necessary to consider the nature, quality, and fate of this allochthonous material.


ORIGIN OF ESTUARINE ECOSYSTEMS

A context for the role of detritus may be established by examining the possible origins of estuarine ecosystems.  Siever (1977) presents a useful and interesting description of early conditions of the earth.  The earth condensed out of the primordial dust cloud about 4.5 billion years ago.  After a major episode of stabilization lasting some 0.2 billion years, weathering processes began under a primitive atmosphere of water, methane, carbon dioxide, carbon monoxide, ammonia, hydrogen, hydrogen sulfide, and sulfur dioxide. These weathering processes produced erosion, sedimentation, and landforms similar to those observable on the earth today.  Along the coasts, near the mouths of rivers, there were shallow areas containing sedimentary accumulations that are characteristic of modern estuaries.  These sedimentary structures (deltas, sand bars, beaches, intertidal flats) differed from their modern counterparts only by the relatively lower content of clay-sized sediment and by the presence of flocculated silica granules (Siever, 1977).  As these early estuaries were forming, simple organic molecules were also forming and precipitating from the atmosphere (Keosian, 1968). Finally, about 3.8 billion years ago, the first organisms appeared (Schopf, 1975).

These first organisms were anaerobic, heterotrophic microorganisms depending on a continual supply of abiogenic (i.e., not formed by organisms) organic matter.  If these abiogenic organic compounds can be considered analogs of what is today defined as detritus, and we can think of no reason why they should not be, then the first food chains must have been detritus-based.

The initial diversification of life on the planet was essentially an invasion, and, as Hutchinson (1959) and others have pointed out, during such invasions, adaptive radiation occurs rather quickly. An important selective process at that time must have been, as today, the acquisition and disposition of food resources. The food supply rate (i.e., abiogenic organic compounds) to the early ecosystem is not known, but if modern microbes can be used as examples, it is safe to assume the uptake of these compounds was rapid. Heterotrophic organisms utilize carbon compounds for two functions: 1) as raw material for synthesis of new tissue, and 2) as energy sources to power both synthesis of new tissue and acquisition of more carbon. Any metabolic, morphological, or behavioral innovation developed by early heterotrophic microbes that would increase the efficiency with which organic compounds were acquired and metabolized would have conferred a selective advantage.

Many innovations are possible, including photophosphorylation, oxidative phosphorylation, and phagotrophy. A possible historical sequence of these metabolic milestones is presented in Figure 2, which is compiled from various sources (e.g., Olson, 1970; Schopf, 1975; and Schwartz and Dayhoff, 1978), as is the following discussion.

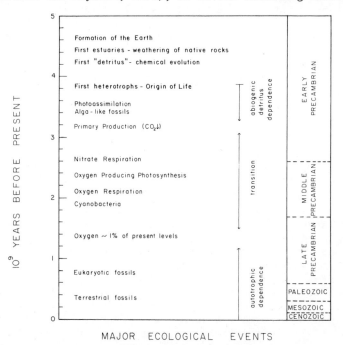

Figure 2. Possible sequence of major events in the history of ecosystems on the Earth show the early origin of detritus-based food webs.

The earliest known fossil-like structures range in age from 3.1 to 3.4 billion years but the exact biological nature of these objects is uncertain. Definite algal stromatolites have been dated as old as 3.1 billion years, setting a lower limit on the origin of photosynthesis (Schopf, 1975). For biochemical reasons, photoassimilation of organic compounds (i.e., the use of solar energy to power the uptake of organic compounds) can be considered to be the ancestral energy generating mechanism for true photosynthetic fixation of carbon dioxide (Olson, 1970). The appearance of true primary productivity about 3.2 billion years ago allowed food chains to diversify and gradually become less dependent on organic detritus. Fossils closely resembling modern cyanobacteria date from about 2.4 billion years ago, suggesting the earliest date for oxygen-producing photosynthesis. From that time, the atmosphere gradually increased in oxygen content and became fully oxidizing, as a consequence, about 1.2 billion years ago. The advent of an oxygenic atmosphere completed the gradual shift of the world ecosystem from dependence on abiogenic organic matter to dependence on photosynthesis.

The use of nitrate as an electron acceptor in respiration probably evolved from electron transport mechanisms used in photosynthesis (Olson, 1970). This form of respiration involves oxidative phosphorylation, is comparable in efficiency of energy production to modern oxygen respiration (Hadjipetrou and Stouthamer, 1965) and preceded it by a considerable period of time. The fossil record is not yet sufficient to provide any information on morphological specializations for phagotrophy. Siever (1977) has suggested that there may have been precipitated organic compounds in shallow parts of early Precambrian estuaries. Exploitation of these particulate residues would have been advantageous to any organisms capable of using them, and it is an easy step from phagotrophic uptake of organic precipitates to predation on other organisms. Both types of feeding exploit preformed, concentrated packages of food.

For a considerable fraction of the history of the earth the ecosystems were anaerobic, heterotrophic, and dependent on supplies of organic "detritus." During this rather long period of time, considerable diversification and metabolic sophistication of consumer species must have taken place. There is no reason to doubt that there were well-developed food chains involving a variety of microbial assimilators, as well as several levels of predators. Anoxic zones occur near the surface of organic particles (Jannasch, 1960; 1978) and below the oxidized surface sediments in most contemporary estuaries. In sediments, these zones support a well developed community of anaerobic organisms, including many metazoans (gastrotrichs, turbellarians, gnathostomulids, nematodes), as well as the usual collection of bacteria and ciliates (Fenchel and Riedl, 1970). *Clostridum* spp. are common in anaerobic sediments (Matches and Liston, 1973) and are considered by Schwartz and Dayhoff (1978) to be representative of an ancestral prokaryotic genotype contemporary in age with the origin of photosynthesis. The anoxic zone has been dubbed the "thiobios" by Boaden (1975, 1977), who concludes that metazoans, as we know them today, had their origins as primitive, benthic, interstitial anaerobes which exploited ready supplies of allochthonous organic matter by absorptive feeding. Contemporary anoxic ecosystems may be ideal models in which to study organization of primordial ecosystems.

THE NATURE OF ESTUARINE ECOSYSTEMS

These lines of evidence suggest that detrital food chains have been central processes in estuarine ecosystems since early Precambrian times. Contemporary studies of estuaries indicate that microbial processes are still important. From this point of view, photosynthetic production of organic matter is a relatively recent, secondary development. This statement is not meant to imply that photosynthetic processes are not important. Primary production is, by definition, the only way in which "new" organic carbon is introduced into the world ecosystem. Ecosystems have dual supports: 1) heterotrophic processing of organic detritus, and 2) photosynthetic fixation of inorganic carbon.

There is a further dimension to the dualistic aspect of estuarine ecosystems. Estuaries receive inputs from both oceanic and terrestrial sources. Estuarine circulation introduces nutrients from the ocean into the intertidal and deeper euphotic zones. Rivers introduce organic matter, sediments, and nutrients. Productivity of the estuarine ecosystem is prescribed by a number of interacting processes as illustrated in Fig. 3. Primary production increases the biomass of plant material available to consumers (the "mixed trophic level" of Odum and Heald, 1975). Microbial reprocessing ensures that detritus resulting from non-grazed plant material is also available to consumers. Imports of material from watershed and ocean stimulate both autotrophic and heterotrophic processes. Finally, an estuary retains material to an extent which is determined by its geomorphological characteristics (Mann, 1975; Naiman and Sibert, 1978). Thus total estuarine productivity is a result of balances between autotrophy and heterotrophy and between import and export. Similar balances may occur in lake ecosystems (Likens, 1972), and may be a general property of ecosystems.

CONCLUSIONS

We have tried to show how ecologists' perception of the role of detritus in the structure and function of ecosystems has changed since the turn of the century. In the 1940's and 1950's a combination of theory and technique shifted interest away from detritus-based food webs, and as a result, significant advances in our understanding of the role of primary production in food chains occurred. Work in the 1960's and 1970's, partly as a result of methodological improvements, returned our attention to detritus-based systems. It is now clear that ecosystems derive their total productivity from two sources: photosynthesis and microbial processing of non-grazed plant material.

The central role of detritus in modern ecosystems can be traced back through the history of life on Earth. The first ecosystems were undoubtedly powered by abiogenic synthesis of organic compounds under a reducing atmosphere. Possibly as the rate of consumption began to approach the rate of supply, extensive biochemical radiation occurred which included photosynthetic, respiratory, and phagotrophic processes.  Thus, abiogenic supplies became augmented and replaced by detritus derived from photosynthesis and were more efficiently exploited.  Abiogenic production of organic matter did not survive the advent of oxygen-evolving photosynthesis.  At that time, the world ecosystem became fully dependent on photosynthesis for production of organic matter.  Nevertheless, the remnants of ancestral detritus dependency persist, and the productivity of contemporary ecosystems has dual sources.

Figure 3.  Major flows of organic carbon and nutrients in estuarine ecosystems.  Emphasis is given to the nature of the ecosystem boundary, to the several inputs and to the dual support of consumer levels by both producer and decomposer levels.

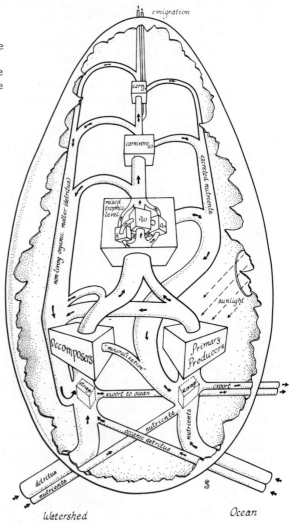

The performance of estuarine ecosystems is currently under
evaluation (e.g., Woodwell *et al.*, 1977; Valiela *et al.*, 1978), and
the 1980's will certainly be a period when properties of whole systems emerge.  We suggest that the ability of an estuarine ecosystem
to retain its production is such a property.  Recent work on estuaries has shown that there is ample detritus (both particulate and
dissolved), that the microbial communities are capable of rapidly
metabolizing it, and that estuaries may be either net importers or
net exporters.  Similar conclusions can be reached for primary producers.  Mann (1975) and Naiman and Sibert (1978) have discussed
the geomorphological features of estuaries which may contribute to
their productivity.  Tyler and Seliger (1978) have discussed dynamic features of estuaries which enable maintenance of productivity.
When interplay between geomorphology, circulation, and biological
processes is investigated from the point of view of analysing total
system performance, the basis for estuarine productivity will be
more understandable and a powerful tool will be provided for the
management of estuaries as productive systems.


ACKNOWLEDGMENTS

We thank Drs. C. D. McAllister, P. A. Breen, H. W. Jannasch,
J. E. Hobbie, and B. Peterson for their stimulating discussions of
ideas contained in this article, and for critically reading the manuscript.  Susan Kerr produced the figures.

REFERENCES

Allee, W. C. *et al.*, 1949.  Principles of Animal Ecology.  Saunders.
Blegvad, H.  1914.  Food and conditions of nourishment among the
    communities of invertebrate animals found on or in the sea bottom
    in Danish waters.  Rept. Danish Biol. Sta. 22:  41-78.
Boaden, P. J. S.  1975.  Anaerobiosis, meiofauna and early metazoan
    evolution.  Zool. Scr. 4:  21-24.
Boaden, P. J. S.  1977.  Thiobiotic facts and fancies (aspects of
    the distribution and evolution of anaerobic meiofauna).  Mikro-
    fauna Meeresboden. 61:  45-63.
Clarke, G. L., and S. S. Gellis.  1935.  The nutrition of copepods
    in relation to the food-cycle in the sea.  Biol. Bull. 68:  231-
    246.
Fenchel, T. M., and C. B. Jorgensen.  1977.  Detritus food chains
    in aquatic ecosystems:  The role of bacteria.  Advances in micro-
    bial ecology. 1:  1-58.
Fenchel, T. M., and R. J. Riedl.  1970.  The sulfide system:  a new
    biotic community underneath the oxidized layer of marine sand
    bottoms.  Mar. Biol. 7:  255-268.
Hadjipetrou, L. P., and A. H. Stouthamer.  1965.  Energy production
    during nitrate respiration by *Aerobacter aerogenes*.  J. Gen. Mi-
    crobiol. 38:  29-34.
Holm-Hansen, O., and C. R. Booth.  1966.  The measurement of adeno-
    sine triphosphate in the ocean and its ecological significance.
    Limnol. Oceanogr. 11:  510-519.

Hutchinson, G. E.   1959.   Homage to Santa Rosalia or why are there
  so many kinds of animals?   Amer. Nat. 63:   145-159.

Jannasch, H. W.   1960.   Versuche uber denitrifikation und die
  verfugbar-keit des saverstoffes in wasser und schlamm.   Arch.
  Hydrobiol. 56:   355-369.

Jannasch, H. W.   1978.   Microorganisms and their aquatic environment, pp. 17-24.   *In* W. E. Krumbein (ed.), Environmental Biogeochemistry.

Keosian, J.   1968.   The origin of life.   Reinhold, 120 pp.

Kuhn, T. S.   1970.   The structure of scientific revolutions.
  (Second edition).   University of Chicago Press.   210 pp.

Likens, G. E.   1972.   Eutrophication and aquatic ecosystems, pp. 3-
  13.   *In* Nutrients and Eutrophication.   ASLO Spec. Symp., Vol. I.

Lindeman, R. L.   1942.   The trophic-dynamic aspect of ecology.
  Ecology. 23:   399-418.

MacGinitie, G. E.   1932.   The role of bacteria as food for bottom
  animals.   Science. 76:   490.

MacGinitie, G. E.   1935.   Ecological aspects of a California marine
  estuary.   Am. Midl. Nat. 16:   629-765.

Mann, K. H.   1972a.   Introductory remarks.   Mem. Ist. Ital. Idrobiol., 29 Suppl.:   13-16.

Mann, K. H.   1972b.   Macrophyte production and detritus food chains
  in coastal waters.   Mem. Inst. Ital. Idrobiol. Suppl. 29:   353-
  384.

Mann, K. H.   1975.   Relationship between morphometry and biological
  functioning in three coastal inlets of Nova Scotia, pp. 634-644.
  *In* L. E. Cronin (ed.), Estuarine Research, V. 1.   Academic Press,
  738 pp.

Matches, J. R., and J. Liston.   1973.   Methods and techniques for
  the isolation and testing of clostridia from estuarine environments, pp. 345-362.   *In* L. H. Stevenson and R. R. Colwell (eds.),
  Estuarine Microbial Ecology.   Univ. of South Carolina Press,
  Columbia, 536 pp.

Naiman, R. J., and J. R. Sibert.   1978.   Transport of nutrients and
  carbon from the Nanaimo River to its estuary.   Limnol. Oceanogr.
  23:   1183-1193.

Naiman, R. J., and J. R. Sibert.   1979.   Detritus and juvenile
  salmon production in the Nanaimo Estuary.   III.   Importance of
  detrital carbon to the estuarine ecosystem.   J. Fish. Res. Board
  Can. 36:   504-520.

Odum, E. P.   1953.   Fundamentals of ecology.   1st ed.   Saunders,
  384 pp.

Odum, E. P.   1959.   Fundamentals of ecology.   2nd ed.   Saunders,
  546 pp.

Odum, E. P.   1971.   Fundamentals of ecology.   3rd ed. Saunders,
  574 pp.

Odum, H. T.   1956.   Primary production in flowing waters.   Limnol.
  Oceanogr. 1:   102-117.

Odum, W. E., and E. J. Heald.   1975.   The detritus-based food web
  of an estuarine mangrove community, pp. 265-286.   *In* L. E.
  Cronin (ed.), Estuarine Research, Vol. 1.   Academic Press, 738
  pp.

Olson, J. M.   1970.   The evolution of photosynthesis.   Science.
  168:   438-446.

Parsons, T. R., and J. D. H. Strickland. 1962. On the production
    of particulate organic carbon by heterotrophic processes in sea
    water. Deep-Sea Res. 8: 211-222.

Pomeroy, L. R. 1974. The ocean's food web, a changing paradigm.
    Bioscience. 24: 499-504.

Richards, F. A., and T. G. Thompson. 1952. The estimation and
    characterization by pigment analysis. II. A spectrophotometric
    method for the estimation of plankton pigments. J. Mar. Res.
    11: 156-172.

Schopf, J. W. 1975. The age of microscopic life. Endeavour. 34:
    51-58.

Schwartz, R. M., and M. O. Dayhoff. 1978. Origins of prokaryotes,
    eukaryotes, mitochondria, and chloroplasts. Science. 199: 395-
    403.

Shelford, V. E. 1929. Laboratory and field ecology. Williams and
    Wilkins, Baltimore, 608 pp.

Siever, R. 1977. Early precambrian weathering and sedimentation:
    an impressionistic view, pp. 13-23. *In* C. Ponnamperuma (ed.),
    Chemical Evolution of the Early Precambrian. Academic Press,
    221 pp.

Steeman-Nielsen, E. 1952. The use of radioactive carbon ($C^{14}$) for
    measurement of organic production in the sea. J. Cons. Int.
    Explor. Mer. 49: 309-328.

Tenore, K. R. 1977. Food chain pathways in detrital feeding ben-
    thic communities: a review with new observations on sediment
    resuspension and detrital recycling, pp. 37-53. *In* B. C. Coull
    (ed.), Ecology of Marine Benthos. Univ. of South Carolina Press,
    Columbia.

Tyler, M. A., and H. H. Seliger. 1978. Annual subsurface transport
    of a red tide dinoflagellate to its bloom area: Water circula-
    tion patterns and organisms distributions in the Chesapeake Bay.
    Limnol. Oceanogr. 23: 227-246.

Wetzel, R. G. 1975. Limnology. Saunders. 743 pp.

Valiela, I., J. M. Teal, S. Volkman, D. Shafer, and E. J. Carpenter.
    1978. Nutrient and particulate fluxes in a salt marsh ecosystem:
    Tidal exchanges and inputs by precipitation and groundwater.
    Limnol. Oceanogr. 23: 798-812.

Woodwell, G. M., D. E. Whitney, C. A. S. Hall, and R. A. Houghton.
    1977. The Flax Pond ecosystem study: Exchanges of carbon in
    water between a salt marsh and Long Island Sound. Limnol.
    Oceanogr. 22: 833-838.

# A Review of Trophic Factors Affecting Secondary Production of Deposit-Feeders

Kenneth R. Tenore
Donald L. Rice

ABSTRACT: This paper reviews factors (detrital
source and related biochemical composition, and
processes of leaching and decomposition) that
affect detrital nutritional value and availability. This paper further summarizes results of
laboratory studies on the effect of interactions
of micro, meio and macroconsumers that affect detrital availability and production of macrodetritivores. Specifically, we review factors affecting the ratio P:O, "production efficiency", i.e.,
net production of macroconsumer to detrital oxidation of the total system. This is a measure of
the amount of resource expended in food chain
transfer from detritus to macroconsumer. A model
is presented that illustrates interactions of the
detrital trophic complex.

Although many data are available on food chain dynamics of grazing (herbivore) food chains in marine systems, trophic mechanisms that might regulate production of detrital-based food chains have not been studied as thoroughly (see Wiegert and Owen, 1971). Due to our limited knowledge of trophic factors affecting food chain transfer efficiencies of deposit-feeding detritivores, there are few unifying principals to assist in understanding secondary production of detrital-based systems. Historically, geologic studies of sediment bioturbation and interest in microbial changes associated with coprophagy and plant decomposition have provided extensive information on aspects of detritus formation and reworking. Less information is available on the effects of macroconsumer bioenergetics and food chain transfer efficiencies on detrital utilization. We wish to limit our review to factors affecting the nutritional value of detritus and possible bioenergetic interactions of microbes, meiobenthos and macroconsumers. Such factors could change what we might call "production efficiency" by determining the amount of resource expended in food chain transfer of detritus to macroconsumer(s) production.

We arbitrarily recognize three phases affecting detrital utilization: 1) mechanical breakdown of particles, 2) orthochemical processes (i.e., leaching and sorption), and 3) nutritional composition, trophic efficiences, and trophic transfer. Unlike the food resources of herbivores and carnivores, "aging" profoundly affects the nutritional availability of organic detritus. These phases assume varying degrees of importance during the time when organic particles first enter the detrital pool until they are completely mineralized.

The mechanical breakdown of large recognizable particles of plant material or fecal pellets into amorphous material is a function of mechanical activity of physical processes such as action, sediment abrasion, and bioturbation by meio and macrofauna. This activity is particularly important in the early phase (days to weeks) of detrital decomposition but continues throughout the residence time of detritus in an ecosystem. The effect of particle size on detrital utilization has been reviewed previously (Levinton *et al.*, 1971; Hargrave, 1976; Rhoads *et al.*, 1977) and in this volume (Levinton, 1980; Yingst and Rhoads, 1980) and will not be considered further here.

ORTHOCHEMICAL PROCESSES ASSOCIATED WITH LEACHING AND DECOMPOSITION

Orthochemical (i.e., leaching and sorption) processes alter not only the bulk composition of detritus but also the availability of its constituents to detritivores. To date, most work on the chemistry of aging detritus has arisen from investigations of terrestrial and fresh-water systems. Recent well-balanced reviews on plant decomposition in terrestrial ecosystems (Singh and Gupta, 1977) and in fresh water (Saunders, 1976) are available. Two major symposium volumes on detritus dynamics (Melchiorri-Santolini and Hopton, 1972; Anderson and Macfadyen, 1976) have been published within the past eight years and offer a plethora of information on

various aspects of detritus chemistry in marine and non-marine environments.

The decompositional and nutritional chemistry of aging detritus
depends largely on the chemical composition of the source material.
Leachable and hydrolyzable components, intractable structural materials (e.g., lignin), nitrogen content, and phenolic residues are
implicated as important factors affecting decomposition rate
(Broadfoot and Pierre, 1939; Kaushik and Hynes, 1971; Nykvist,
1959; Waksman and Tenney, 1928; Suberkropp *et al.*, 1976; King and
Heath, 1967; Singh and Gupta, 1977). Readily soluble organic and
inorganic components are lost early in the aging process. When
these substances are present in substantial quantities, virtually
all of the mass loss in the first few days may be ascribed to simple leaching of ash (Nykvist, 1959; Crossley and Hogland, 1962;
Boyd, 1970; King and Heath, 1967; Kucera, 1959) and export of soluble organic material (Nykvist, 1959). Fresh *Zostera* leaf tissue
loses a greater proportion of its organic content during decomposition than does dead tissue, and about 80% of the organic matter lost
during aging of *Zostera* is due to simple leaching (Harrison and
Mann, 1975). Although initial losses may result from simple leaching, autolysis contributes to the pool of soluble and hydrolyzable
substances, possibly as much as 60% of the initial weight of marine
macroalgae (Khailov and Burkalova, 1969), 40% of fresh-water vascular plants (Otsuki and Wetzel, 1974), and 30% of deciduous leaf
litter (Cummins, 1974).

In general, algal detritus is more susceptible to simple leaching than is vascular plant detritus. Algae are commonly high in nitrogen, soluble organics, and ash. For example, after one hour of
leaching in sterile seawater, detritus derived from the red seaweed
*Gracilaria* and *Hypnea* lost approximately one-half of their dry mass
as soluble ash, carbohydrate, and protein while detritus derived
from standing dead leaves of *Spartina* and *Thalassia* lost only 1%
and 5-10% respectively (Rice, 1979). Leaching results in potential
loss of assimilable carbohydrates and nitrogen from the algal material to subsequent trophic levels; 50% of the protein and almost all
of the simple carbohydrates are quickly exported. The general rule-
of-thumb that high-nitrogen tissue has a rapid initial rate of decomposition seems to hold when algae and vascular plant materials
are compared, although for the most part the form of organic nitrogen (protein, lignin, chitin) should be considered. For example,
the nitrogen content of different forest leaf litter is an undepend-
able index of decomposability (Melin, 1930; Daubenmire and Prusso,
1963).

The initial composition of detritus affects both later decomposition and its nutritive value to consumers. Complex carbohydrates and intractable lignins are generally unavailable to detritivores that do not possess carbohydrases or xylases. In such
cases "aging", i.e., breakdown and transformation into available
substances over periods of weeks or months, by microbes, is necessary to improve food quality. Even after leaching, algal detritus
is processed more easily by microbes than grass or leaf detritus;
consequently, algal detritus is more quickly and easily assimilated
by macroconsumers (Tenore, 1977b, Tenore and Hanson, 1980).

Plant leaf tissue and marine macroalgae, particularly the brown seaweeds, contain considerable quantities of water-soluble and alkali- or alcohol-extractable polyhydroxy benzene derivatives commonly called polyphenols. Along with carbohydrates, polyphenols are probably the major soluble component of leaf litter (Suberkropp *et al.*, 1976). Dark condensation products of these substances from dead leaves often impart a deep amber color to lagoon and river waters. Polyphenolic exudates from brown algae may color coastal waters a dull yellow ("Gelbstoff") and create thick surface foams during turbulence. Polyphenols, like nonstructural carbohydrates and hydrolyzable proteins, are lost rapidly during the initial stage of decomposition and then more slowly as they polymerize into intractable materials. These substances, when present in high concentrations, may be antibiotic (Prakash *et al.*, 1972; McLachlan and Cragie, 1964; Conover and Sieburth, 1963; Sieburth and Conover, 1965) and toxic to marine larvae (Conover and Sieburth, 1965; Sieburth and Jensen, 1969). They readily form strong complexes with proteins and carbohydrates (Sieburth and Jensen, 1969). Phenolic compounds in leaf litter correlate inversely with the efficiency of microbial degradation (King and Heath, 1967; Satchell and Lowe, 1967) and might render plant material unpalatable (Stout *et al.*, 1976, Feeney and Bostock, 1968) or toxic (Feeney, 1970) to macroconsumers. Thus fresh detritus, even though it is high in available organic nitrogen and chemical energy, may become an available food for microbes and macrodetritivores only after antibiotic and unpalatable phenols are removed. If this is true, then from an ecological viewpoint leaching of detritus is not so much an unavoidable event in the scheme of things as it is a necessary prelude to detrital utilization.


"AGING" AND NUTRITIONAL VALUE OF DETRITUS TO MACROCONSUMERS

Historically, the aging of detritus has been linked to a build-up on detrital particles of microbial biomass that supposedly was the actual food for deposit feeders. As described in the coprophagic concept of detrital nutrition, deposit feeders reingested (their) fecal pellets on which microbes had grown. The microbes were stripped from the fecal pellets in the gut of the deposit feeder but the fecal pellet was egested essentially unchanged. These pellets could then be recolonized by microbes (Newell, 1965; Frankenberg and Smith, 1967). Because fecal pellets (and many other detrital sources) are low in nitrogen, aging is viewed essentially as "protein enrichment" by microbes (Fenchel, 1969, 1972; Mann, 1972). Aging improved nutritional quality by lowering the carbon/nitrogen ratio (Mann, 1972). This is considered especially necessary for detritus originally low in nitrogen, such as *Thalassia* (Zieman, 1975), *Zostera* (Harrison and Mann, 1975), *Juncus* and *Rhizophora* (Newell, 1973; Fell and Master, 1973; Odum and Heald, 1975; Fell *et al.*, 1975), and *Spartina* (Odum and de la Cruz, 1967). Emphasis on the concept of "enrichment" seems an over-simplification of the role of aging and microbial decomposition of detritus

and should be reviewed in light of recent information.  Unquestion-
ably, at least for nitrogen-poor detritus, microbial production
adds to the nutritional quality of detritus by increasing the organic nitrogen pool.  However, the decompositional role of microbes
(bacteria and fungi) that depolymerize complex organic materials
into caloric substrates utilizable by macroconsumers should not be
de-emphasized (Meyers, 1968; Parkinson, 1975; Tenore *et al.*, 1979).
Also, many types of detritus contain organic substances that are
themselves readily available to macroconsumers (Tenore, 1980).
"Available" algal detritus from *Gracilaria* is incorporated by *Capitella capitata* to a greater extent than structurally complex "unavailable" *Spartina* detritus (Tenore *et al.*, 1979).  Thus, energy
and organic nitrogen from detritus similar to *Gracilaria* may be
available to the detritivore directly, or from microbial biomass
*per se* or microbially processed substrate.

    Furthermore, increases of organic nitrogen with aging reported
in the literature might not be due to microbial biomass increases.
There is little doubt that aging lowers carbon/nitrogen ratios.
There is less evidence that this is the cause of a concomitant increase in nutritional value to macroconsumers.  Direct counts generally have not yielded the high numbers of bacteria necessary to
account for protein enrichment suggested by measurements of organic
nitrogen nor can microbial production account for estimated energy
requirements of macroconsumers (Christian and Wetzel, 1978; Cammen
*et al.*, 1978).  There is some evidence that nitrogen enrichment in
aging detritus may be due to the accumulation of extracellular organic nitrogen (Paerl, 1978; Sieburth, 1976; Rice, 1979; Hobbie and
Lee, 1980).  Microbes secrete exoenzymes that render otherwise intractable material assimilable to macroconsumers.  Natural phenols complex with proteins in leaf material (Davies *et al.*, 1964; Feeney
and Bostock, 1968; Feeney, 1970) and there is evidence of protein
complexation with marine algal polyphenols (Sieburth and Jensen,
1969).  Suberkropp *et al.*, (1976) suggested that part of the nitrogen build-up in leaf litter results from condensation products of
microbial proteins and plant phenolics.  In decomposition studies
of marine macrophyte detritus, Rice (1979) found that detritus derived from *Spartina* and the brown macroalga *Spatoglossum schroderi*
showed slight increases in total nitrogen and significant absolute
losses of protein over 150 days.  *Rhizophora* detritus showed significant increases of both nitrogen and protein.  The absolute
content of phenolic compounds decreased in *Spartina* and *Spatoglossum*; phenols were continuously removed throughout the aging.  On
the other hand, *Rhizophora* detritus, which quickly lost its enor-
mous amount of tannins early in the leaching stage, exhibited a constant level of apparently intractable phenols (lignins and "pseudo-
lignins") throughout the remainder of the experiment.  Rice's interpretation was that microbially produced extra-cellular proteins
combine with the highly reactive phenols and are exported from or
retained in the detritus depending upon the mobility of the phenols.

    The degree of protein-phenolic complexation should have a
direct bearing upon its availability to macroconsumers.  Although
lignous nitrogen is probably not available to detritivores (Feeney,
1970), at least some part of the bound protein is probably assimi-

lable.  If the total nitrogen content (detritus particles plus microbial) provides an adequate nitrogen source to a macroconsumer,
then caloric availability stimulated by microbial action upon the
detritus particles should limit secondary production (Tenore, 1980).
The nutritional chemistry of aging detritus is complex, and there
is every reason to believe that we must look in new directions to
solve questions posed by yesterday's answers.

## FACTORS AFFECTING DETRITAL UTILIZATION BY MACROCONSUMERS

"Observation collects whatever nature offers, whereas experimentation takes from nature whatever it requires." (Ivan Petrovich
Pavlov).

Despite the admitted drawbacks and limitations of laboratory
studies, properly-designed experiments can reduce the complexity of
the functioning of detrital systems to testable hypotheses about
specific sub-processes and permit a degree of manipulation, control of variables and measurement of factors difficult, if not impossible under field conditions.  Information thus gained can be
used to point out the possible relative importance of different regulatory processes that influence detrital availability and utilization and indicate patterns that will provide a common basis for comparing natural systems.

There is growing information gained from laboratory studies on
the effect of quantity and quality (biochemical composition) of detritus on detrital availability and utilization.  Detrital source,
amount, and state of decomposition can affect nutritional quality
and its utilization by benthic deposit-feeders (Tenore, 1977a and b;
Tenore *et al.*, 1977 and 1979; Tenore, 1980).  The population carrying capacity (K) of *Capitella* correlates with even such crude indices as ration levels of detrital organic nitrogen and "available"
caloric content.[1]  In our review of the chemistry of detrital aging,
we saw that the biochemical composition of detritus changes due to:
1) leaching of substances that inhibit microbial growth or macrobenthic feeding and 2) nutritional enrichment of detritus by either microbial biomass itself or transformation products produced by microbes.  These changes affect not only gross organic nitrogen-calor-
ic content but also relative composition of biochemical suites.
For instance, fatty acid composition changes as detritus ages
(Johnson and Calder, 1973).  Detritus ages at different rates depending on biochemical composition and resistance to decay.  Thus
parts of a detrital pool will become available to macroconsumers
at different times.  This spreads food availability for macroconsumers over time.

There are limited data on the exact role of micro and meiobenthos on detrital availability to macroconsumers.  The presence of
meiofauna increases both detrital oxidation and utilization by the
polychaetes *Nephtys incisa*, *Nereis succinea* and *Capitella capitata*
(Tenore *et al.*, 1977; Briggs *et al.*, 1979; Tenore and Tietjen, un-

---

[1]"Available caloric content" was defined arbitrarily as that
portion of calories hydrolyzed by 1N HCl for 6 hrs at *ca* 20°C.

published data), but the mechanisms of the interactions are undoc-
umented.  For instance, we know little of the effect of meio and
macro benthic grazing on the growth and production of microbial
populations, due to the lack of techniques to properly measure microbial production and activity (see Hanson, 1980).

Biotic interactions can affect the efficiency of macroconsumer
production in detrital systems at a given trophic level.  In any
trophic level there is metabolic partitioning of food assimilated
into that used for metabolic activity and net production by organ-
ism(s) (Fig. 1a).  However, when dealing with detrital-based food
chains, we must realize that a simple trophic-dynamics model that
represents a single detritivore trophic level is misleading; in reality there is a "detrital trophic complex" comprised of microbial
populations and/or meiobenthos that exhibit an array of trophisms
(i.e., nutritional sources) and feeding interactions (Fig. 1b).
For example, Hanson (1980) pointed out the danger of using data on
substrate-specific metabolic rates to represent total microbial activity on aging detritus.  The microbial community associated with
detritus is composed of populations that use many different substrates in their heterotrophic and autotrophic activity.  In addition, protozoans and meiofauna surely form a food web of complex
interactions that ultimately affect nutrition for macroconsumers.
We cannot attempt to delineate these interactions but wish to make
the point that due to the composite nature of detrital trophic dynamics, there are many sources of energy expenditure.  When summed,
these represent the total metabolic expenditure in the detrital
trophic complex.  It is the sum of these oxidations that we will
use to measure the cost of net production of detritivore(s) in the
ratio P:O[2].  The ratio P:O, i.e., net production of macroconsumer
(P) to detrital oxidation by the detrital trophic complex (O) reflects the relative amounts of the resource that are conserved and
expended in food chain transfer from detritus to macroconsumer.

Experimental studies of detrital utilization by deposit-feeding
polychaetes in laboratory microcosms show that P:O is affected by:
1) the presence of ciliates and meiobenthos; 2) detrital source and
amount and 3) length of detrital aging (Tenore *et al.*, 1977; Briggs
*et al.*, 1979; Tenore and Hanson, 1980) (Table 1).  Drawing on limited data, the addition of mixed meiofauna or only nematodes to microcosms containing the polychaetes *Nephthys incisa* and *Capitella
capitata* significantly increased the oxidation rate of detritus and
incorporation by macroconsumers; the effect on P:O varied (Table 1).
Mixed meiofauna and the nematode *Rhabdites* increased P:O; the nematode *Diplolaimella* decreased P:O, perhaps due to the high individual
(species) metabolic cost of this nematode (Tietjen, personal communication).  The P:O of microcosms containing *Nereis succinea* were
quite high (>0.50) and increased (0.75) when the ciliate *Aspidis-
ca sp* was added.  These last data should be used cautiously because

---

[2]This ratio should not be confused with the NR/NP "respiratory
coefficient" of Lindeman nor of maintenance metabolism-net production relations (see Grodzinski *et al.*, 1975).  In these situations
the measures of metabolic cost are those of the macro-organism
alone.

the microcosm oxidation values are low compared to all previous
data.  However, the difference of the P:O ratios with and without
the ciliate present gives a relative measure of change.  These few
initial data suggest a possible fruitful approach to delineating the
role of micro and meiobenthos in macroconsumer production of detrital-based food chains.

    Results of utilization rates of aging detritus derived from
different sources (periphyton, *Gracilaria*, *Spartina*), albeit based
on few detrital sources, suggest that: a) overall, the more rapidly
degradable the detritus, the higher the P:O and b) with time of
aging (at least for the time spans observed) P:O increased for all
three detritus types (Tenore and Hanson, 1980).  Concomitant changes
in microbial efficiency of detrital utilization were suggested as
the cause of the observed changes.  The nutritional value of a given
detritus, a function of both qualitative and quantitative differences, could also affect the production efficiency of detrital systems.  Some of our data indicate that in microcosms containing *Capitella*, P:O decreased from 0.15 to 0.08 when the food was tripled.
*Capitella* is an opportunistic species usually found in environments
with high food levels.  The trophic efficiences (assimilation, gross
growth, etc.) of this polychaete and changes in its grazing on bacteria might cause the observed change.

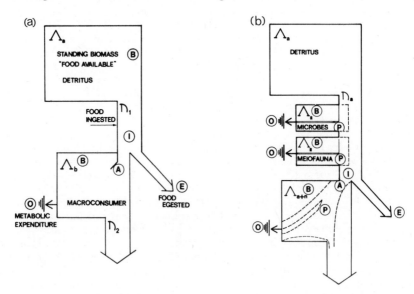

    Figure 1.  Representation of trophic dynamics
of detrital food chains:  (a) the traditional simple diagram of detritus-macroconsumer; (b) a diagram of the detrital "trophic complex", showing
intermediary role of microbes and meiofauna and
bioenergetic partitioning of resources at the macroconsumer level.

Table 1. The effect of ciliates and meiofauna on production efficiency of detritus food chains.

| Meiofauna | Oxidation of Microcosm (MG Detritus/Day/Microcosm) | | Incorporation (or Production) by Macroconsumer (MG Detritus/Total Dry Wt Worm Population) | | P:O[1] | |
|---|---|---|---|---|---|---|
| | Absent | Present | Absent | Present | Absent | Present |
| **Experimental Animals** | | | | | | |
| Mixed Meiofauna and *Nephthys Incisa* on *Zostera* Detritus[2] | 43 | 84 | 0.5 | 1.3 | 0.01 | 0.02 |
| *Rhabditis* and *Capitella capitata* on *Gracilaria* Detritus[3] | 32 | 49 | 0.5 | 1.4 | 0.02 | 0.03 |
| *Diplolaimella* and *Capitella capitata* on *Gracilaria* Detritus[4] | 46 | 79 | 2.6 | 3.0 | 0.06 | 0.04 |
| *Aspidisca* and *Nereis Succinea* on *Gracilaria* Detritus[5] | 14 | 15 | 7.1 | 11.6 | 0.52 | 0.75 |

[1] P:O is the ratio of net production (i.e. net incorporation of detritus) of the macroconsumer to total microcosm detrital oxidation.

[2] Tenore et al., 1977

[3] and [4] Tenore and Tietjen, unpublished data

[5] Briggs et al., 1979

The results-to-date in our studies of factors affecting the secondary production of macroconsumers in detrital-based food chains in microcosms are incorporated into a conceptual model (Fig. 2).

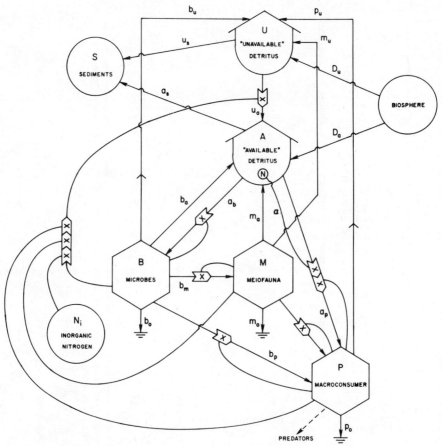

Figure 2.  A model of trophic factors affecting production of macroconsumers in detrital-based food chains.

Explanation for Figure 2.

Reservoirs

U = "unavailable" detritus
A = "available" detritus
B = microbes
M = meiofauna
P = secondary production
$N_i$ = inorganic nitrogen
S = losses from detritus ecosystem

Specific Transfer Rates

Lower case letters are specific transfer rates from one reservoir to another; subscripts indicate reservoir to which energy is transferred. For example, $u_a$ is the specific rate of transfer of unavailable detritus energy to the pool of available detritus. The lower case Greek letters ($\beta$, $\mu$, and $\pi$) below are rate constants.

$D_u$ and $D_a$ are gross rates of unavailable and available (respectively) detritus input from outside the detritus-based system.

$\alpha$ is a factor indicating whether detritus organic nitrogen or available caloric content is the limiting factor to macroconsumer production.

$$\alpha = \text{constant} \times (N) \times (\text{Avail. Calories}) \text{ when } N < N^*$$
$$\alpha = \text{constant} \times (\text{Avail Calories}) \text{ when } N \geq N^*$$

where N is nitrogen content of the detritus and $N^*$ is the critical level of nitrogen (100 mg $N \cdot m^{-2} \cdot day^{-1}$)

## Rate Equations

Unavailable pool
$$\frac{dU}{dt} = D_u + Mm_u + Bb_u + Pp_u - U(\beta BMPN_i + u_s)$$

Available pool
$$\frac{dA}{dt} = D_a + BMPN_i U + Mm_a - A(\pi_a P + a_b + a_s)$$

Microbes
$$\frac{dB}{dt} = B(Aa_b - b_a - b_o - M\mu - P\pi_b)$$

Meiofauna
$$\frac{dM}{dt} = M(B\mu - m_a - m_o - P\pi_m)$$

Secondary Production
$$\frac{dP}{dt} = P(\pi_a \alpha A + B\pi_b + Mm_m - p_u - p_o)$$

Oxidation
$$\frac{dO}{dt} = Bb_o + Mm_o + Pp_o$$

---

The pools of "unavailable" and "available" detritus are maintained by input ($D_u$, $D_a$) from the biosphere outside the detritus system as well as within the system by recycling ($P_u$, $b_u$, $M\mu$). Only "available" detritus (A) enters the macroconsumer level, either directly or indirectly as microbial biomass. Unavailable substrates (U) must be microbially processed, the specific rate of conversion to available material ($U_a$) being enhanced by inorganic nitrogen and by meio- and macrofaunal activity (the three adjacent "work gates"). Bacteria are consumed by grazing meiofauna (M) which in turn may (or may not) be consumed by the detritivore. Non-living organic material from these organisms may be returned to the detrital pool (U and A), although this model assumes that biodeposits of the macroconsumer are initially "unavailable". Because either nitrogen content or available caloric content of detritus may be the important limiting factor in secondary production, the factor $\alpha$ at low nitrogen levels ($< 100$ mg $N \cdot m^{-2} \cdot day^{-1}$) is a function of both nitrogen and available caloric content and at high nitrogen levels ($> 100$ mg $N \cdot m^{-2} \cdot day^{-1}$) is a function of available caloric content (and more than likely other nutritional requirements). Because detritus with low nitrogen content such as that derived from vascular aquatic plants is also usually low in available caloric content, the nutritional quality of such detritus is influenced both by nitrogen and available caloric content. Because the nitrogen input to coastal regions reported in the literature are $< 50$ mg $N \cdot m^{-2} \cdot day^{-1}$, this interaction of nitrogen-available caloric content probably controls the food value of detritus.

Although microbial processing increases the nutritional value of detritus, there are concomitant oxidative processes ($b_o, m_o$) which diminish the overall quantity available to the macroconsumer. If 0 is the sum of oxidative losses, the secondary production efficiency of the system (in the absence of predation) is given by:

$$SPE = \frac{\dot{P}}{\dot{P} + \dot{S} + \dot{O}} \qquad (\dot{P} = \frac{dP}{dt}; \quad \dot{O} = \frac{dO}{dt}; \quad \dot{S} = \frac{dS}{dt})$$

where $\dot{S}$ is the rate of loss (e.g. by burial) from the detritus ecosystem. Neglecting non-trophic losses for simplicity:

$$SPE = \frac{\dot{P}/\dot{O}}{\dot{P}/\dot{O} + 1} = \frac{1}{1 + \dot{O}/\dot{P}}$$

which essentially states that some portion of available energy must be expended (via microbes and meiofauna) in the production of nutritious detrital food. Easily-decomposed detritus may be more available to macroconsumers, but it is also more rapidly mineralized by the processes (B and M). Decay-resistant detritus may require months of aging before it becomes nutritionally beneficial, but long-term temporal resource partitioning adds stability to a detritus-based ecosystem. A "dynamic" P/O ratio for the model

$$\frac{dP}{dO} = \frac{P(\pi_a \alpha A + B\pi_b + M\pi_m - P_u - P_o)}{Bb_o + Mm_o + Pp_o}$$

demonstrates, for example, that there is a trade-off between P, B, and M in the numerator and denominator (between secondary production and total oxidation) and that the density of macroconsumers (P) increases the efficiency of the system. The other rate equations may be manipulated to illustrate the interrelationship of important factors such as microbial activity, "availability", and oxidation, but we leave that for another day.

ACKNOWLEDGMENTS

This research has been supported by the Oceanography Section of the National Science Foundation, Grant No. OCE78-25862.

REFERENCES

Anderson, J. M., and A. Macfadyen (eds.). 1976. The role of terrestrial and aquatic organisms in decomposition processes. The
17th Symposium of the British Ecological Society, April, 1975.
Blackwell Scientific Publ., Oxford.

Boyd, C. E. 1970. Chemical analyses of some vascular aquatic
plants. Arch. Hydrobiol. 67: 78-85.

Briggs, K. B., K. R. Tenore, and R. B. Hanson. 1979. The role of
microfauna in detrital utilization by the polychaete, *Nereis succinea* (Frey and Leuckart). J. exp. mar. Biol. Ecol. 36: 225-
234.

Broadfoot, W. M., and W. H. Pierre. 1939. Forest soil studies:
I. Relation of rate of decomposition of tree leaves to their
acid-base balance and other chemical properties. Soil Sci. 48:
329-348.

Cammen, L., P. Rublee, and J. E. Hobbie. 1978. The significance
of microbial carbon in the nutrition of the polychaete *Nereis
succinea* and other aquatic deposit feeders. UNC Sea Grant, Publication UNC-57-78-12, Raleigh, N. C.

Christian, R. R., and R. L. Wetzel. 1978. Interaction between substrate, microbes and consumers of *Spartina* detritus in estuaries,
pp. 93-144. *In* M. L. Wiley (ed.). Estuarine Interactions. Academic Press, N.C.

Conover, J. T., and J. M. Sieburth. 1963. Effect of *Sargassum* distribution on its epibiota and antibacterial activity. Bot. Mar.
6: 147-157.

Conover, J. T., and J. M. Sieburth. 1965. Effect of tannins se-
creted from Phaeophyta on planktonic animal survival in tide
pools, pp. 99-100. *In* E. G. Young and J. L. McLachlan (eds.),
Proc. 5th Int. Seaweed Symposium, Halifax, N.S. Pergamon Press,
Oxford.

Crossley, D. A., and M. P. Hogland. 1962. A litter bag method for
the study of microarthropods inhabiting leaf litter. Ecology.
43: 571-573.

Cummins, K. W. 1974. Structure and function in stream ecosystems.
Bio-Science. 24: 631-641.

Davies, R. I., C. B. Coulson, and D. A. Lewis. 1964. Polyphenols
in plant, humus and soil. III. Stabilization of gelatin by
polyphenol tanning. J. Soil. Sci. 15: 299-309.

Daubenmire, R., and D. C. Prusso. 1963. Studies on the decomposition rates of the litter. Ecology. 44: 589-592.

Feeney, P. 1970. Seasonal changes in oak leaf tannins and nutrients as a cause of spring feeding by winter moth caterpillars.
Ecology. 51: 565-581.

Feeney, P., and H. Bostock. 1968. Seasonal changes in the tannin
content of oak leaves. Phytochemistry. 7: 871-880.

Fell, J., and I. M. Master. 1973. Fungi associated with the degradation of mangrove (*Rhizophora mangle* L.) leaves in south
Florida, pp. 445-466. *In* L. H. Stevenson and R. R. Colwell
(eds.). Estuarine Microbial Ecology. Univ. of South Carolina
Press, Columbia.

Fell, J. W., R. C. Cefalu, I. M. Master, and A. S. Tallman. 1975.
Microbial activities in the mangrove (*Rhizophora mangle* L.) leaf

detrital system, pp. 661-679. *In* Proc. Int. Symp. Biol. and
Manag. of Mangroves, Hawaii, 1974.

Fenchel, T.  1969.  The ecology of marine microbenthos.  IV.  Structure and function of the benthic ecosystem, its chemical and
physical factors and the microfauna communities with special reference to ciliated protozoa.  Ophelia. 6:  1-182.

Fenchel, T.  1972.  Aspects of decomposer food chains in marine
benthos.  Verh. Deut. Zool. Gesell. 65:  14-20.

Frankenberg, D., and K. L. Smith, Jr.  1967.  Coprophagy in marine
animals.  Limnol. Oceanogr. 12:  443-450.

Hanson, R. B.  1980.  Microbial activities in trophic interaction
and detrital decomposition, pp. 347-358.  *In* K. R. Tenore and
B. C. Coull (eds.).  Marine Benthic Dynamics.  Univ. of South
Carolina Press, Columbia.

Hargrave, B. T.  1976.  The central role of invertebrate feces in
sediment decomposition, pp. 301-321.  *In* J. M. Anderson and A.
Macfadyen (eds.).  The role of terrestrial and aquatic organisms
in decomposition processes.  17th Symp. Brit. Ecol. Soc. Blackwell.

Harrison, P. D., and K. H. Mann.  1975.  Detritus formation from
eelgrass (*Zostera marina*):  The relative effects of fragmentation
leaching and decay.  Limnol. Oceanogr. 20:  924-934.

Hobbie, J. H., and C. Lee.  1980.  Microbial production of extra-
cellular material:  importance in benthic ecology.  pp. 341-346.
*In* K. R. Tenore and B. C. Coull (eds.).  Marine Benthic Dynamics.
Univ. of South Carolina Press, Columbia.

Johnson, R. W., and J. M. Calder.  1973.  Early diagenesis of fatty
acids and hydrocarbons in a salt marsh environment.  Geochin.
Cosmochum. Acta. 37:  1943-1955.

Kaushik, N. K., and H. B. N. Hynes.  1971.  The fate of dead leaves
that fall into streams.  Arch. Hydrobiol. 68:  465-515.

Khailov, K. M., and Z. P. Burkalova.  1969.  Release of dissolved
organic production to inshore communities.  Limnol. Oceanog. 14:
521-527.

King, H. G. C., and G. W. Heath.  1967.  The chemical analysis of
small samples of leaf material and the relationship between the
disappearance and composition of leaves.  Pedobiologia. 7:  192-
197.

Kucera, C. L.  1959.  Weathering characteristics of deciduous leaf
litter.  Ecology. 40:  485-487.

Levinton, J. S., G. R. Lopex, H. Heidemannlassen, and U. Rahn.
1971.  Feedback and structure in deposit-feeding marine benthic
communities, pp. 409-416.  *In* B. F. Keegan, P. O. O'Ceidigh,
P. J. S. Boaden (eds.), Biology of Benthic Organisms.  Pergamen
Press, N.Y.

Levinton, J. S.  1980.  Particle feeding by deposit-feeders:  models,
data and a prospectus.  pp. 423-439.  *In* K. R. Tenore and B. C.
Coull (eds.).  Marine Benthic Dynamics.  Univ. of South Carolina
Press, Columbia.

Mann, K. H.  1969.  The dynamics of aquatic ecosystems.  Adv. Ecol.
Res. 6:  1-31.

Mann, K. H.  1972.  Macroscopic production and detritus food chains
in coastal areas.  Mem. Ist. Ital. Hydrobiol. 29(Suppl.):  353-
382.

McLachlan, J., and J. S. Cragie.  1964.  Algal inhibition by yellow
    ultraviolet-absorbing substances from *Fucus vesiculosus*.  Can.
    J. Bot. 42:  287-292.

Melchiorri-Santolini, U., and J. W. Hopton (eds.).  1972.  Detritus
    and its role in aquatic ecosystems.  Mem. Ist. Ital. Idrobiol.
    29(Suppl.).

Melin, E.  1930.  Biological decomposition of some types of litter
    from North American forests.  Ecology. 11:  72-101.

Meyers, S. P.  1968.  Degradative activities of filamentous marine
    fungi, pp. 594-599.  *In* A. H. Walter and J. J. Elphick (eds.),
    Biodeteriation of Materials.  Microbiology and Allied Aspects.
    Proceedings of the First International Biodeterioration Sympo-
    sium.  Elsevier, New York.

Newell, R.  1965.  The role of detritus in the nutrition of two
    marine deposit feeders, the prosobranch *Hydrobia ulvae* and the
    bivalve *Macoma balthica*.  Proc. Zool. Soc. Lond. 144:  25-45.

Newell, S. Y.  1973.  Succession and role of fungi in the degrada-
    tion of mangrove seedlings, pp. 467-480.  *In* L. H. Stevenson and
    R. R. Colwell (eds.), Estuarine Microbial Ecology.  Univ. of
    South Carolina Press, Columbia, S.C.

Nykvist, N.  1959.  Leaching and decomposition of litter.  Oikos
    10:  190-224.

Odum, W. E., and E. J. Heald.  1975.  The detritus food web of the
    estuarine mangrove community, pp. 265-286.  *In* L. E. Cronin
    (ed.).  Estuarine Research, Vol. 1.  Academic Press, N.Y.

Odum, E. P., and A. A. de la Cruz.  1967.  Particulate organic de-
    tritus in a Georgia salt marsh-estuarine ecosystem, pp. 383-388.
    *In* G. H. Lauff (ed.), Estuaries.  Amer. Assoc. Adv. Sci. Publ.
    No. 83.

Otsuki, A., and R. G. Wetzel.  1974.  Release of dissolved organic
    matter by autolysis of submerged macrophyte, *Scirpus subtermi-
    nalis*.  Limnol. Oceanogr. 19:  842-845.

Paerl, H.  1978.  Microbial organic carbon recovery in aquatic eco-
    systems.  Limnol. Oceanogr. 23:  927-935.

Parkinson, P.  1975.  Terrestrial decomposition, pp. 55-60.  *In*
    Productivity of World Ecosystems.  Natl. Acad. Sci., Washington.

Prakash, A., A. Jensen, and M. A. Rashid.  1972.  Humic sub-
    stances and aquatic productivity, pp. 259-268.  *In* D. Povoledo
    and H. L. Golterman (eds.), Humic Substances.  Centre for Agri-
    cultural Publishing and Documentation, Wageningen, The Nether-
    lands.

Rhoads, D., R. C. Aller, and M. Goldhaber.  1977.  The influence of
    colonizing benthos on physical properties and chemical diagenesis
    of the estuarine seafloor.  pp. 113-138.  *In* B. Coull (ed.).
    Ecology of Marine Benthos.  Univ. of South Carolina Press,
    Columbia.

Rice, D. L.  1979.  Trace element chemistry of aging marine detritus
    derived from coastal macrophytes.  Ph.D. Thesis, Georgia Insti-
    tute of Technology, Atlanta, Georgia.

Saunders, G. W.  1976.  Decomposition in freshwater, pp. 341-373.
    *In* J. M. Anderson and A. Macfadyen (eds.).  The Role of Terres-
    trial and Aquatic Organisms in Decomposition Processes.  Black-
    well Scientific Publ., Oxford.

Satchell, J. E., and D. G. Lowe.  1967.  Selection of leaf litter

by *Lumbricus terrestris*, pp. 102-119. *In* O. Graff and J. E.
    Satchell (eds.), Progress in Soil Biology. Braunschweig, Vieweg,
    and John, Amsterdam.
Sieburth, J. M. 1976. Bacterial substrates and production in
    marine ecosystems. Ann. Rev. Ecol. Syst. 7: 259-285.
Sieburth, J. M., and J. T. Conover. 1965. *Sargassum* tannin, an
    antibiotic which retards fouling. Nature. 208: 52-53.
Sieburth, J. M., and A. Jensen. 1969. Studies on algal substances
    in the sea. II. The formation of Gelbstoff (humic material) by
    exudates of Phaeophyta. J. Exp. Mar. Biol. Ecol. 3: 275-289.
Singh, J. S., and S. R. Gupta. 1977. Plant decomposition and soil
    respiration in terrestrial ecosystems. Bot. Rev. 43: 449-528.
Stout, J. D., K. R. Tate, and L. F. Molloy. 1976. Decomposition
    processes in New Zealand soils with particular respect to rates
    and pathways of plant degradation, pp. 97-144. *In* J. M. Anderson
    and A. Macfadyen (eds.). The Role of Terrestrial and Aquatic Or-
    ganisms in Decomposition Processes. Blackwell Scientific Publ.,
    Oxford.
Suberkropp, K., G. L. Godshalk, and M. J. Kug. 1976. Changes in
    the chemical composition of leaves during processing in a wood-
    land stream. Ecology. 57: 720-727.
Tenore, K. R. 1977a. Growth of *Capitella capitata* on various
    levels of detritus derived from different sources. Limnol.
    Oceanogr. 22(5): 936-941.
Tenore, K. R. 1977b. Differential availability of aged detritus
    from different sources to the polychaete, *Capitella capitata*.
    Mar. Biol. 44: 51-55.
Tenore, K. R. 1980. The effect of organic nitrogen and caloric
    content on detrital utilization by the deposit-feeding poly-
    chaete, *Capitella capitata*. Submitted to Estuarine and Coastal
    Marine Science.
Tenore, K. R., R. B. Hanson. 1980. Availability of different de-
    tritus with aging to a polychaete macroconsumer *Capitella capi-
    tata*. Submitted to Limnology and Oceanography.
Tenore, K. R., J. H. Tietjen, and J. J. Lee. 1977. Effect of
    meiofauna on incorporation of aged eelgrass, *Zostera marina*,
    detritus, by the polychaete *Nephthys incisa*. J. Fish. Res. Bd.
    Canada. 34: 563-567.
Tenore, K. R., R. B. Hanson, B. E. Dornseif, and C. N. Wiederhold.
    1979. The effect of organic nitrogen supplement on the utiliza-
    tion of different sources of detritus. Limnol. Oceanogr. 84:
    350-355.
Waksman, S. A., and F. G. Tenney. 1928. Composition of natural
    organic materials and their decomposition in the soil. III.
    The influence of nature of plant upon the rapidity of its decom-
    position. Soil Sci. 26: 155-171.
Wiegert, R., and D. F. Owen. 1971. Trophic structure, available
    resources and population density in terrestrial vs. aquatic eco-
    systems. J. Theoret. Biol. 30: 69-81.
Yingst, J., and D. Rhoads. 1980. pp. 407-421. *In* K. R. Tenore
    and B. C. Coull (eds.). Marine Benthic Dynamics. Univ. South
    Carolina Press, Columbia.
Zieman, J. C. 1975. Quantitative and dynamic aspects of the ecol-
    ogy of turtle grass, *Thalassia testudinum*, pp. 541-562. *In* L. E.
    Cronin (ed.), Estuarine Research, V. 1, Academic Press.

# Microbial Production of Extracellular Material: Importance in Benthic Ecology

J. E. Hobbie
Cindy Lee

ABSTRACT:   It has long been assumed that microbes
make up most of the food of detritivores.  While
this is reasonable for small animals, such as
protozoans and nematodes, there is some doubt
that microbes are abundant enough in sediments to
nourish a non-selective feeder.  The hypothesis
is presented here that the extracellular mucopolysaccharides of microbes are more abundant than the
microbes themselves and provide the majority of
the food for many benthic animals.  The evidence
for this comes from studies of the attachment of
bacteria to particles, from measures of microbial
biomass and nitrogen in sediments and litter, and
from indications of the accumulation of microbial
slimes in situations where water moves past surfaces (e.g., incubation chambers in the ocean,
rocks in streams).  The source of the mucopolysaccharide could be the dissolved organic matter
of the water.

INTRODUCTION

It is generally believed that microbes make up most of the food of detritivores. When detritus is eaten by an animal, the microbes are stripped off and the bulk of the detritus passes through the animals unscathed (Fenchel, 1970). Microbes are also believed to make up most of the increase in nitrogen which occurs in detritus as it ages; this added nitrogen changes detritus from a poor to a good quality food for animals (Newell, 1965; Fenchel, 1970; Mann, 1972).

These beliefs must now be re-evaluated because the actual biomass of bacteria and fungi has recently been measured in sediments and detritus of a number of habitats. These measurements show that microbial biomass accounts for only a small fraction of the carbon necessary to feed detritivores and for only a small fraction of the nitrogen in aged detritus or sediments. Several lines of evidence lead us to suggest that extracellular particulate matter produced by microbes may make up most of the food of detritivores and may also account for the relative increase in nitrogen in detritus during decomposition.

DISCREPANCIES IN AMOUNTS OF NITROGEN AND ENERGY

Decomposition of vascular plant tissues, including those of aquatic macrophytes, results in changes in the relative and absolute amounts of carbon and nitrogen present in detritus. With increasing decay, the concentration of carbon decreases while the nitrogen concentration increases (Teal, 1962; Odum *et al.*, 1973; Kaushik and Hynes, 1971). Although some of the increase in nitrogen relative to carbon may be merely the effect of carbon being preferentially lost, under some conditions there is an absolute increase in the amount of detrital nitrogen. Traditionally, much of the increase in nitrogen was thought to be from protein-N present in microbes colonizing the detritus. It was thought that the microbial formation of protein from the relatively cellulose-rich macrophyte improves the nutritional value of the detritus to secondary consumers (Newell, 1965; Odum *et al.*, 1973; Fenchel, 1970).

Rublee *et al.* (1978) investigated the contribution of bacterial nitrogen to detritus formed by the decomposition of *Spartina alterniflora*. They found that bacterial biomass (determined by acridine orange epifluorescence) could account for only a small percentage of the detrital nitrogen (3.3%) or carbon (0.5%). They concluded that other microorganisms such as fungi, algae, or protozoans must be the source of the nitrogen if the nitrogen is contained in the microbes. Indeed evidence does point to the microbes as the source of the nitrogen. For example, in a fertilized stream microcosm, Howarth and Fisher (1976) found that the increase in detrital (maple-leaf) nitrogen was associated with an increase in microbial activity. Lee *et al.* (in press) estimated the contribution of nitrogen from both bacteria (by acridine orange direct counting) and fungi (by fungal sterol composition) to detritus formed by decomposing *S. alterniflora*. Fungal and bacterial biomass were about equal during decomposition and the contribution of nitrogen from these microorganisms was low. Helfrich and Hobbie (personal communication) found similar results by directly counting fungal biomass. Lee *et al.* (in prep.) suggested that the excess nitrogen formed may be due to the accumulation of microbial exudates rich in nitrogen, such as mucopolysaccharides or glycoproteins.

There have also been several studies on microbial contribution to the energy requirements of detritivores. Protozoans survive well by consuming bacteria in pond sediments (Fenchel, 1975). But protozoans are small enough that they can select the bacteria from the mass of detritus. Larger animals may not be as selective. Measurements of the bacterial content of the detrital food of a few animals shows that the bacterial biomass alone cannot provide the energy the animals require. For example, chironomid larvae living in the sediment of an Alaskan tundra pool could obtain only a small percentage of their energy needs from microbes (Mozley and Butler, in prep.). In another study, Cammen *et al.* (1978) concluded that microbes provided only 26-45% of the energy requirements of a *Nereis succinea* population in a salt marsh. In contrast, Wetzel (1976) believed that microbial carbon could meet the energy needs of the snail *Nassarius obsoletus*. Again, part of the difference may be due to selectivity. The worm *Nereis* is unable to preferentially select bacteria as food while *Nassarius* may feed only on surfaces which are very rich in microbes.

## MICROBIALLY MEDIATED FORMATION OF PARTICULATE ORGANIC MATTER

Microbes produce extracellular polymer fibers with which they attach to surfaces (Marshall, 1973; Costerton *et al.*, 1978). This is shown graphically in SEM pictures of bacteria and their associated fibers on a detrital particle from Lake Tahoe (Paerl, 1974). Paerl suggested the potential use of these capsular and fibrillar materials as a food source. Geesey *et al.* (1978) reported that the mucilage-producing algae on rock surfaces in streams may be as important as the bacteria in producing polysaccharide material; the mass of extracellular particulate matter may be 4-5 times greater than the living biomass (Costerton, personal communication).

Microbes appear to be capable of removing dissolved organic matter from solution and transforming it to particulate matter. This removal is particularly effective when microbes are attached to a surface and in contact with moving waters. The first hints that this removal might be occurring came from experiments in which particulate matter was formed when filtered, but not sterile, seawater was bubbled with air (Barber, 1966). The same type of particles are formed in streams, and many have bacteria associated with them (Lush and Hynes, 1973). Even better evidence for the importance of this removal process is the work of Paerl (1978) in Lake Tahoe. In these experiments particulate matter was incubated in darkness in the lake, enclosed in chambers made from dialysis membrane. After 20 days, the amount of particulate organic matter increased sixfold over sterile controls while the microbial biomass, as ATP, changed very little. Paerl concluded from these measure-

ments and from SEM observations that a small, living bacterial population could remove dissolved organic matter from the lake water
and produce a large quantity of particulate materials by extracellular secretion and cellular death.

    Mucopolysaccharides and Their Role

    Mucopolysaccharides are a complex and ill-defined group of mac-
romolecules which contain polymeric sugar and amino sugar residues.
The amino sugars, usually hexosamines, contain N-acyl, N-methyl, or
sulfamido groups.  Frequently, mucopolysaccharides are so closely
associated with proteins (and sometimes inorganic salts) that it is
difficult to separate the two.  For this reason, it has been hard
to determine whether or not there is actually covalent bonding between the polysaccharides and the protein, as is the case with glycoproteins.  Because mucopolysaccharides are so diverse in their chemical components, they are not easily characterized as a group.  A
few examples may illustrate this diversity.  The most abundant mucopolysaccharide, chitin, is composed of a linear array of β-linked
N-acetyl-D-glucosamine units and contains about 7% nitrogen by
weight.  It is found as a structural component in the cell walls of
fungi, insects, crustacea and other organisms.  Chitin is very resistant to chemical attack and is decomposed in nature only by specific microorganisms.  Another mucopolysaccharide, fucomucin, is
found in human epithelial mucus (Odin, 1958).  It contains N-acetyl-
glucosamine and galactosamine as well as the non-amino sugars,
fucose and galactose.  The amino sugars make up about half of the
sugar content, so fucomucin contains about 3-4% nitrogen.
    Mucopolysaccharides (also called glucosaminoglycans) and the
structurally related glycoproteins occur in a wide variety of polymeric forms and serve in many different biological roles.  Extracellular polymeric fiber capsules serve to aid bacteria in attach-
ing to surfaces, as discussed in Section 2.  Bacteria also synthesize extracellular polysaccharides which diffuse away from the
cells in liquid medium or form a mucoid layer around organisms
growing on an agar surface (Hepper, 1975).  The function of this
non-capsular polysaccharide is not clear.  The extracellular poly-
mers surrounding aerobic bacteria are polysaccharides which frequently contain amino sugars (Hepper, 1975; Cagle, 1975).
Costerton *et al.* (1978) and Hepper (1975) discuss the resistance
of this polymeric material to microbial enzyme attack, a factor
which suggests to us the possibility of localized accumulation in
a detrital system.  Perhaps the presence of compounds such as these
might help explain the presence of resistant amino sugars observed
in marine sediments (Mopper and Degens, 1972).  A mucus-like material thought to be a bacterial exudate has been observed in sediments of Long Island Sound and may be responsible for binding and
stabilizing the sediments (Rhoads *et al.*, 1977).
    The resistance of the polymeric material to microbial enzyme
attack does not rule out its use by animals as food.  This use
would explain many of the discrepancies in energy budgets such as
the *Nereis* and chironomid examples already mentioned.  J. J. Lee
(in press) reports on the ability of nematodes which feed on bacteria to produce mucopolysaccharide hydrolases.  Other nematodes

which do not eat bacteria did not have this enzyme. Lee suggested that the nematode enzyme may be important in the breakdown of bacterial capsules and slime layers.

In conclusion, we have presented the hypothesis that the extracellular particulate matter produced by microbes may be an important food source for detritivores. If true, the hypothesis would explain discrepancies between energy needed by detritivores and the microbial food available. It would also explain differences between the amount of nitrogen accumulating in decomposing detritus and the nitrogen actually present inside the microbes. Thus, extra-cellular particulate matter must be viewed as a potentially important food source for benthic animals.

## ACKNOWLEDGMENTS

We have benefitted from discussion with R. Howarth and appreciate his comments on this manuscript. We also thank S. Henrichs, F. Lipschultz, B. Peterson and R. Naiman. C. Lee acknowledges support from the National Science Foundation, OCE 77-26180. J. Hobbie is supported by National Science Foundation, DPP 77-23879 and DEB 76-83877. This is contribution number 4400 from the Woods Hole Oceanographic Institution.

## REFERENCES

Barber, R. T. 1966. Interaction of bubbles and bacteria in the formation of organic aggregates in seawater. Nature. 211: 257-258.

Cagle, G. C. 1975. Fine structure and distribution of extracellular polymer surrounding selected aerobic bacteria. Can. J. Microbiol. 21: 395-408.

Cammen, L., P. Rublee, and J. E. Hobbie. 1978. The significance of microbial carbon in the nutrition of the polychaete *Nereis . succinea* and other aquatic deposit feeders. UNC Sea Grant, UNC 57-78-12, Raleigh, N. C.

Costerton, J. W., G. C. Geesey, and K.-J. Cheng. 1978. How bacteria stick. Scientific American. 238: 86-95.

Fenchel, T. 1970. Studies on the decomposition of organic detritus derived from the turtle grass *Thalassia testudinum*. Limnol. Oceanogr. 15: 14-20.

Fenchel, T. 1975. The quantitative importance of the benthic microfauna of an arctic tundra pond. Hydrobiologia. 46: 445-464.

Geesey, G. G., R. Mutch, J. W. Costerton, and R. B. Green. 1978. Sessile bacteria: an important component of the microbial population in small mountain streams. Limnol. Oceanogr. 23: 1214-1223.

Hepper, C. M. 1975. Extracellular polysaccharides of soil bacteria, pp. 93-110. *In* N. Walker (ed.), Soil Microbiology. Wiley.

Howarth, R. W., and S. Fisher. 1976. Carbon, nitrogen and phosphorus dynamics during leaf decay in nutrient-enriched stream microecosystems. Freshwater Biology. 6: 221-228.

Kaushik, N. K., and H. B. N. Hynes. 1971. The fate of dead leaves that fall into streams. Arch. Hydrobiol. 68: 465-515.

Lee, C., R. W. Howarth, and B. L. Howes (in prep.). Sterols in de-
    composing *Spartina alterniflora* and use of ergosterol in esti-
    mating the contribution of fungi to detrital nitrogen. Limnol.
    Oceanogr.

Lee, J. J. (in press). A conceptual model of marine detrital de-
    composition and the organisms associated with the process. *In*
    Advances in Aquatic Microbiology, Vol. 2. Academic Press.

Lush, D., and H. B. N. Hynes. 1973. The formation of particles in
    freshwater leachates of dead leaves. Limnol. Oceanogr. 18: 968-
    977.

Mann, K. H. 1972. Macrophyte production and detritus food chains
    in coastal waters. Mem. Ist. Ital. Idrobiol. 29 (Suppl.): 353-
    383.

Marshall, K. C. 1973. Mechanism of adhesion of marine bacteria to
    surfaces, pp. 625-632. *In* R. F. Acker, B. R. Brown, J. R.
    DePalma and W. P. Iverson (eds.), Proc. 34th Intern. Congress
    Mar. Corrosion Fouling. Northwestern University Press, Evanston,
    Illinois.

Mopper, K., and E. T. Degens. 1972. Aspects of the biogeochemis-
    try of carbohydrates and proteins in aquatic environments.
    Woods Hole Oceanographic Institution Technical Report 72-68.
    105 pp.

Newell, R. 1965. The role of detritus in the nutrition of two
    marine deposit feeders, the prosobranch *Hydrobia ulvae* and the
    bivalve *Macoma baltica*. Proc. Zool. Soc., London. 144: 25-45.

Odin, L. 1958. Mucopolysaccharides of epithelial mucus, pp. 234-
    244. *In* M. O'Conner (ed.), Chemistry and Biology of Mucopoly-
    saccharides. Little, Brown and Co., Boston.

Odum, W. E., J. C. Zieman, and E. J. Heald. 1973. The importance
    of vascular plant detritus to estuaries, pp. 91-135. *In* R. H.
    Chabreck (ed.), Proceedings of the Coastal Marsh and Estuary
    Management Symposium. L.S.U. Div. of Continuing Education,
    Baton Rouge.

Paerl, H. 1974. Bacterial uptake of dissolved organic matter in
    relation to detrital aggregation in marine and freshwater sys-
    tems. Limnol. Oceanogr. 19: 966-972.

Paerl, H. 1978. Microbial organic carbon recovery in aquatic eco-
    systems. Limnol. Oceanogr. 23: 927-935.

Rhoads, D. C., R. C. Aller, and M. B. Goldhaber. 1977. The influ-
    ence of colonizing benthos on physical properties and chemical
    diagenesis of the estuarine sea floor, pp. 113-138. *In* B. C.
    Coull (ed.), Ecology of Marine Benthos. University of South
    Carolina Press, Columbia, S.C.

Rublee, R. P., L. M. Cammen, and J. E. Hobbie. 1978. Bacteria in
    a North Carolina salt marsh: standing crop and importance in
    the decomposition of *Spartina alterniflora*. UNC Sea Grant Pub-
    lication 78-11. 80 pp.

Teal, J. M. 1962. Energy flow in the salt marsh ecosystem of
    Georgia. Ecology. 43: 614-624.

Wetzel, R. L. 1976. Carbon resources of a benthic salt marsh in-
    vertebrate *Nassarius obsoletus* Bay (Mollusca: Nassariidae), pp.
    293-308. *In* M. Wiley (ed.), Estuarine Processes, v. 2. Academ-
    ic Press.

# Measuring Microbial Activity
# to Assess Detrital Decay and Utilization

Roger B. Hanson

ABSTRACT: Classical and newly developed techniques to measure metabolic activities and microbial biomass should be used with caution when
studying detrital decay and utilization. For example, glucose uptake is too process orientated
and does not assess total microbial activity. We
found that flux measurements of organic substrates
do not estimate "total" microbial activity because
of nutrient equilibrium in the cell, substrate metabolic partitioning between growth and maintenance, and microbial succession and physiological
activity during detritus degradation. If nutrient
substrates are used to evaluate metabolic activity
or growth, they should be utilized by a large proportion of the microbes and synthesized into cellular components with low turnover. In addition,
measurements of total adenylate are preferred over
adenosine triphosphate (ATP) because of the rapid
turnover of ATP and physiological control. Adenylate energy charge ratio as an index of growth
rate potential in benthic systems remains questionable.

INTRODUCTION

Many methods, employing isotope and non-isotope tracers, are available to measure specific microbial processes, but all are not suitable to investigate microbial activity in benthic systems. Most radioisotopic techniques address functional and physiological aspects of specific microbial processes that may not be correlated or tightly coupled to total growth or biomass production. Many radio-labelled compounds that are taken up by microbes undergo metabolic partitioning for growth and maintenance. Furthermore, this metabolic partitioning of the labelled material is influenced by the presence of other metabolites in the system (Crawford *et al.*, 1974). Oxygen uptake has been the most extensively used non-radiolabel technique to assess microbial activity (Smith, 1974; Pamatmat, 1975). But many systems are dominated by anaerobic processes and such measurements are thus conservative estimates of benthic microbial activity (Pamatmat, 1975, 1980). Electron transfer system (ETS) measurements (Pamatmat and Skjoldal, 1974; Christensen and Packard, 1977) have been used to correct this underestimation because it measures aerobic and anaerobic electron transport systems. However, one must use this technique cautiously in benthic systems because of chemical reduction of the tetrazolium dye and the absence of ETS in fermentation. Although other methods, such as lipid synthesis, changes in ATP, etc., are available to provide adequate analyses of microbial activities (White *et al.*, 1977; Paul and Johnson, 1977; Sieburth *et al.*, 1977), most fall short of obtaining information on "total" microbial activity, growth rate, and production.

One method of assessing microbial growth rates uses $^3$H-thymidine as a tracer for nucleic acid synthesis (Brock, 1971). However, $^3$H-thymidine uptake is not solely utilized in DNA synthesis because it can be metabolized and incorporated in other cellular components, and steps have been taken to extract and separate DNA from cells (Tobin and Anthony, 1978). Still, this method only estimates relative growth rate because thymidine pools *in situ* and *in vivo* are unknown. Another method measuring the growth state of microbes is the adenylate energy charge ratio (AEC). Chapman *et al.* (1971) developed this concept to follow the physiological state of axenic cultures, and Wiebe and Bancroft (1975) suggested that it might be used in natural systems. But the usefulness of this ratio for mixed microbial communities is questionable, and it might be restricted to the study of growth state of single populations or individual organisms.

This paper gives the results of using glucose uptake velocities and adenylate energy charge ratio measurements to estimate microbial activity in microcosms containing the polychaete, *Capitella capitata*, feeding on detritus differing qualitatively and quantitatively.

MATERIALS AND METHODS

Microcosms were layered with clean fine sand (< 0.3 μm) and detritus aged for different periods of time. *Capitella* (75 individ-

uals) were added and the microcosms sealed with airtight lids. Seawater, 0.45 μm filtered, was metered through the chambers at a rate
of 35 liters/chamber per day and the experiments were run for four
days.

We looked at changes in glucose uptake as a result of qualitative and quantitative differences in detritus. Differences in the
detritus included effects of nitrogen content, aging, and quantity:

1) Detrital Availability With Nitrogen Supplement. Different
organic nitrogen concentrations (0 to 6% nitrogen by dry wt.) were
obtained by supplementing carbon-14 labelled *Gracilaria* and *Spartina*
detritus with a casein-gelatin (6:1) mix. The results on macroconsumer utilization and total microbial biomass (total adenylates)
are given in Tenore *et al.* (1979).

2) Detrital Availability With Aging. Labelled detrital stocks
were aged in 7 liter chambers with continuous inorganic nutrient
additions ($NH_4$-N, 145 μmole/l) at a rate of 9 liters/day. Periphyton and *Gracilaria* detritus was harvested at day 5 and then at 10
day intervals; the decay resistant *Spartina* was collected at day 60
and then at 20-30 day intervals and added to microcosms containing
the polychaete *Capitella capitata*. Further details and data on worm
utilization and microbial biomass are given in Tenore and Hanson
(1980).

3) Detrital Concentration. Four concentrations of three day-
aged *Gracilaria* detritus were added to microcosms. Details of experimental design are given in Tenore *et al.* (1979).

In all these experiments the chambers were opened after four days
and sediment samples taken for microbial biomass and glucose uptake
activity. Uptake of $^3$H-glucose was measured in sediment slurries
(Hanson and Gardner, 1978; Briggs *et al.*, 1979). Glucose was measured by a modified fluoremetric technique (Hanson and Snyder,
1979). Adenylates were determined by converting AMP and ADP to ATP
enzymatically and measuring ATP with a firefly lantern tail extract
(Tenore *et al.*, 1979).


RESULTS

Effect of Nitrogen Supplement

The relative rate of glucose uptake decreased with organic nitrogen supplements of *Gracilaria* and *Spartina* detritus (Fig. 1).
However, relative "growth states" (AEC) for the most part were relatively constant at the organic nitrogen concentrations investigated
(Fig. 2). The drop in AEC ratio seen for one of the *Spartina* experiments was probably related to low water flow and the onset of
anaerobiosis. Glucose concentrations (70 to 95 $μg \cdot l^{-1}$) in the pore
water were not significantly different among the various organic nitrogen supplements or between the two detrital stocks.


Effect of Detrital Concentration

Microbial activity (glucose uptake) increased at low *Gracilaria*
detrital rations but leveled off at relatively high detrital rations

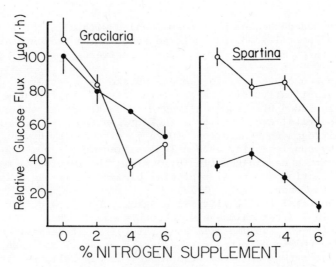

Figure 1.   Relative glucose flux associated with detritus derived from *Gracilaria foliifera* and *Spartina alterniflora* supplemented with casein-gelatin mix.   The value for duplicate experiments (0 and ●) are means of six samples ± S.E.

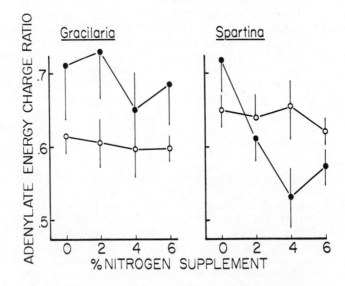

Figure 2.   Adenylate energy charge ratio of the microbial community in chambers with *Spartina* and *Gracilaria* detritus supplemented with organic nitrogen.   Replication and sample number as in Figure 1.

(Fig. 3a).   However, microbial biomass increased linearly (Fig. 3a) and the "growth state" was approximately constant over all detrital rations (Fig. 3b).   Glucose concentrations were relatively high (10-20 $\mu g \cdot l^{-1}$) at low detrital rations, but at high detrital rations glucose concentrations graduatlly decreased to less than 5 $\mu g \cdot l^{-1}$.

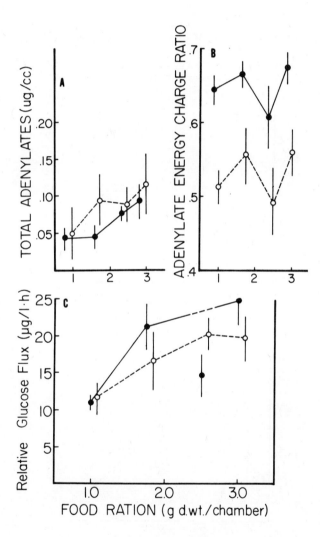

Figure 3.   Total adenylates (a), adenylate energy charge ratio (b) and relative glucose flux (c) in chambers with *Gracilaria* detritus at various rations.   Replication and sample number as in Figure 1.

Effect of Detrital Availability With Aging.

In microcosms that received detritus derived from different
sources and aged for different lengths of time, glucose uptake was
maximum after *Gracilaria* and periphyton detritus had aged for 30
days, whereas the uptake activities decreased for *Spartina* detritus
aged for 60 to 220 days (Fig. 4b). However, the growth state decreased over the same period for *Gracilaria* and periphyton detritus.
The "growth state" of microbial communities in the *Spartina* chambers
remained relatively constant over the experimental period (Fig. 5).
Glucose concentrations varied for the different ages of *Gracilaria*
and periphyton detritus, whereas, from day 60 to day 220 concentrations in the *Spartina* chambers remained at a relatively low level
(Fig. 4a).

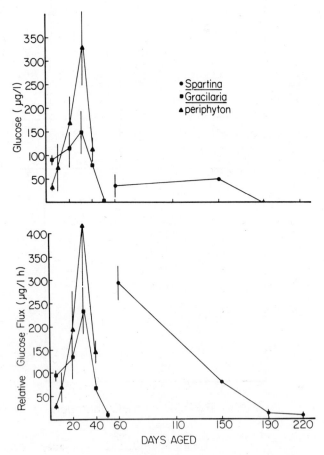

Figure 4. Glucose concentration and relative glucose flux in chambers with aged *Spartina*, *Gracilaria* and periphyton detritus. Values
are means of replicate experiments ± S.E. (six
samples per chamber).

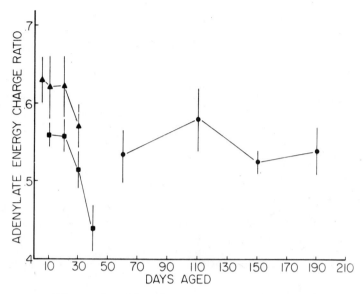

Figure 5. Adenylate energy charge ratio
in chambers with differentially aged detritus
from various sources. Replication, sample number and symbols as in Figure 4.

DISCUSSION

A great deal of the research in detrital food webs has centered
around decomposition and enrichment of the detritus by microorganisms, but greater effort is required to couple microbial activities
with macroconsumer production. The methods employed here may not
be suitable for measuring detrital trophic interactions. Microorganisms play a central role in transforming detritus into microbial
protein (either cellular or extracellular excretions) that supports
macrodetritivore growth. Yet there have been some questions as to
how much of the macroconsumer's diet depends on microorganisms associated with detritus (Hargrave, 1970a; Yingst, 1976), sequestered
organic nitrogen due to chemical adsorption (Rice, 1979), mucopolysaccharides of microbial origin (Paerl, 1974, 1975; Hobbie and Lee,
1980), and detritus *per se* (i.e., detritus devoid of microbes (Baker
and Bradman, 1976; Harrison, 1977; Tenore and Rice, 1980). Most of
the contradictions reported are related to previous preconditioning
of the detritus, i.e., leaching, particle size, age and source
(Gosselink and Kirby, 1974; Harrison and Mann, 1975; Tenore and
Hanson, 1980), feeding nature of the detritivore, and number of detritivores in the system (Hargrave, 1970b; Tenore, 1975, 1976;
Hanlon and Anderson, 1979).

Methods are available to investigate specific microbial activities but few are adequate to estimate total metabolic partitioning (growth and maintenance) of the total microbes associated with detritus. Our results on glucose flux did not coincide with other measures of metabolic activity (microbial and macroconsumer) such as $O_2$, $CO_2$, and microbial biomass measurements (Tenore and Hanson, 1980). In addition, the net incorporation rate of a labelled detrital-complex by *Capitella* correlated with microbial biomass (total adenylates) and microcosm oxygen consumption rate, and not glucose uptake.

Microbiologists have used glucose uptake velocities to measure microbial activities with the understanding that only a fraction of the microbes may utilize glucose. Biologists untrained in microbiology should be made aware that the uptake velocities of glucose, as well as other organics, should not be used with the intent of measuring total microbial activities during detrital aging or to determine the effect of detritivores on total microbial activities. Radiolabelled glucose flux measurements assess a functional attribute of only a part of total metabolic activity. Therefore, microbial activity measurements using radiolabelled organics should be restricted to studies of nutrient cycling for the following reasons. Nutrient assimilation may not be closely coupled to growth processes because of external and internal fluctuations in nutrient pools, and the equilibrium kinetics of the label within the cell (i.e., the allocation of materials to cell maintenance, energy production, and biosynthesis of cellular components for growth). (Hanson and Wiebe, 1977; Button, 1978). Furthermore, nutrient flux measurements in decomposition studies are complicated by the temporal succession of microbes utilizing and producing the nutrients. For example, depending on the source of detritus, the end products of decomposition (e.g., glucose) may not be released at a constant rate (see Fig. 4). Therefore, without knowledge of the natural substrate concentration, specific activity of the added substrate may be drastically altered and the level of activity of the microbes misinterpreted.

Adenylate energy charge (AEC) ratio is frequently calculated from total adenylate (ATP, ADP and AMP) concentrations with little understanding as to the meaning or relevance of the ratio. In benthic systems, the meaning or relevance is almost certainly misleading as to growth state (cell, population, community, or system). For example, soil microbes possess a vast number of metabolic processes that are regulated by many edaphic and biotic factors. In addition, nutrient concentration and juxtaposition of metabolic processes dictate whether certain processes operate near optimum rates. AEC ratio is influenced by the proportion of enzymes involved in ATP regeneration and ATP utilization at all levels of organization (i.e., cell, population, etc.). Thus, it could be misleading to estimate by AEC ratio the growth state of the microbial community in a natural system based on the relation of AEC ratio to growth state of laboratory cultures. AEC ratio has limited application to natural systems. Because the ratio $[\text{ATP} + \frac{1}{2} \text{ADP} \div (\text{ATP} + \text{ADP} + \text{AMP})]$ is written in terms of chemical energy, high-energy anhydride phosphate bonds (Atkinson, 1977), it provides a suitable measure of the energy stored or available for work in the system. In other words, when the ratio is relatively high (>0.6), it is reasonable to be-lieve that the system is charged to provide energy for work and metabolically active.  Therefore, the AEC ratio is an instantaneous measure of the stored chemical energy and metabolic potential and not "growth state" *per se* of the system.

Unlike ATP, which has an established ATP:carbon ratio of 250:1, total adenylate has no such factor.  However, we choose to measure total adenylate instead of ATP as an indicator of microbial biomass.  Theoretically, ATP concentrations may fluctuate whereas the adenylate concentrations remain relatively constant and only change with an increase or decrease in microbial biomass.  Even though ATP is generally much greater than either ADP or AMP in the system, at times ADP and AMP concentrations are substantial.  There-fore, in decomposition and grazing studies, estimates of ADP and AMP concentrations may be important and may further facilitate a better understanding of energy allocation in the system.

Oxygen consumption is used most frequently to measure micro-bial activity in microcosms (Hanlon and Anderson, 1979).  However, benthic production is still difficult to determine from such mea-surements.  For example, the "respiratory quotient" (RQ) that many investigators use may vary substantially.  We have found "RQ" values ranging from 0.55:1 to 60:1 (Tenore and Hanson, 1980).  These re-sults suggest that a more direct measure of microbial growth is re-quired.  A technique that appears promising is DNA synthesis (Tobin and Anthony, 1978), even though a few problems have not been re-solved (e.g., thymidine concentrations within the cell).

In summary, this paper is not meant for most microbial ecolo-gists but for the ecologist interested in incorporating microbiolog-ical techniques in his studies of decomposition rates and utiliza-tion of detritus.  Workers in these allied fields, therefore, should consider adopting the least specific measurement to determine total microbial activity or productivity.  In this regard, we recommend oxygen consumption to monitor oxidative metabolism, total adenylates to quantitate microbial biomass, and possibly DNA synthesis to esti-mate production rates.

## ACKNOWLEDGMENTS

This research was supported by the National Science Foundation, Grant Number OCE77-19944 and done in collaboration with Dr. Kenneth R. Tenore (NSF Grant OCE7825862).  We thank Donna Paiva for typing and Dan McIntosh for drafting.

## REFERENCES

Atkinson, D. E.  1977.  Cellular energy metabolism and its regula-tion.  Academic Press, New York, 312 pp.

Baker, J. H., and L. A. Bradman.  1976.  The role of bacteria in the nutrition of aquatic detritivores.  Oecologia. 24:  95-104.

Briggs, K. B., K. R. Tenore, and R. B. Hanson.  1979.  The role of microfauna in detrital utilization by the polychaete *Nereis suc-*

*cinea* (Frey and Lueckart). J. exp. mar. Biol. Ecol. 36: 225–234.

Brock, T. D. 1971. Microbial growth rate in nature. Bact. Rev. 35: 39–58.

Button, D. K. 1978. On the theory of control of microbial growth kinetics by limiting nutrient concentrations. Deep-Sea Res. 25: 1163–1177.

Chapman, A. G., L. Fall, and D. E. Atkinson. 1971. Adenylate energy charge in *Escherichia coli* during growth and starvation. J. Bacteriol. 108: 1072–1086.

Christensen, J. P., and T. T. Packard. 1977. Sediment metabolism from the northwest African upwelling system. Deep-Sea Res. 24: 331–343.

Crawford, C. C., J. E. Hobbie, and K. L. Webb. 1974. The utilization of dissolved free amino acids by estuarine microorganisms. Ecology. 55: 551–563.

Gosselink, J. G., and C. J. Kirby. 1974. Decomposition of salt marsh grass *Spartina alterniflora* Loisel. Limnol. Oceanogr. 19: 825–832.

Hanlon, R. D. G., and J. M. Anderson. 1979. The effects of *Collembola* grazing on microbial activity in decomposing leaf litter. Oecologia. 38: 93–99.

Hanson, R. B., and W. J. Wiebe. 1977. Heterotrophic activity associated with particulate size fractions in a *Spartina alterniflora* salt marsh estuary, Sapelo Island, Georgia, and the continental shelf waters. Mar. Biol. 42: 321–330.

Hanson, R. B., and W. S. Gardner. 1978. Uptake and metabolism of two amino acids by anaerobic microorganisms in four diverse salt-marsh soils. Mar. Biol. 46: 101–107.

Hanson, R. B., and J. Snyder. 1979. Enzymatic determination of glucose in marine environments. Improvement and note of caution. Mar. Chem. 7: 353–362.

Hargrave, B. T. 1970a. The utilization of benthic microflora by *Hyalella azteca* (Amphipoda). J. Anim. Ecol. 39: 427–437.

Hargrave, B. T. 1970b. The effect of a deposit-feeding amphipod on the metabolism of benthic microflora. Limnol. Oceanogr. 15: 21–30.

Harrison, P. G. 1977. Decomposition of macrophyte detritus in seawater: effects of grazing by amphipods. Oikos. 28: 165–169.

Harrison, P. G., and K. H. Mann. 1975. Detritus formation from eelgrass (*Zostera marina* L.): the relative effects of fragmentation, leaching, and decay. Limnol. Oceanogr. 20: 924–934.

Hobbie, J. E., and C. Lee. 1980. Microbial production of extracellular material: Importance in benthic ecology. pp. 341–346. *In* K. R. Tenore and B. C. Coull (eds.), Marine Benthic Dynamics. Univ. of South Carolina Press, Columbia.

Paerl, H. W. 1974. Bacterial uptake of dissolved organic matter in relation to detrital aggregation in marine and freshwater systems. Limnol. Oceanogr. 19: 966–972.

Paerl, H. W. 1975. Microbial attachment to particles in marine and freshwater ecosystems. Microbial Ecol. 2: 73–83.

Pamatmat, M. M. 1975. *In situ* metabolism of benthic communities. Cah. Biol. Mar. 16: 613–633.

Pamatmat, M. M. 1980. Facultative anaerobiosis of benthos. *In* K. R. Tenore and B. Coull (eds.), Marine Benthic Dynamics. University of South Carolina Press, Columbia.

Pamatmat, M., and H. R. Skjoldal. 1974. Dehydrogenase activity and adenosine triphosphate concentration of marine sediments in Lindaspollene, Norway. Sarsia. 56: 1-11.

Paul, E. A., and R. L. Johnson. 1977. Microscopic counting and adenosine 5'-triphosphate measurement in determining microbial growth in soils. Appl. Environ. Microbiol. 34: 263-269.

Rice, D. L. 1979. The trace metal chemistry of aging marine detritus derived from coastal macrophytes. Ph.D. Dissertation, 144 pp. Georgia Inst. of Technology.

Sieburth, J. McN., K. M. Johnson, C. M. Burney, and D. M. Lavoie. 1977. Estimation of *in situ* rates of heterotrophy using diurnal changes in dissolved organic matter and growth rates of pico-plankton in diffusion cultures. Helgol. wiss. Meeresunt. 30: 565-574.

Smith, K., Jr. 1974. Oxygen demands of San Diego Trough sediments: an *in situ* study. Limnol. Oceanogr. 19: 939-944.

Tenore, K. R. 1975. Detrital utilization by the polychaete, *Capitella capitata*. J. Mar. Res. 33: 261-274.

Tenore, K. R. 1976. Food chain pathways in detrital feeding benthic communities: a review, with new observations on sediment resuspension and detrital recycling. pp. 37-53. *In* B. C. Coull (ed.), Ecology of Marine Benthos, Univ. of South Carolina Press, Columbia.

Tenore, K. R., and R. B. Hanson. 1980. Temporal resource partitioning: differential availability of aging detritus derived from various sources. Limnol. Oceanogr. (in press).

Tenore, K. R., and D. L. Rice. 1980. A review of trophic factors affecting secondary production of deposit feeders. *In* K. R. Tenore and B. Coull (eds.), Marine Benthic Dynamics. University of South Carolina Press, Columbia.

Tenore, K. R., R. B. Hanson, B. Dornseif, and C. Wiederhold. 1979. The effect of organic nitrogen supplement on the utilization of different sources of detritus. Limnol. Oceanogr. 24: 350-355.

Tobin, R. S., and D. H. J. Anthony. 1978. Tritiated thymidine incorporation as a measure of microbial activity in lake sediments. Limnol. Oceanogr. 23: 161-165.

Wiebe, W. J., and K. Bancroft. 1975. Use of the adenylate energy charge ratio to measure growth state of natural microbial communities. Proc. Nat. Acad. Sci. USA. 72: 2112-2115.

White, D. C., R. J. Bobbie, S. J. Morrison, D. K. Oosterhof, C. W. Taylor, D. A. Meeter. 1977. Determination of microbial activity of estuarine detritus by relative rates of lipid biosynthesis. Limnol. Oceanogr. 22: 1089-1099.

Yingst, J. Y. 1976. The utilization of organic matter in shallow marine sediments by an epibenthic deposit-feeding Holothurian. J. exp. mar. Biol. Ecol. 23: 55-69.

# Laboratory Model of the Potential Role of Fungi in the Decomposition of Red Mangrove (Rhizophora mangle) Leaf Litter

Jack W. Fell
I. M. Master
S. Y. Newell

ABSTRACT: Field studies of quantitative changes in carbon, nitrogen and dry weight of decomposing mangrove leaves at different stations and seasons demonstrated similar net weight and carbon losses but varying changes (0 to 78% increase) in nitrogen content. A laboratory model examined the field responses. This model consisted of 15 o/oo seawater, 161 µg-at P/L and senescent mangrove leaves. Testing included an inoculum of the fungus *Phytophthora*, $HgCl_2$ sterilized controls and levels of inorganic nitrogen, $(NH_4)_2 SO_4$, from 25 to 18,000 µg-at N/L. Because of variation between leaves, a split leaf technique was developed in which one half of the leaf was used to determine the original carbon, nitrogen and dry weight of the other half that was subjected to experimental decay. Nitrogen immobilization required the presence of microorganisms, such as *Phytophthora*, and the addition of an exogenous nitrogen source with optimal levels in excess of 400 µg-at N/L. The final amount of nitrogen in the leaf was directly related to the original

amount; however, change in nitrogen was inversely related to original nitrogen.  Field nitrogen
responses could be attributed to this inverse relationship in conjunction with the level of available nitrogen.  Carbon and dry weight losses increased with the presence of the fungus; neither
carbon nor dry weight were affected by addition of
nitrogen.


INTRODUCTION

   Nitrogen immobilization occurs during the decomposition of
plant litter in forests (Gosz *et al.*, 1973) and freshwater systems
(Boyd, 1971; Hunter, 1976; Iverson, 1973) as well as estuarine and
marine habitats (Odum *et al.*, 1973; Hunter, 1976).  This nutrient
enrichment is considered to be mediated by microbes (Thayer *et al.*,
1977; Gosselink and Kirby, 1974) and to result in a primary food
source for detritivores that in turn support commercial species of
fish and crustaceans (Fenchel, 1977; Bärlocher and Kendrick, 1975;
Tenore *et al.*, 1977).  Fungi are associated with leaf litter
(Hudson, 1971) and are presumed to be one of the primary agents of
decomposition and enrichment (Willoughby, 1974; Bamforth, 1977).
Lower fungi, such as members of the genus *Phytophthora* (Fuller and
Poyton, 1964; Höhnk, 1956; Siepman, 1959) occur in marine environments, particularly in association with decomposing plant litter
(Anastasiou and Churchland, 1969; Fell and Master, 1975), although
the role of these fungi has not been determined.  In a laboratory
study we assessed the role of fungi and inorganic nitrogen in caus-
ing the changes in nitrogen immobilization, and carbon and dry
weight losses of field decomposed red mangrove (*Rhizophora mangle*)
litter.


METHODS AND MATERIALS

   Field Studies

   The decomposition of *R. mangle* leaves was studied at two loca-
tions:  one in a fringe mangrove forest along the west coast of Card
Sound (CS) which is 45 km south of Miami, and the other in the
Fahkahatchee Strand (Fahk), which is in Collier County on Florida's
southwest coast.  The Fahk station differed from the station at Card
Sound by the presence of more wave action.  There were two sampling
periods at Card Sound:  winter (WCS) Nov. 1973-March 1974, and summer (SCS) July-Oct. 1974, and one sampling period in Fahk:  summer,
June-Aug. 1974.
   Yellow senescent leaves with a nearly complete abscission layer
were collected and put in nylon 1 m x 30 cm No. 5 mesh bags with 50
leaves/bag, a limit which prevented excessive packing.  Weights were
pinned to the bags to keep them in place on the sediment, and the
bags secured to the roots of mangrove trees.  The bags of leaves were
submerged at the sampling site on the day of collection.  Collections

of 30 leaves were made at CS after each of the periods of 0,2,4,6
and 15 weeks. At Fahk, 25 leaves were collected at each of 0,2,4
and 6 weeks. The studies were terminated (after 6 weeks at Fahk
and 15 weeks at CS) when the leaves reached a point of near total
disintegration, such that laboratory processing could not be used.
One of the purposes of the program was to examine the carbon and nitrogen changes that take place within the leaf due to fungal activities. Generally in leaf litter decomposition processes, fungi penetrate the internal layers of the leaves, whereas the bacteria are
confined to the leaf surfaces (Harley, 1971). There is a complex
community that inhabits the leaf surface, including bacteria, meiofauna and diatoms; within this matrix there is an accumulation of
detrital particles. This surface layer was removed by gently rubbing the leaves by hand under running tap water. A 21 mm diameter
disc was cut with a cork borer from the center of the leaf with the
mid-rib as the center line. The remainder of the leaves and discs
were dried to a constant weight, both of which were recorded. Details of the field study will be presented by Newell *et al.* (ms, in
prep.)


Experimental Procedures

Leaf decomposition was examined in the laboratory in shaken
flasks with 15 o/oo seawater that was changed at 24 h intervals to
eliminate buildup of potentially toxic compounds, such as tannins
that leach from the leaves, and to maintain added inorganic nitrogen. Senescent leaves were picked immediately before each experiment. For identification purposes, the leaves were selected so that
each flask contained a similar set of five different sizes of leaves.
The leaves were measured, split down the midrib and each half
weighed; reference halves (left) were oven dried at $104^{\circ}C$ for three
days and reweighed. Experimental halves (right) were sealed in
plastic bags and UV sterilized by placing the bags at a distance of
10-15 cm from short wave UV lamps (254 nm) at $4^{\circ}C$ for 48 hours on
each side of the leaf. We selected UV sterilization because the
high heat of autoclaving altered the organic compounds of the
leaves. This was evident from the altered color of the leaves and
the increase in rate of leaching of solubles from the leaves after
autoclaving. The effect of UV treatment appeared to be limited to
the surface of the leaf and was particularly effective in the elimination of the bacterial flora which inhabits the leaf surfaces.
The efficiency of the treatment was tested by cutting 40 leaf discs
from UV treated leaves and placing the discs in individual test
tubes with 5.0 ml of seawater containing 0.1% yeast extract and 1%
glucose. After one week of incubation at room temperature, bacteria were not detected in any of the tubes. However, after 48 h
fungi grew out of the internal layers of the leaf discs.
One liter flasks contained Gulf Stream seawater diluted to 15
o/oo and filter-sterilized (0.45 µm Millepore; 500 ml/flask); additions included 161 µg-at P/L ($KH_2PO_4$) and nitrogen as $(NH_4)_2SO_4$
at levels of 25, 75, 125, 133, 400, 1200, 1428, 3600, 7139, 10709,

14279 and 17849 µg-at N/L. Under natural estuarine conditions, there are a variety of organic and inorganic nitrogen compounds in low concentrations in the water that are available to the fungi. The lab model was not designed to reproduce this nitrogen complex, but rather to determine the quantity of nitrogen required for nitrogen immobilization in the field. Therefore the most energetically efficient nitrogen source was used. Controls included no nitrogen (O-N), O-N + $HgCl_2$ at 1:10,000 w/v, and 1428 µg-at N + $HgCl_2$. Gulf Stream water contained 2.0 µg-at particulate organic N/L and 0.5 µg-at $NO_3$ - N/L (Roman, pers. comm.). Each flask contained five leaf-halves, and there were four replicate flasks/treatment. Although mangrove sediments are basically anaerobic, the surface of the sediment, containing the leaf litter, is aerobic with a continuous flow of water due to tidal action. To maintain an aerobic condition, the flasks were placed on a reciprocating shaker (58 cycles/min at an amplitude of 5 cm) at $25^{\circ}C$. Interest was in the initial 1-3 weeks of decomposition; therefore, *Phytophthora* spp. fungi which are prevalent during the early stages of decomposition (Fell and Master, 1975) were used as the inocula. Due to the different growth requirements of the fungi, the inocula consisted of *Phytophthora spinosa* in agar blocks, and *P. vesicula* on sesame seeds (Fell and Master, 1975). A heavy growth of *Phytophthora* was observed on the leaves. However, it is presumed that some of the indigenous fungi were also active. Time period for the tests was 10 days. At termination, the leaves were washed under running tap water by gently rubbing the fungal-detrital layer off by hand. Leaves were dried at $104^{\circ}C$ for three days and reweighed.

## Analytical Procedures

Fresh and dry weights were measured to the nearest milligram. Nitrogen and carbon percentages of leaves were obtained with a Perkin-Elmer Model 240 Elemental Analyzer after drying ($104^{\circ}C$) and grinding in a Wiley mill (No. 60 mesh sieve). Surface areas of leaves were measured with a Hayashi Denko AAM-5 area meter.

## Statistical Procedures

The original fresh weight, dry weight, C and N contents of mangrove leaves were variable; for example, in one collection (n = 35), DW:FW ratio of leaves ranged from 0.16 to 0.45. Therefore, the split leaf technique was used in which the original levels in reference halves of the leaves were used to base changes in the experimental halves. To determine the reliability of the technique, both halves of individual leaves were compared for C:N, C%, N%, and DW:FW. In all instances linear regressions demonstrated a high degree of correlation between leaf halves for each test factor (n = 35, C:N sample correlation coefficient (r) = 0.96, F = 373, slope (b) = 1.01; C% r = 0.94, F = 281, b = 0.71; N% r = 0.95, F = 339, b = 0.95; DW:FW r = 0.95, F = 345, b = 0.79).
The data analyzed included the original variables:  fresh and

dry weight, surface area, carbon and nitrogen content (%) of the left and right halves of the leaves. The information generated included:

1) predicted dry weight (PDW) of the experimental right half (R) prior to decomposition was calculated from the dry weight to fresh weight ratio (DW:FW) of the reference left half (L) multiplied by the fresh weight (FW) of the right half: $PDW\text{-}R = \dfrac{DW\text{-}L}{FW\text{-}L} \times FW\text{-}R$

2) weight loss (%WT LS) of the right half is the difference between the predicted and final dry weights expressed as a percentage:

$$\%WT\ LS = \frac{(PDW\text{-}R) - (DW\text{-}R)}{(PDW\text{-}R)} \times 100$$

3) to avoid relative increases in N% due to weight loss, the nitrogen percentages are based on the quantity of nitrogen in grams per gram dry weight of the senescent leaves before decomposition:

$$Original\ N\% = \frac{N\text{-}L}{DW\text{-}L} \times 100$$

$$Final\ N\% = \frac{N\text{-}R}{PDW\text{-}R} \times 100$$

4) final carbon (%): derived in the same manner as final N%.
5) change in nitrogen ($\Delta$N%) from the original expressed as a percentage:

$$\Delta N\% = \frac{(Final\ N) - (Orig.\ N)}{Orig.\ N} \times 100$$

$$= \frac{(N\%\text{-}R)\ (DW\text{-}R) - (N\%\text{-}L)\ (PDW\text{-}R)}{(N\%\text{-}L)\ (PDW\text{-}R)} \times 100$$

6) change in carbon ($\Delta$C%): derived in the same manner as $\Delta$N%.
7) change in C:N ($\Delta$C:N): derived in the same manner as $\Delta$N% and $\Delta$C%, also expressed as percent change.

Key parameters (%WT LS, $\Delta$N%, $\Delta$C%, final N%, final C%, $\Delta$C:N) were tested for correlation with the original variables by stepwise multiple regression. Significant correlations resulted in the use of that original variable as a covariate in analysis of covariance (Nie *et al.*, 1975). Analysis of covariance (ANCOVA), utilizing the appropriate covariates, was used on the key parameters to ascertain the effect of N treatment levels. The Newman-Keuls multiple range test (Zar, 1974) was performed on the ANCOVA adjusted means to determine significant differences. Regressions were tested, in a stepwise procedure, with linear and polynomial models (Zar, 1974) to the third power.

RESULTS

Field Studies

Leaves at the two stations decomposed at different rates (Figs. 1–3); Fahk leaves degraded most rapidly due to wave activity. Dry weight loss at Fahk was 60% in six weeks (Fig. 1), as compared to 37% at summer Card Sound (SCS) and 45% at winter Card (WCS) at six weeks. While rates of dry weight loss differed, total dry weight losses at 15 weeks (60% Fahk, 58% SCS, 59% WCS) were approximately the same. Carbon losses at 15 weeks (57% Fahk, 50% SCS, 51% WCS) reflected weight losses.

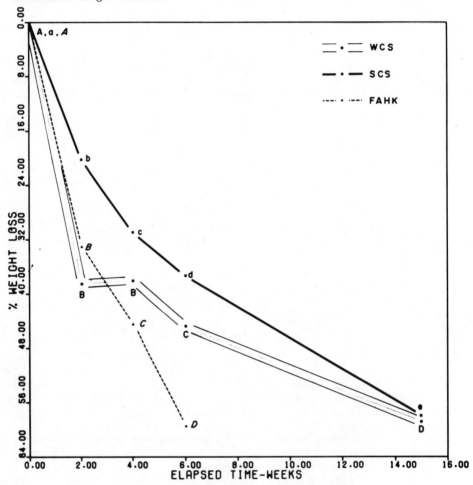

Figure 1. Weight loss as a percentage of original dry weight.

Explanation for Figures 1–3: Changes in *Rhizophora mangle* leaves in litter bags. WCS = Winter Card Sound; SCS = Summer Card Sound; FAHK = Fahkahatchee Strand. Points with different letters on the same line are significantly different at the 0.05 α level.

ΔN% (Fig. 2) showed more variation than the other factors measured. The amount of nitrogen in the senescent leaves (Fig. 3, elapsed time = 0) was the same for both CS collections (Fig. 3) while the quantity of nitrogen in the Fahk leaves was lower. During the 15 weeks at WCS, ΔN% (Fig. 2) did not show a consistent pattern of change. There was a loss during the first four weeks with a 14% increase between the fourth and sixth weeks followed by another loss. In contrast, SCS ΔN% values moved consistently upward, gaining 33% by week 15. The Fahk leaves had a ΔN of 78% after four weeks, reaching the same final N% (Fig. 3) as the SCS leaves.

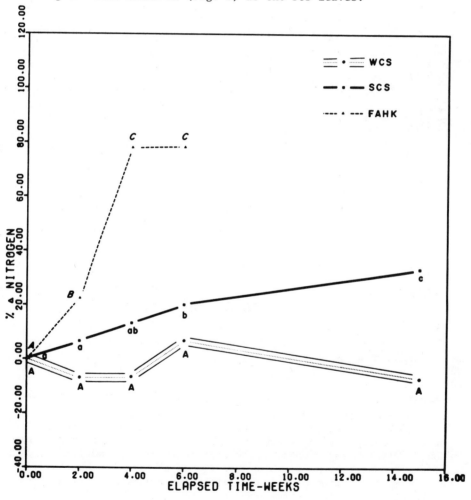

Figure 2. Change in nitrogen content as a percentage of original nitrogen.

Laboratory Model

The laboratory model disclosed several characteristics of mangrove leaf litter decomposition: the effect of fungi on carbon and weight losses and N immobilization, the requirement for exogenous

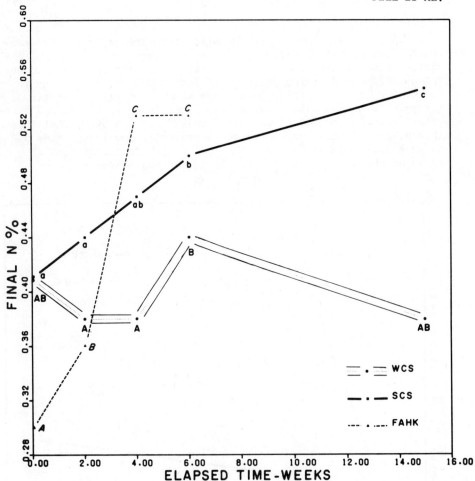

Figure 3. Nitrogen content as a percentage of original dry weight.

dissolved N, the relationship of original N level of a leaf to ΔN%, and the relationship of DW:FW to weight loss.

Weight and carbon losses (Table 1) were significantly less in sterile controls than treatments with fungi. The increased carbon loss indicated that fungal activities contribute an additional 3.0% of the C outflow from leaching (or 15% of the total C outflow). A considerable portion of this fungal carbon outflow is probably due to respiration. In the absence of fungi, ΔN% did not increase and C:N remained high.

The addition of inorganic nitrogen to the system did not affect % weight loss or ΔC%, but increased levels of inorganic nitrogen are directly related to ΔN% and inversely related to ΔC:N. The lack of a statistically demonstrable effect of added inorganic nitrogen on weight and C losses may be a result of the incubation period (10 days), as direct effects of nitrogen additions to weight loss have been reported for 5-week decomposition studies of leaves in a fresh water stream (Howarth and Fisher, 1976).

Table 1. Effects of dissolved nitrogen concentration and fungal presence on key changes during decomposition of *Rhizophora mangle* leaves inoculated with *Phytophthora* in a laboratory flask experiment. Nitrogen as $(NH_4)_2 SO_4$; Hg = sterile controls with $HgCl_2$ at 368 µg-at Hg/L. Bars indicate treatments that are not significantly ($\alpha = 0.05$) different. n = 350 for each figure.

| | | $(NH_4)_2SO_4$ Concentrations – µg-at N/L | | | |
|---|---|---|---|---|---|
| | 0+Hg | 0 | 25–75 | 125–133 | 400–17850 |
| % Weight loss | 25.6 | 29.2 | 30.0 | 31.9 | 30.7 |
| Δ C% | −16.8 | −19.8 | −20.8 | −22.0 | −19.9 |
| Δ N% | − 5.0 | − 1.4 | +13.5 | +21.0 | +43.3 |
| C:N | 133.9 | 120.5 | 107.0 | 98.8 | 80.5 |
| Δ C:N% | −10.0 | −17.5 | −28.3 | −34.2 | −42.6 |

There is a direct correlation of final N% with original nitrogen %, which differs with the level of inorganic nitrogen added to the system (Fig. 4). This relationship of final N% to original N% is often reported in the literature (e.g., Fenchel and Harrison, 1975). The relationship of ΔN% and original N% (Fig. 5) suggests that at low test levels of inorganic nitrogen (0-N and 25 µg-at N/L) there was little, if any, ΔN% at original N levels below 0.40%; leaves with greater than 0.40% N lost nitrogen under these conditions. At high levels of inorganic nitrogen (> 400 µg-at N/L), there was an inverse relationship ($r = -0.70$) of ΔN% to original N%. For example, with these higher levels of inorganic nitrogen (Fig. 5), leaves with original N% of 0.20 increased in N by 75%. This leveled out in the vicinity of 0.32 original N% to 30% ΔN. In a separate test (n = 5), a control (Hg + 1428 µg-at N/L) determined that there was no abiotic immobilization of nitrogen (ΔN = -4.2%, ΔC = -10.3%, ΔC:N = -6.2%, WT LS = 16.3%).

As an additional point of information there is an inverse relationship between weight loss and initial dry wt:fresh wt (n = 280, r = -0.53, F = 109, b = -65.2). The accuracy of the computation can be increased by the use of leaf surface area: i.e., when weight loss is standardized by surface area, the inverse relationship with DW:FW became stronger (n = 80, r = -0.71, F = 81, b = -79.5). This relationship may be due to a higher fiber content in the high ratio leaves, a possiblilty suggested by a direct relationship of original C% to DW:FW (r = 0.93, F = 880, n = 360, b = 49.98).

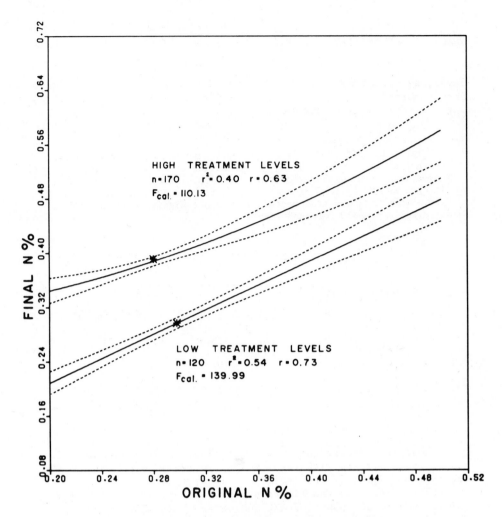

Figure 4. Relationship of the final to original nitrogen %; low treatment levels included 0 and 25 μg-at N/L; high treatment levels included 400 to 17,850 μg-at N/L.

DISCUSSION

Use of laboratory data to interpret the field responses has limitations, since a complex system has been reduced to a single exogenous nitrogen source and one type of microorganism. Assuming these limitations, various principles can be considered. The laboratory model demonstrates that nitrogen immobilization can be the result of fungal activities in the presence of exogenous nitrogen.

Explanation for Figures 4 and 5: n = sample size; $r^2$ = sample coefficient of determination; r = sample correlation coefficient; all $F_{cal}$ values are significant at the 0.05 α level; * = $\bar{X}$, $\bar{Y}$; dotted lines are the ± 95% confidence belts for the regression line; each regression model is the best fit.

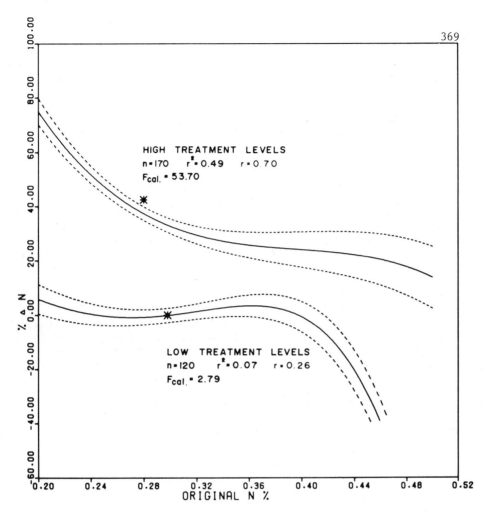

Figure 5. Relationship of percent change in nitrogen
to original N%; low treatment levels included 0 and 25 µg-at
N/L; high treatment levels included 400 to 17,850 µg-at N/L.

While species of *Phytophthora* can immobilize nitrogen, the field nitrogen immobilization is most likely the result of the activities
of a variety of fungi, bacteria and microalgae. The requirement for
exogenous nitrogen in the field is evident from the laboratory results. While the differences in rates of nitrogen immobilization
at Fahk and SCS can be attributed to environmental conditions, the
lack of nitrogen immobilization at WCS may be due to insufficient
exogenous nitrogen. Measurement of standing levels of nitrogen (0.2
to 3.0 µg-at $NO_3$ N/1) in waters of Card Sound (Gerchakov *et al.*,
1973) suggests that ambient nitrogen is below requirement levels.
Therefore, additional sources of nitrogen must be considered. One
possible source is $N_2$ fixation which takes place on decomposing mangrove leaves (Gotto and Taylor, 1976) as well as on the roots and
associated sediments (Kimball and Teas, 1974; Zuberer and Silver,
1978). Jones and Stewart (1969) demonstrated that marine occurring
fungi are capable of utilizing extracellular nitrogen that has been

fixed and liberated by cyanobacteria. While $N_2$ fixation takes place during most of the year, Capone (1978) showed a reduction in the $N_2$ fixation rate at winter water temperatures of $20^{\circ}C$, the condition during the WCS study.

By use of laboratory and field data it is possible to predict the $\Delta N$ and final N for a given level of original N and for a given level of added N. The laboratory model confirmed the difference in $\Delta N\%$ observed at SCS and Fahk, i.e., leaves with low original N level result in a higher $\Delta N\%$. The field data predicts that final N (given sufficient exogenous nitrogen) will be the same regardless of original N content. This was not demonstrated in the laboratory model because the model was restricted to the first 10 days of decay, at which point < 50% of the total $\Delta N\%$ has taken place. The model should be expanded to a longer time period with the fungi, such as *Fusarium* sp. and *Lulworthia* sp., that are associated with later stages of decay.

The prediction that final N will be the same regardless of original N suggests two concepts. First, in high original N content leaves there is a factor, in addition to exogenous nitrogen, that limits the level of nitrogen immobilization. The factor may be available carbon. This is suggested by the high fiber content of high original N leaves as indicated by the direct relationship of original C to DW:FW and a direct relationship of original C to original N (n = 350, r = 0.53, F = 45). Second, due to fungal nitrogen immobilization, the nitrogen contribution to the ecosystem from litter with low original N levels will be the same quantity as the contribution from litter with high original N levels, a point worth considering when evaluating the relative roles of mangrove forests.

## ACKNOWLEDGMENTS

This research was supported by the Department of Energy, Contract No. EY 765053081.

## REFERENCES

Anastasiou, C. J., and L. M. Churchland. 1969. Fungi on decaying leaves in marine habitats. Can. J. Bot. 47: 251-257.

Bamforth, S. S. 1977. Litter and soils as fresh water ecosystems. *In* Aquatic Microbial Communities. Cairns, J. (ed.). Garland Publ. Co., pp. 243-256.

Bärlocher, F., and B. Kendrick. 1975. Leaf-conditioning by microorganisms. Oecologia (Berl.). 20: 359-362.

Boyd, C. E. 1971. The dynamics of dry matter and chemical substances in a *Juncus effusus* population. Am. Midl. Nat. 86: 28-45.

Capone, D. G. 1978. Dinitrogen fixation in subtropical seagrass and macroalgal communities. 114 pp. Ph.D. Dissertation, Univ. of Miami, Coral Gables, Florida.

Fell, J. W., and I. M. Master. 1975. Phycomycetes (*Phytophthora* spp. nov. and *Pythium* sp. nov.) associated with degrading mangrove (*Rhizophora mangle*) leaves. Can. J. Bot. 53(24): 2908-2922.

Fenchel, T. 1977. Aspects of the decomposition of seagrasses. *In* Seagrass Ecosystems. A Scientific Perspective. McRoy, C. D. and C. Helfferich (eds.), New York: Marcel Dekker. Mar. Sci. Vol. 4: pp. 123-146.

Fenchel, T., and P. Harrison. 1975. The significance of bacterial grazing and mineral cycling for the decomposition of particulate detritus. *In* The Role of Terrestrial and Aquatic Organisms in Decomposition Processes. Anderson, J. M. and A. Macfadyen (eds.), London: Blackwell Sci. Publ. pp. 285-299.

Fuller, M. S., and R. O. Poyton. 1964. A new technique for the isolation of aquatic fungi. Bioscience. 14(9): 45-46.

Gerchakov, S. M., D. A. Segar, and R. D. Stearns. 1973. Chemical and hydrological investigations in the vicinity of a thermal discharge into a tropical marine estuary. *In* Radionuclides in Ecosystems, Proc. Third Nat. Symp. Radio-ecology, May 10-12, 1971, Oak Ridge, Tenn., D. J. Nelson (ed.), Vol. 1, pp. 603-618. USAEC CONF-710501-P1.

Gosselink, J. G., and C. J. Kirby. 1974. Decomposition of salt marsh grass, *Spartina alterniflora* Loisel. Limnol. and Oceanogr. 19(5): 825-832.

Gosz, J. R., G. E. Likens, and F. H. Bormann. 1973. Nutrient release from decomposing leaf and branch litter in the Hubbard Brook Forest, New Hampshire. Ecological Monographs. 43: 173-191.

Gotto, J. W., and B. F. Taylor. 1976. $N_2$ fixation associated with decaying leaves of the red mangrove (*Rhizophora mangle*). Applied and Environ. Microbiol. 31(5): 781-783.

Harley, J. L. 1971. Fungi in ecosystems. J. Appl. Ecol. 8(3): 627-642.

Höhnk, W. 1956. Studien zur Brack- und Seewassermykologie. VI. Über die pilzliche Besiedlung verschieden salziger submerser Standorte. Veröffentl. Inst. Meeresforsch. Bremerhaven. 4: 67-110.

Howarth, R. W., and S. G. Fisher. 1976. Carbon, nitrogen, and phosphorous dynamics during leaf decay in nutrient-enriched stream ecosystems. Freshwater Biol. 6: 221-228.

Hudson, H. J. 1971. The development of the saprophytic fungal flora as leaves senesce and fall. *In* Ecology of Leaf Surface Microorganisms. Preece, T. F. and C. H. Dickinson (eds.). London: Academic Press. pp. 447-455.

Hunter, R. D. 1976. Changes in carbon and nitrogen content during decomposition of three macrophytes in freshwater and marine environments. Hydrobiologia. 51(2): 119-128.

Iversen, T. M. 1973. Decomposition of autumn-shed beech leaves in a spring brook and its significance to the fauna. Arch. Hydrobiol. 72(3): 305-312.

Jones, K., and W. D. P. Stewart. 1969. Nitrogen turnover in marine and brackish habitats. IV. Uptake of the extracellular products of the nitrogen-fixing alga *Calothrix scopularum*. J. Mar. Biol. Assoc. U.K. 49: 701-716.

Kimball, M. C., and H. J. Teas. 1974. Nitrogen fixation in mangrove areas of southern Florida. *In* Proc. of Int. Symposium on

Biology and Management of Mangroves, Univ. of Fla.  G. Walsh,
S. Snedaker and H. Teas (eds.).  pp. 654-660.

Nie, N. H., C. H. Hull, J. G. Jenkins, K. Steinbrenner, and D. H.
Bent.  1975.  SPSS.  Statistical package for the social sciences,
Second edition, 675 pp.  N.Y.:  McGraw-Hill Book Co., N.Y.

Odum, W. E., J. C. Zieman, and E. J. Heald.  1973.  Importance of
vascular plant detritus to estuaries.  *In* Proceedings of the
Coastal Marsh and Estuary Management Symposium.  R. H. Chabreck
(ed.), Baton Rouge:  L.S.U. Division of Continuing Education.
pp. 91-114.

Siepmann, R.  1959.  Ein Beitrag zur saprophytischen Pilzflora des
Wattes der Wesermündung.  I.  Systematischer Teil.  Veröffentl.
Inst. Meeresforsch. Bremerhaven. 6:  213-281.

Tenore, K. R., J. H. Tietjen, and J. J. Lee.  1977.  Effect of meiofauna on incorporation of aged eelgrass detritus by the polychaete, *Nephthys incisa*.  J. Fish. Res. Bd. Canada. 34:  563-567.

Thayer, G. W., D. W. Engel, and M. W. Lacroix.  1977.  Seasonal distribution and changes in the nutritive quality of living, dead
and detrital fractions of *Zostera marina* L.  J. Exp. Mar. Biol.
Ecol. 30(2):  109-127.

Willoughby, L. G.  1974.  Decomposition of litter in fresh water.
*In* Biology of Plant Litter Decomposition.  II.  Dickinson, C. H.
and G. J. F. Pugh (eds.).  London: Academic Press.  pp. 659-681.

Zar, J. H.  1974.  Biostatistical Analysis.  Englewood Cliffs, N.J.:
Prentice-Hall.  620 pp.

Zuberer, D. A., and W. S. Silver.  1978.  Biological dinitrogen fixation (acetylene reduction) associated with Florida mangroves.
Appl. Environ. Microbiol. 35(3):  567-575.

# Interactions of Bacteria, Microalgae, and Copepods in a Detritus Microcosm: Through a Flask Darkly

Paul Garth Harrison
Brenda J. Harrison

ABSTRACT: Freshly collected dead leaves of eel-grass (*Zostera marina* L.) had more soluble organics, ash, and carbon but less soluble nitrogen than did leaves aged five months in nature. Increased nitrogen content (total and soluble) in aged detritus was attributed to the activity of colonizing microorganisms. As detritus aged in batch cultures, bacterial densities decreased from $100 \times 10^3$ cells$\cdot$mm$^{-2}$ to $20\text{-}40 \times 10^3$ and microalgal densities increased from 0 to $100\cdot$mm$^{-2}$. In the dark, the microbial decomposition rate was 0.5% of detrital dry weight$\cdot$day$^{-1}$. In the light, weight loss from the *Zostera* leaves equalled weight gain by attached microalgae; therefore, detrital dry weight was constant. Bacteria and algae may compete for nutrients in detritus systems: enrichment with minerals increased bacterial abundance when detritus was plentiful but algae increased when detritus was scarce. Selective grazing by harpacticoid copepods affected this competition. In dark cultures, with only bacteria as a food source, grazing decreased bacterial activity, but copepods selectively grazed algae in light cultures and bacterial activity increased.

INTRODUCTION

    Detritus consisting of dead leaves of eelgrass *Zostera marina*
L. and the associated microorganisms (bacteria, fungi, protozoa,
and algae) is an important, long-lasting source of organic matter
in many temperate, near-shore marine ecosystems (Petersen and
Jensen, 1911; Mann, 1972; Kikuchi and Pérès, 1977).  The roles of
bacteria and protozoa are reviewed by Fenchel and Harrison (1976)
and Fenchel and Jørgensen (1977).  Little is known about fungi on
eelgrass.  Despite the abundance of microalgae on macrophyte detritus (Fenchel, 1970; Rho and Gunner, 1978) and the potential for competition for minerals between algae and bacteria (Johannes, 1965;
Thayer, 1974), the algae have not been studied.  Bacteria and algae
may be grazed by harpacticoid copepods which in turn are fed upon
by macroconsumers (Kaczynski *et al.*, 1973; Sibert *et al.*, 1977;
Roland, 1978), but the role of copepods in detrital-based systems
is unclear (McIntyre, 1969; Coull, 1973; Heinle *et al.*, 1977).  In
this paper we present new data on the relative abundances and activities of bacteria and microalgae on eelgrass detritus and we explore
the response of the microflora to chemical alteration of the detritus (by extraction and enrichment with minerals) and to grazing by
harpacticoid copepods.
    The experiments reported here used small batch cultures stocked
with a "natural" assemblage of organisms.  Much of our knowledge of
eelgrass decay and utilization is based on extrapolation from such
microcosm studies (Fenchel and Harrison, 1976; Fenchel and Jørgensen, 1977).  The need for replication requires accurate weighing of
pre-dried samples of detritus, but the act of drying alters the
physical and chemical nature of the plant material.  For this reason
our microcosms most closely resemble the extensive near-shore areas
where leaves previously deposited on shore by storms and dried there
at low tide are later washed back into the sea.  Microcosms which
are essentially closed to the movements of minerals and organisms
may not reflect all field conditions, but they do enable us to ma-
nipulate parameters not easily controlled in the field (Levandowsky,
1977; Ringelberg and Kersting, 1978) and thus can contribute to our
knowledge of factors regulating detritus-based food chains.


MATERIALS AND METHODS

    In Boundary Bay south of Vancouver, B.C., leaves of eelgrass
(*Zostera marina* L.) were collected in early November when they were
dead but not yet decomposed (=fresh leaves) and again in March after
they had aged five months in nature (=aged leaves).  Leaves were
rinsed to remove algae and salt, oven dried at 60°C for 24 h, and
ground in a Wiley Mill to pass a #40 mesh (200-400 µm particles).
Some particles were then extracted for 30 min with boiling 80%
aqueous methanol to remove most soluble materials.
    Batch cultures consisted of 250 ml flasks containing 100 ml
0.45 µm-filtered seawater (24°/oo) and 50, 100, 250, or (in grazing
experiments) 1000 mg dry wt of eelgrass particles (whole or extracted).  Cultures were inoculated with a drop of slurried sediment from the collection site containing bacteria, protozoa, and,

in some treatments, microalgae. Some cultures were enriched with
the mineral medium of Nordby and Hoxmark (1972) which included 13
mg $NO_3^--N \cdot 1^{-1}$ and 3.5 mg $PO_4^{3-}-P \cdot 1^{-1}$; 5.7 mg $SiO_4^{3-}-Si \cdot 1^{-1}$ was
added. All cultures were incubated with a 12-h night at 10°C and
either a 12-h day at 20°C under a combination of high output day-
light fluorescent and 40 W incandescent lamps that yielded 400 mi-
croeinsteins$\cdot m^{-2} \cdot sec^{-1}$ or 12-h in the dark at 18°C. Cultures were
shaken one hour of every three on a rotary shaker at 175 RPM to
increase gas diffusion between water and air.

After incubation for 2-10 weeks, dry weight was determined by
filtering the flask contents·onto pre-weighed 0.45 μm membrane filters, rinsing first with seawater to avoid damage to cells and then
briefly with distilled water to remove salts, drying 24 h at 60°C,
and reweighing. Organic content was determined by weight loss after
ashing 4 h at 500°C. Carbon and nitrogen analyses were performed
by Canadian Microanalytical Service (Vancouver, Canada) on a Carlo-
Erba CHN analyzer.

Algal cells on the detritus were counted using a Leitz Dialux
microscope with an epifluorescence unit including a broad band blue
excitation filter. We could identify different algae from the characteristic shapes of their red autofluorescing chloroplasts. Ten
haphazardly selected fields of 0.04 $mm^2$ were counted in each of two
cultures for each treatment to correct for some of the variability
due to patchy distribution of cells. Bacteria were counted on
samples stained with acridine orange; 10 fields of 64 $μm^2$ from each
replicate flask were counted. Because of problems with a pale
greenish background and the uncertainty as to the colour of living
cells, we counted all bacterium-size cells which fluoresced yellow-
ish green to green (Francisco *et al.*, 1973; Hobbie *et al.*, 1977).

For grazing studies, fifteen individuals of each of two species
of harpacticoid copepods, *Tachidius triangularis* and *Halectinosoma*
sp., were added to duplicate detritus cultures (1 g initial dry wt
of fresh or aged detritus) after they had incubated in light or dark
two weeks. The grazing community consisted of five gravid females
and ten non-gravid adults (males and females) of each species, numbers typical of densities in nature. These cultures and duplicate
controls (no grazers) were incubated a further four weeks in light
or dark. The flask contents were then fixed with 4% glutaraldehyde,
and the animals were removed and counted.

Metabolic activities of autotrophs and heterotrophs were measured using uptake of [14]carbon before grazers were added and four
weeks later before the samples were fixed. For autotrophy, two subsamples of detritus (10-30 mg dry wt) from each of the grazed and
ungrazed flasks were transferred to individual 50 ml vials with 20
ml 0.22 μm-filtered seawater. The four samples from each duplicated
treatment (e.g., two flasks with grazed fresh detritus) were treated
as follows: 1 ml of $NaH^{14}CO_3^-$ (10μCi) was added to each and one
sample (control) was filtered immediately; the other three vials
were incubated in the light as above for 1 h. For all vials the
contents were filtered onto pre-weighed 0.45 μm membranes, the vials
were rinsed with seawater and the rinsings were filtered, and the
filters were rinsed briefly with distilled water to remove salts.
The filters were dried at 60°C and reweighed. The whole sample with
filter was transferred to a scintillation vial with 10 ml Aquasol-2

and 3 ml distilled water to form a gel to suspend the particles.
Samples were counted in a Nuclear Chicago Unilux-IIA scintillation
counter using an external standard channel ratios method to determine efficiency of counting. Data were expressed as disintegrations
per minute (DPM) per mg dry weight of detritus corrected for surface
adsorption and loss of label during drying in controls. Further
calculations of actual rates of photosynthesis followed the methods
of Strickland and Parsons (1968). To obtain an index of net heterotrophic uptake, uniformly labeled $^{14}$C-glycine (0.1μCi) was added.
The vials were incubated 1 h in darkness (Hall *et al.*, 1972) and
treated as above.


RESULTS

    Comparison of dead leaves collected in November (fresh) with
those collected in March (aged) showed higher organic and nitrogen
contents in aged leaves, but fresh leaves had higher soluble fractions of all components except nitrogen (Fig. 1). Aged leaves
lacked soluble ash. The extract from aged leaves, although only 10%
of the dry weight, was all organic and rich in nitrogen (Table 1).
After extraction fresh and aged leaves had similar chemical compositions (Table 1). Fresh leaves lost 36% of their dry weight on
extraction, but the material was nitrogen-poor (Table 1). After
extraction, fresh and aged leaves had similar chemical compositions
(Table 1).

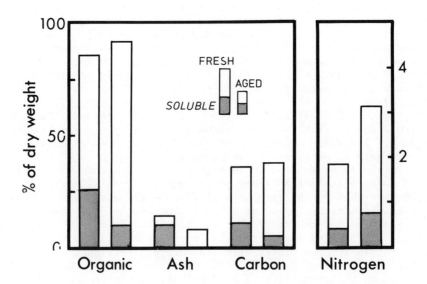

    Figure 1. Chemical composition of freshly collected
dead eelgrass leaves and leaves aged in nature five months.
Soluble fraction equals methanol extract. Data are means
of duplicate analyses.

Table 1. Composition of methanol extract and residual fractions of eelgrass detritus collected in November (fresh) and after five months in nature (aged). Data are means of duplicate analyses.

| Component | % of Total Dry Wt. | | % of Extract or Residual Fraction Dry Wt. | |
|---|---|---|---|---|
| | Fresh | Aged | Fresh | Aged |
| **Extract** | | | | |
| Total | 36 | 10 | 100 | 100 |
| Organic | 26 | 10 | 72 | 100 |
| Carbon | 11 | 5 | 30 | 52 |
| Nitrogen | 0.4 | 0.7 | 1.1 | 7.4 |
| **Residual** | | | | |
| Total | 64 | 90 | 100 | 100 |
| Organic | 60 | 81 | 94 | 90 |
| Carbon | 26 | 33 | 40 | 37 |
| Nitrogen | 1.5 | 2.4 | 2.3 | 2.7 |

Epifluorescence counts of attached bacteria and measurements of net $^{14}$C-glycine uptake varied in a similar way. After incubation for two weeks, fresh detritus had $100 \times 10^3$ cells·mm$^{-2}$ with an uptake rate of $10^3$ DPM·mg dry wt$^{-1}$·hr$^{-1}$ whereas aged detritus had only $20\text{--}40 \times 10^3$ cells·mm$^{-2}$ and an uptake rate of 200 DPM·mg dry wt$^{-1}$·hr$^{-1}$. In addition, fresh detritus supported a two-week bloom of suspended bacteria. In nutrient enriched cultures, aged detritus lost 46% of its dry weight in 10 weeks in the dark (Table 2). Since leaching (as evidenced by the extraction) could account for a loss of up to 10%, the heterotrophic decay rate was at least 0.5% of dry weight per day.

Particles of aged leaves incubated in the light showed no long-term change in dry weight although the percent of carbon and nitrogen declined (Table 2). Compared to the combined losses due to leaching and microbial decay, algal growth increased carbon 58% and nitrogen 33%. The gain of dry weight was proportional to the numbers of algal cells counted with epifluorescence (Fig. 2). In short-term experiments the algae fixed 150-200 µg carbon·g dry detritus$^{-1}$·hr$^{-1}$, which accounts for the measured long-term changes in weight assuming constant rates of photosynthesis for 12 h a day. As cultures aged, the initially dominant pennate diatoms (mainly *Cylindrotheca* (*Nitzschia*) *closterium*) gave way to non-motile and flagellated prasinophytes (*Pyramimonas* sp.) and later (after six weeks) to unidentified filamentous blue-green algae.

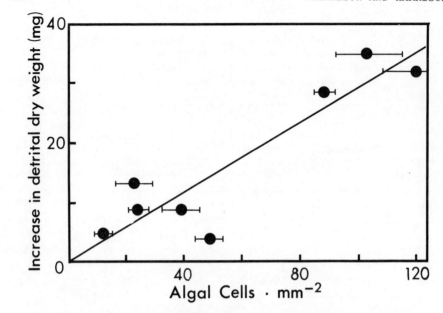

Figure 2.  Increase in dry weight over dark controls
of particles of aged eelgrass detritus incubated two weeks
on a 12:12 light:dark cycle, showing the relationship with
number of algal cells (mean $\pm$ SE; n = 20), 90% of which were
diatoms.  Line shows calculated linear regression with $r^2$ =
0.8.

Table 2.  Composition of aged detritus
initially, after extraction with aqueous methanol, and after incubation for 70 days in dark
and light.  Data are means of three replicates;
coefficients of variation were 2–4%.

| | | Treatment | | |
|---|---|---|---|---|
| Calculation | Initial | Extracted | Incubation | |
| | | | Dark | Light |
| % of Remaining Dry wt. as: | | | | |
| Carbon | 38.2 | 36.7 | 37.8 | 32.2 |
| Nitrogen | 3.2 | 2.7 | 2.6 | 2.0 |
| % Loss of: | | | | |
| Dry Wt. | – | 10.0 | 46.0 | 0.0 |
| Carbon | – | 13.6 | 46.6 | 15.7 |
| Nitrogen | – | 25.0 | 56.2 | 37.5 |

In cultures not nutrient enriched, highest bacterial and lowest algal counts were found consistently on fresh detritus, giving a ratio of over 2,000 bacteria per algal cell (Table 3). The ratio decreased when the detritus was extracted or aged. When nutrients were added, either bacteria or algae responded with population increases while the other remained stable or declined. The response depended on the type and amount of detritus but all enriched cultures had similar ratios of bacteria to algae (Table 3). When the concentration of detritus was low, enrichment with minerals led to large increases in the number of algal cells and concomitant increases in total dry weight (Fig. 3). When detritus was more plentiful, and bacteria were at least potentially more abundant, enrichment led to smaller increases in weight, i.e., greater rates of decay.

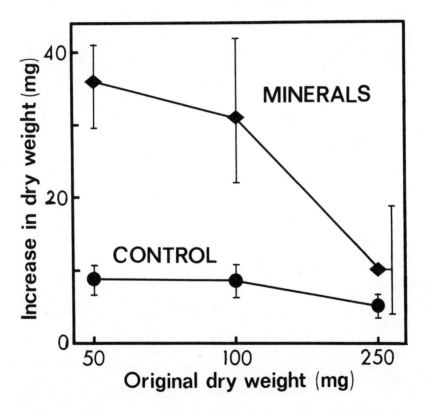

Figure 3. Increases in dry weight of particles of aged eelgrass detritus after three weeks incubation on a 12:12 light:dark cycle, showing the effects of increasing initial densities of detritus and of the addition of a complete mineral enrichment. Data are means ± SE; n = 3.

Table 3.  Ratios of attached bacteria to algae (cell numbers) after incubation of eelgrass detritus for two weeks, showing the effects of the nature of the detritus and of the addition of a complete mineral enrichment.  Data used were means of ten counts on each of two replicate cultures; coefficients of variation were less than 7%.

| Detritus | Bacteria : Algae | |
|----------|---------|-----------|
|          | Control | +Nutrients |
| Fresh            | 2040 | 755 |
| Fresh, Extracted | 140  | 760 |
| Aged             | 360  | 740 |
| Aged, Extracted  | 570  | 830 |

The two copepod grazers swim just above the detritus surface, descending to burrow into the top centimeter: *Halectinosoma* sp. scrapes microflora from particles and *Tachidius triangularis* picks cells from packets of flocculent detritus.  We commonly saw algal remains in the guts of preserved individuals of both species. After one month, *T. triangularis* was found only in cultures with abundant algae, and *Halectinosoma* sp., although present in all cultures, was common only when algae were plentiful.  The rate of uptake of inorganic $^{14}$C was much greater in control cultures (without copepods) after four weeks in the light than at the start of the period, but the increase was much smaller when copepods were present (Fig. 4, autotrophy).  Grazing in the light increased the rate of uptake of organic $^{14}$C (Fig. 4, heterotrophy, light).  In cultures incubated in darkness, however, copepods decreased heterotrophic activity.


DISCUSSION

The results of this study concerning loss of soluble materials from, and microbial decomposition of, detritus broadly confirm earlier reports, but they deserve brief consideration before the interactions of bacteria, algae, and copepods are discussed.  Leaching and autolysis from dead eelgrass leaves once were considered to remove all soluble material soon after leaves died, but Harrison and Mann (1975a) showed that leaching potentially can occur for many months, albeit at a low rate, if cell walls remain intact.  Although the extraction procedure used here was harsher than natural leaching, the results suggest that soluble ash may leach particularly quickly but that even leaves aged in nature five months can lose soluble organics.  The rapid loss of ash, including minerals needed by bacteria, may be one reason for the slow rates of decay reported here and elsewhere for eelgrass (Harrison and Mann, 1975b; Fenchel, 1977a), but other factors, such as the availability of the organic

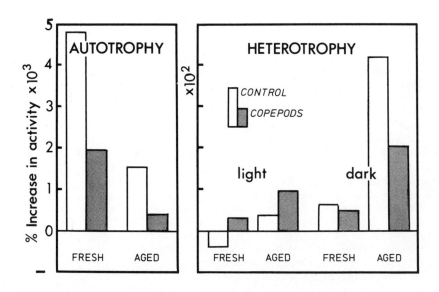

Figure 4.  Changes in autotrophic and heterotrophic activities in a four-week period in cultures of aged eelgrass
detritus showing the effect of the harpacticoid copepods
*Tachidius triangularis* and *Halectinosoma* sp.

material to decomposers are important (Tenore *et al.*, 1979).  The
large amount of nitrogen, total and soluble, in aged leaves most
likely reflected the buildup of attached microbial debris, including
attachment molecules and extracellular enzymes (Hobbie and Lee,
1980; Tenore and Rice, 1980), but about 30% of the carbon extracted
came from the leaves.

We have established that microalgae attached to eelgrass detritus are a potentially important source of organic matter for macroconsumers.  The densities of microalgae reported here are in the
range Fenchel (1970) found on turtle grass (*Thalassia*) detritus in
nature.  The long-term increases in biomass due to primary production balanced the losses due to leaching and microbial decomposition
(Table 2; Fig. 2).  Short-term measurements of algal activity ($^{14}$C-
fixation) were equivalent to rates of respiration of bacteria on
marsh grass detritus reported by Gallagher and Pfeiffer (1977).

In our experiments, when inorganic nutrients were scarce, algae
reached maximal abundances when bacteria were low and vice versa.
Either group of organisms could have produced substances inhibitory
to the other (Sieburth, 1964; Berland *et al.*, 1972; Moebus, 1972;
Delucca and McCracken, 1977).  Alternatively, the organisms may have
competed for the available organic or inorganic nutrients.  Littoral algae such as the main diatom in this study, *Cylindrotheca closterium*, can use dissolved organics (Lewin and Lewin, 1960; Droop,
1974) as can bacteria.  Inorganic nutrients such as nitrate and
phosphate are essential to algae, but bacteria on eelgrass detritus
also need an exogenous supply of these ions because of their low

concentrations in the detritus (Fenchel, 1977a). Competition with bacteria may explain the replacement of pennate diatoms by green unicells on aging detritus; a similar trend was observed in a chemostat when the specific nutrient flux decreased, corresponding in our experiments to a decreasing supply of free nutrients (Turpin and Harrison, 1979).

Our grazing experiments showed that *Halectinosoma* sp. decreased net heterotrophic uptake of glucose in cultures with no algae. Uptake of one substrate cannot be interpreted as a measure of total heterotrophic activity (Hanson, 1980), but in a relative way our data indicate that *Halectinosoma* sp., although an algal feeder, can graze bacteria. *Tachidius triangularis* appeared to require algal food. Heterotrophic activity in the light flasks was stimulated by the grazers, but we conclude that the effect was not a direct positive feedback since similar stimulation did not occur in the dark. Because bacteria and algae are likely to compete for nutrients, a negative feedback on the algae by selective grazing has a positive effect on the bacteria. Such negative feedback loops may contribute significantly to the overall stability of detritus systems (Odum, 1969). Copepods might stimulate microfloral activity by the combined effects of mechanical breakdown of detritus which frees new surfaces for exploitation, stirring which aerates the system, and nutrient regeneration (McIntyre, 1969; Coull, 1973; Tenore *et al.*, 1977; Gerlach, 1978). In our experiments any direct stimulatory effect of the copepods was completely masked by their grazing pressure. The copepods did not appear to fragment detrital particles, and mechanical shaking of the cultures promoted much higher rates of gas exchange than would stirring by the copepods. These effects could be pronounced, however, in many natural systems where sediments are not disturbed by waves or where other grazers are active.

Harpacticoid copepods as grazers show complex effects on their microfloral foods, some of which resemble the effects proposed for smaller (protozoan) and larger (amphipod) grazers (Fenchel, 1970; Hargrave, 1970; Fenchel, 1977b; Fenchel and Jørgensen, 1977; Lopez *et al.*, 1977). In general, grazers prevent microbial populations from senescing (Johannes, 1965; Christian and Wetzel, 1978). As our data suggest, copepods often are selective grazers and hence may control the intricate interrelationships among microbial populations in detritus.

ACKNOWLEDGMENTS

Financial support was provided by an operating grant to PGH and a postgraduate scholarship to BJH both from the Natural Sciences and Engineering Research Council of Canada and by an operating grant to PGH from the Natural, Applied, and Health Sciences Grants Committee of U.B.C.

We thank Dr. K. R. Tenore and several anonymous reviewers for their guidance in improving the final manuscript.

REFERENCES

Berland, B. R., D. J. Bonin, and S. Y. Maestrini. 1972. Étude des relation algues-bactéries du milieu marin: possibilité d'inhibition des algues par les bactéries. Tethys. 4: 339-348.

Christian, R. R., and R. L. Wetzel. 1978. Interaction between substrate, microbes, and consumers of *Spartina* detritus in estuaries, pp. 93-114. *In* M. L. Wiley (ed.), Estuarine Interactions. Academic Press.

Coull, B. C. 1973. Estuarine meiofauna: a review: trophic relationships and microbial interactions, pp. 499-512. *In* L. H. Stevensen and R. R. Colwell (eds.), Estuarine Microbial Ecology. Univ. of South Carolina Press, Columbia.

Delucca, R., and M. D. McCracken. 1977. Observations on interactions between naturally-collected bacteria and several species of algae. Hydrobiologia. 55: 71-75.

Droop, M. R. 1974. Heterotrophy of carbon, pp. 530-559. *In* W. D. P. Stewart (ed.), Algal Physiology and Biochemistry. Univ. of California.

Fenchel, T. 1970. Studies of the decomposition of organic detritus derived from the turtle grass *Thalassia testudinum*. Limnol. Oceanogr. 15: 14-20.

Fenchel, T. 1977a. Aspects of the decomposition of seagrasses, pp. 123-145. *In* C. P. McRoy and C. Helfferich (eds.), Seagrass Ecosystems. Marcel Dekker.

Fenchel, T. 1977b. The significance of bactivorous protozoa in the microbial community of detrital particles, pp. 529-544. *In* J. Cairns (ed.), Aquatic Microbial Communities. Garland.

Fenchel, T., and P. Harrison. 1976. The significance of bacterial grazing and mineral cycling for the decomposition of particulate detritus, pp. 285-299. *In* J. M. Anderson and A. Macfadyen (eds.), The Role of Terrestrial and Aquatic Organisms in Decomposition Processes. Blackwell.

Fenchel, T., and B. B. Jørgensen. 1977. Detritus food chains of aquatic ecosystems: the role of bacteria, pp. 1-58. *In* M. Alexander (ed.), Advances in Microbial Ecology, V. 1. Plenum.

Francisco, D. E., R. A. Mah, and A. C. Rabin. 1973. Acridine orange-epifluorescence technique for counting bacteria in natural waters. Trans. Amer. Micros. Soc. 92: 416-421.

Gallagher, J. L., and W. J. Pfeiffer. 1977. Aquatic metabolism of the communities associated with attached dead shoots of salt marsh plants. Limnol. Oceanogr. 22: 562-565.

Gerlach, S. 1978. Food-chain relationships in subtidal silty sand marine sediments and the role of meiofauna in stimulating bacterial productivity. Oecologia (Berlin). 33: 55-69.

Hall, K. J., P. M. Kleiber, and I. Yesaki. 1972. Heterotrophic uptake of organic solutes by microorganisms in the sediment. Mem. Ist. Ital. Idrobiol. 29 Suppl.: 441-471.

Hanson, R. B. 1980. Microbial activities in trophic interaction and detrital decomposition. *In* K. R. Tenore and B. C. Coull (eds.), Marine Benthic Dynamics. Univ. of South Carolina Press, Columbia.

Hargrave, B. T. 1970. The effect of a deposit-feeding amphipod on the metabolism of benthic microflora. Limnol. Oceanogr. 15: 21-30.

Harrison, P. G., and K. H. Mann. 1975a. Chemical changes during the seasonal cycle of growth and decay in eelgrass (*Zostera marina* L.) on the Atlantic coast of Canada. J. Fish. Res. Bd.

Can. 32:  615-621.

Harrison, P. G., and K. H. Mann.  1975b.  Detritus formation from
    eelgrass (*Zostera marina* L.):  the relative effects of fragmen-
    tation, leaching, and decay.  Limnol. Oceanogr. 20:  924-934.

Heinle, D. R., R. P. Harris, J. F. Ustach, and D. A. Flemer.  1977.
    Detritus as food for estuarine copepods.  Mar. Biol. 40:  341-
    353.

Hobbie, J. E., R. J. Daley, and S. Jasper.  1977.  Use of nucleopore
    filters for counting bacteria by fluorescence microscopy.  Appl.
    Environ. Microbiol. 33:  1225-1228.

Hobbie, J. E., and C. Lee.  1980.  Microbial production of extra
    cellular material:  importance in benthic ecology.  *In* K. R.
    Tenore and B. C. Coull (eds.), Marine Benthic Ecology.  Univ. of
    South Carolina Press, Columbia.

Johannes, R. E.  1965.  Influence of marine protozoa on nutrient re-
    generation.  Limnol. Oceanogr. 10:  434-442.

Kaczynski, V. W., R. J. Feller, J. Clayton, and R. J. Gerke.  1973.
    Trophic analysis of juvenile pink and chum salmon in Puget Sound.
    J. Fish. Res. Bd. Can. 30: 1003-1008.

Kikuchi, T., and J. M. Pérès.  1977.  Consumer ecology of seagrass
    beds, pp. 147-193.  *In* C. P. McRoy and C. Helfferich (eds.),
    Seagrass Ecosystems.  Marcel Dekker.

Levandowsky, M.  1977.  Multispecies cultures and microcosms, pp.
    1399-1458.  *In* O. Kinne (ed.), Marine Ecology, V. 3(2).  John
    Wiley and Sons.

Lewin, J. C., and R. A. Lewin.  1960.  Auxotrophy and heterotrophy
    in marine littoral diatoms.  Can. J. Microbiol. 6:  127-134.

Lopez, G. R., J. S. Levinton, and L. B. Slobodkin.  1977.  The
    effect of grazing by the detritivore *Orchestia grillus* on *Spartina*
    litter and its associated microbial community.  Oecologia
    (Berlin). 30:  111-127.

Mann, K. H.  1972.  Macrophyte production and detritus food chains
    in coastal waters.  Mem. Ist. Ital. Idrobiol. 29 Suppl.:  353-
    383.

McIntyre, A. D.  1969.  Ecology of marine meiobenthos.  Biol. Rev.
    Cambridge Philos. Soc. 44:  245-290.

Moebus, K.  1972.  Seasonal changes in antibacterial activity of
    North Sea water.  Mar. Biol. 13:  1-13.

Nordby, Ø., and R. C. Hoxmark.  1972.  Changes in cellular param-
    eters during synchronous meiosis in *Ulva mutabilis* Føyn.  Exp.
    Cell Res. 75:  321-328.

Odum, E. P.  1969.  The strategy of ecosystem development.  Science.
    164:  262-270.

Petersen, C. G. J., and P. B. Jensen.  1911.  Valuation of the sea.
    I.  Animal life of the sea-bottom, its food and quantity.  Rep.
    Danish Biol. Station. 20:  3-79.

Rho, J., and H. B. Gunner.  1978.  Microfloral response to aquatic
    weed decomposition.  Water Res. 12:  165-170.

Ringelberg, J., and K. Kersting.  1978.  Properties of an aquatic
    microecosystem:  I.  General introduction to the prototypes.
    Arch. Hydrobiol. 83:  47-68.

Roland, W.  1978.  Feeding behavior of the kelp clingfish *Rimicola
    muscarum* residing on the kelp *Macrocystis integrifolia*.  Can. J.
    Zool. 56:  711-712.

Sibert, J., T. J. Brown, M. C. Healey, B. A. Kask, and R. J. Naiman. 1977. Detritus-based food webs: exploitation by juvenile chum salmon (*Oncorhynchus keta*). Science. 196: 649-650.

Sieburth, J. McN. 1964. Antibacterial substances produced by marine algae. Develop. Industrial Microbiol. 5: 124-134.

Strickland, J. D. H., and T. R. Parsons. 1968. A practical handbook of seawater analysis. J. Fish. Res. Bd. Can. 167.

Tenore, K. R., R. B. Hanson, B. E. Dornseif, and C. N. Wiederhold. 1979. The effect of organic nitrogen supplement on the utilization of different sources of detritus. Limnol. Oceanogr. 24: 350-355.

Tenore, K. R., and D. L. Rice. 1980. A review of trophic factors affecting secondary production of deposit-feeders. *In* K. R. Tenore and B. C. Coull (eds.), Marine Benthic Dynamics. Univ. of South Carolina Press, Columbia.

Tenore, K. R., J. H. Tietjen, and J. J. Lee. 1977. Effect of meiofauna on incorporation of aged eelgrass, *Zostera marina*, detritus by the polychaete *Nephthys incisa*. J. Fish. Res. Bd. Can. 34: 563-567.

Thayer, G. W. 1974. Identity and regulation of nutrients limiting phytoplankton production in the shallow estuaries near Beaufort, N.C. Oecologia (Berlin). 14: 75-92.

Turpin, D. H., and P. J. Harrison. 1979. Limiting nutrient patchiness and its role in phytoplankton ecology. J. Exp. Mar. Biol. Ecol. 39: 151-166.

# The Availability of Microorganisms Attached to Sediment as Food for Some Marine Deposit-Feeding Molluscs, with Notes on Microbial Detachment due to the Crystalline Style

G. R. Lopez

ABSTRACT: Feeding experiments with *Ilyanassa obsoleta* (Gastropoda: Nassariidae) indicated that attachment of microorganisms on sand grains affected ingestion rate of the microorganisms but not retention efficiency, while microbial association with silt-clay sediment fraction did not affect either ingestion or retention. Similar experiments with *Bittium varium* (Gastropoda: Cerithiidae) showed that ingestion rate was not affected when microbial attachment to silt-clay was reduced by sonification, but the retention of ingested microorganisms increased.

Experiments with crystalline style extract from *Mytilus edulis* (Bivalvia: Mytilidae) showed that dissolved style material detached bacteria and algae from sediment particles.

Thus the availability of sediment-associated microorganisms for deposit-feeding molluscs might depend upon mode of digestion and whether there are specific adaptations for detachment of microorganisms from sediment particles.

INTRODUCTION

   In a pioneering study on deposit-feeders, Newell (1965) suggested that *Hydrobia ulvae* (Gastropoda; Hydrobiidae) and *Macoma
balthica* (Bivalvia; Tellinidae) digested only living microorganisms
from ingested sediment.  Several studies since then have shown that
microorganisms are usually digested efficiently, while organic debris is inefficiently utilized (Hargrave, 1970; Fenchel, 1970;
Kofoed, 1975; Wetzel, 1976; Yingst, 1976; Lopez *et al.*, 1977;
Cammen *et al.*, 1978), although nutritional value of organic debris
also depends on the parent material, age, and nutrient supply
(Tenore, 1977a, b; Tenore *et al.*, 1979).
   The present study examines some preliminary evidence of the
effect of microbial attachment to sediment and detachment by crystalline style activity on the availability of microorganisms as
food for deposit-feeding molluscs, the snails *Ilyanassa obsoleta*
(= *Nassarius obsoletus*) (Say) (Gastropoda:  Nassariidae) and *Bittium varium* (Say) (Gastropoda:  Cerithiidae).  The purpose of these
experiments was to determine whether alteration by sonification of
microbial attachment onto sediment particles or size of sediment
particle onto which microorganisms were attached would affect ingestion or retention of the ingested microorganisms.
   Sediment-microorganism associations can affect the availability of microorganisms for animals.  The prosobranch *Theodoxus fluviatilis* did not digest algae on smooth surfaces, while algae attached to a rough surface were digested because they were broken
during ingestion (Neumann, 1961).  Grazing by intertidal gastropods
resulted in microflora that was tightly attached to the substrate,
suggesting that it was less available to browsers (Nicotri, 1977).
The hydrobiid gastropod *Hydrobia ventrosa* digested more diatoms
from ingested sediment (particle size < 10 μm) after the sediment
was blended briefly, suggesting that loosened or detached cells were
more available for digestion (Lopez and Levinton, 1978).  Brief
blending most likely did not affect bacterial attachment and thus
did not increase digestion of bacteria (Lopez and Levinton, 1978).
   Several types of sediment-dwelling animals have specific adaptations for removing attached microorganisms from particle surfaces.
The cumacean *Cumella vulgaris* ingested small particles but scraped
larger particles with its mouthparts, a behavior termed "epistrate
browsing" (Wieser, 1956).  The amphipods *Bathyporeia sarsi* and *B.
pilosa* scraped microorganisms from sand grains (Nicholaisen and
Kanneworff, 1969).  This behavior is probably widespread among sand-
dwelling crustaceans.  Deposit-feeding benthic nematodes have been
distinguished from epistrate browsers by buccal structure (Wieser,
1953).
   A recent study of the feeding behavior of the hydrobiid gastropods *Hydrobia ulvae*, *H. ventrosa*, *H. neglecta*, and *Potamopyrgus jen-
kinsi* on a wide size-range of sediment fractions showed that these
snails were capable of feeding on microorganisms attached to sediment particles by swallowing small particles (deposit-feeding) and
by browsing upon particle surfaces (Lopez and Kofoed, 1979). Hydrobiids appear to be adapted to browsing on sand grains small enough
to be swallowed, a process which the authors call "epipsammic browsing."  The snail takes a small particle into the buccal cavity,

scrapes off the attached microorganisms, and then spits out the particle. Hydrobiids are also capable of browsing upon the surfaces of particles too large to be swallowed (Fenchel, 1975; Lopez and Levinton, 1978; Lopez and Kofoed, 1979).

Different microorganisms are differentially available for digestion by deposit-feeders. For example, blue-green algae are sometimes relatively indigestible (Hargrave, 1970; Kofoed, 1975). Strains of bacteria may be differentially available (Wavre and Brinkhurst, 1971; Baker and Bradnam, 1976).

Bacterial abundance in sediment increases with decreasing sediment particle size (Zobell, 1938; Dale, 1974). Abundance of small, epipsammic diatoms also exhibits an inverse relationship with sediment particle size (Lopez and Levinton, 1978). Most bacteria in sediment adhere quite firmly to sediment particles (Wieser and Zech, 1976; Meyer-Reil *et al.*, 1978). Benthic diatoms range from strongly attached, epipsammic species to large, very mobile cells (Round, 1965; Harper, 1969). Motile diatoms are attached to sediment particles by mucus slime even while moving (Harper and Harper, 1967). Many factors affect microbial attachment onto surfaces: turbulence (Meadows and Anderson, 1968); nutrients (Jannasch and Pritchard, 1972; Paerl, 1975); temperature (Fletcher, 1977) and light (Nagata, 1977). Microorganisms on sediment particles are often in highest densities in depressions and crevices, which has been attributed to the effect of physical abrasion (Meadows and Anderson, 1968; Weise and Rheinheimer, 1978).

*I. obsoleta*, one of the most abundant animals on the intertidal mud flats of the Atlantic coast of North America, is a facultative deposit-feeder and scavenger which develops a crystalline style when deposit-feeding and loses it when scavenging (Jenner, 1956; Brown, 1969). *I. obsoleta* assimilates sediment-associated bacteria and algae but not ingested *Spartina* detritus (Wetzel, 1976). *Bittium varium* is one of the most abundant epifaunal species of the *Zostera marina* community (Thayer *et al.*, 1975). Preliminary observations indicated that *B. varium* is a grazer and facultative deposit-feeder.

MATERIALS AND METHODS

Ingestion and Retention of Sediment-Associated Microorganisms by *Ilyanassa obsoleta* and *Bittium varium*

Sediment and animals were collected from Bogue Sound, North Carolina. The experiments were conducted at the Institute of Marine Sciences, University of North Carolina, Morehead City, North Carolina.

Only 7 mm (shell length) specimens of *I. obsoleta* were used in this study; all snails of this size class are juveniles (C. E. Jenner, personal communication). Initial observations indicated that juveniles began feeding under experimental conditions, while adult feeding behavior appeared to be correlated with tidal cycle (Robertson, 1979; Wetzel, 1976). The *B. varium* used were approximately 3 mm in length.

Particle size fractions of sediment were prepared by wet-sieving and decantation.

Sediment was labelled by adding approximately 5 µCi $^{14}$C-glucose
to 6 ml sediment and 20 ml seawater, and incubating for 1-4 h in the
dark at approximately 20°C.  Labelled sediment was then mixed twice
with 20 ml seawater, centrifuging after each mixing.  All seawater
used in the experiments described here was filtered through a 0.22
µm pore-size filter.  A subsample of the labelled sediment was then
sonified for one min in an ultrasonic bath; sonified and unsonified
sediment were both centrifuged for 10 minutes at 4,000 rpm.

A preliminary experiment was conducted to determine whether sonification destroys detached microorganisms.  Two ml of sandy sediment
was washed four times to remove unattached material and then placed
in 30 ml seawater.  This was sonified for 30 sec and the overlying
water containing detached microorganisms was then transferred to
another beaker.  The suspension was then subjected to additional
sonification periods of 0, 30, 90, and 150 sec.  Subsamples were
taken at each period for estimates of bacteria and algae.  Direct
microscopic counts were made using the method of Hobbie *et al.*
(1977) for bacteria and the method of Lopez and Levinton (1978) for ·
algae.  The results (Fig. 1) show that the number of algae decreased
steadily with increasing sonification time (p < 0.01), while there
was no effect of sonification on bacterial counts.

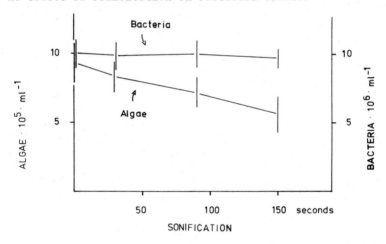

Figure 1.  The effect of sonification on detached
microorganisms.  Vertical bars indicate 95% confidence
intervals.

The results for bacteria are consistent with previously reported use of ultrasonic energy to recover bacteria from a variety of
substrates (Puleo *et al.*, 1967; Zvyagintsev and Galkina, 1967; Weise
and Rheinheimer, 1978).  One minute sonification of sediment probably did not destroy bacteria.

Sediment labelling was probably due to uptake of $^{14}$C-glucose
by bacteria adhering to the surface of sediment particles (see Munro
and Brock, 1968; Hobbie and Rublee, 1977).  In the experiments presented here, I assume that the bacteria detached or loosened by sonification have the same specific activity as those cells remaining

attached during sonification.  Therefore, ingestion of [14]C probably
represents ingestion of [14]C-labelled bacteria.  Retention of [14]C
after several hours of feeding, during which the gut contents were
flushed several times, is a measure of the [14]C-labelled bacteria
that were digested and assimilated.  The amount of [14]C lost from the
sediment during the course of the experiments was usually less than
3% per hour.

     Gut residence time for each species on given sediment fractions
was estimated with sediment stained with methylene blue.

     One day before the initiation of an experiment, animals were
transferred to dishes containing unlabelled sediment of the same size
fraction to be used in the experiment.  The next day animals were
transferred to glass dishes (10.5 cm diameter) containing 50 ml seawater (8 *I. obsoleta* or 80-100 *B. varium* per dish).  After 1 hour a
layer of labelled sediment was added to each dish.  Coarser sand
could be spooned in, but the silt-clay (< 63 µm) sediment was added
by slowly pipetting a thick slurry onto the bottom of each dish.
Little sediment was resuspended in this manner, and the animals did
not appear unduly disturbed.  During the experiment, animals that
crawled onto the walls of the dishes were gently pushed back onto
the sediment.

     Groups of animals were sacrificed after they had fed upon la-
belled sediment for a given period of time (usually 10, 20, or 30
min).  Sampling schedule was arranged so that estimated gut residence time was bracketed by sampling times.  At the last sampling
time, other groups of animals were transferred to dishes containing
unlabelled sediment.  They were allowed to feed for 2-4 h.  This
was usually several times longer than the estimated gut residence
time.   [14]C associated with these animals is presumed to be the
amount retained in the tissue; it equals the amount assimilated
minus losses to respiration and other metabolic losses.  All animals
removed from the dishes with labelled sediment were initially transferred to dishes of seawater before final transfer to unlabelled
sediment or prepared for scintillation counting.  They began crawl-
ing after a short period (usually 1 minute), leaving behind most of
the sediment that had adhered to their feet.

     Animals to be prepared for scintillation counting were killed
with formalin (3% final concentration) within 2 min of removing them
from the sediment, then wiped with tissue paper to remove adhering
material.  The shells of *I. obsoleta* were cracked with a hammer, and
then snails were placed individually in glass scintillation vials
containing 1 ml of NCS tissue solubilizer.  *B. varium* were not
cracked and were placed in groups of 15 individuals in similar vials.
Samples were digested overnight at 45°C, cooled, and then 10 ml OCS
scintillator solution was added.

     Retention efficiency of ingested label was calculated as

$$([14]C \text{ x animal}^{-1}, t_1) \text{ x } ([14]C \text{ x animal}^{-1}, t_2)^{-1} \text{ x } 100,$$

where $t_1$ is time of transfer of animals from the labelled sediment
to the unlabelled sediment, and $t_2$ is the time of final collection
of animals from the unlabelled sediment.  Because of destructive
sampling, parallel groups of animals were used to calculate retention efficiency.

Experiment # 1. To determine the effects of sediment particle size and microbial attachment on microbial availability to *I. obsoleta*, animals were presented with unsonified or sonified: a) natural sediment (unsieved muddy sand) or b) a sediment fraction of < 63 μm, and allowed to feed for 10, 20, or 30 min. After 30 min on [14]C-labelled sediment, two groups of animals were transferred to unlabelled sediment, which were sampled after 2 h and 3 h, respectively. Preliminary observations with sediment stained with methylene blue indicated that *I. obsoleta* passed these sediments through its gut in approximately 20 min. Several feeding attempts with < 63 μm sediment, however, were unsuccessful; the animals did not appear to feed actively on this fraction. The next experiment was conducted to investigate this problem.

Experiment # 2. Experiment 1,b was repeated, except that samples were taken after 15 min and 30 min upon the [14]C-labelled sediment, and retention of [14]C was measured for animals that had fed for 30 min on labelled sediment, then 3 h on unlabelled sediment.

Experiment #3. To determine the effect of microbial attachment to sediment particles too large to be swallowed by *I. obsoleta*, animals were presented with unsonified or sonified 308–1000 μm sediment fraction and allowed to feed for 20, 40, or 60 min. Retention of [14]C label was measured in animals that had fed for 60 minutes on labelled sediment, then for 2 h or 4 h on unlabelled sand. Gut residence time of material scraped from sand grains was approximately 50 min. The animals were not observed to swallow sand grains in this size range.

Experiment #4. To determine the effects of sediment particle size and microbial attachment on microbial availability to *B. varium* (2–3 mm shell length), animals were presented with unsonified or sonified a) < 63 μm sediment or b) 63–125 μm sediment. Ingestion was measured after 15 min and 30 min, while retention was measured in animals that fed on labelled sediment for 30 min, followed by 3 h on unlabelled sediment. Gut residence time was estimated to be approximately 20 min.

## Crystalline Style Experiments

A series of experiments was conducted to determine whether molluscan crystalline styles affect the detachment of microorganisms from sediment particles. These experiments were conducted at Rønbjerg Marine Station, Aarhus University, Denmark, and at Odense University, Odense, Denmark. Crystalline styles were dissected from *Mytilus edulis* (L.) and dissolved in filtered seawater (0.45 μm pore filter), which was then filtered through a 0.45 μm pore filter. For some experiments, a lyophilized extract was prepared. To estimate the number of microorganisms detached from the sediment, direct microscopic counts of bacteria and algae in the water overlying the sediment were made using methods referred to above.

Experiment #1. To determine whether fresh style material affected microbial detachment from fine sand grains, 1 ml of 45–63 μm sediment was mixed with either 1 ml seawater or 1 ml crystalline style extract (c.s.e.) (25 styles dissolved in 5 ml seawater), agitated on a test-tube shaker for 30 sec, allowed to settle for 20 min, and then the overlying water was sampled for direct counts of bacteria and algae detached from the sediment.

Experiment #2. The effect of crystalline style material on microbial detachment from sediment when samples were vigorously agitated was tested by adding 2 ml of < 63 μm sediment to either 8 ml seawater or 8 ml seawater containing 0.5 mg lyophilized c.s.e. x ml$^{-1}$, agitating the sediment suspensions for 15 sec with a test tube shaker, allowing the suspension to settle for 45 minutes, and then repeating agitation and settling. The top 5 ml of the overlying water was sampled for microscopic counts.

Experiment #3. To determine the effect of crystalline style material on microbial detachment from sediment when gently agitated, one ml of < 63 μm sediment was mixed with 4 ml seawater or with 4 ml seawater containing 0.5 mg c.s.e./ml. These samples were then mixed by flicking each test tube 3 times. After they were allowed to stand for 2.5 h, 4 ml of overlying water was sampled for counts of bacteria and algae.

Experiment #4. The effect of concentration of c.s.e. on microbial detachment from sediment was tested by adding 2 ml sediment (muddy sand) to 8 ml seawater containing 0, 0.5, or 0.005 mg lyophilized c.s.e./ml. Because dissolved crystalline style material increases the viscosity of seawater (Kristensen, 1972), a fifth treatment consisting of mixing 2 ml sediment with 7 ml seawater plus 1 ml glycerol was run to test the effect of increased viscosity on microbial detachment. The samples were mixed for 15 sec on a test tube shaker, allowed to settle for 20 min, and then the top 3 ml of overlying water from each sample was taken for counts of detached algae.

RESULTS

Effect of Sediment Sonification and Sediment Particle Size on Ingestion and Retention of Attached Microorganisms by *I. obsoleta* and *B. varium*

Experiment #1. When *I. obsoleta* was presented with muddy sand, it ingested approximately twice as much $^{14}$C from sonified sediment as from unsonified sediment (Fig. 2a) (p < .001). For both sonified and unsonified sediment, ingestion increased with time after 10 min and 20 min ingestion periods (p < .001), but not between 20 and 30 min. The snails that had fed on sonified sediment retained approximately 20% more $^{14}$C-labelled sediment as those that fed on unsonified sediment (Fig. 2a) (p <.001). There was no statistical difference in the amount of $^{14}$C retained between the 2 h and 3 h samples, within treatment of sediment. There was no difference in the retention efficiency of ingested $^{14}$C due to sediment treatment (66.4% unsonified *vs* 45.6% sonified).

When *I. obsoleta* was presented with < 63 μm sediment, neither sonification nor time allowed for feeding affected amount of $^{14}$C in the animals (Fig. 2b). Ingestion (dpm x animal$^{-1}$) was only slightly above background. Sediment sonification did not affect the amount of $^{14}$C retained after feeding for 3 h on unlabelled sediment, nor did it affect retention efficiency (76.7%, unsonified *vs* 71.4%, sonified).

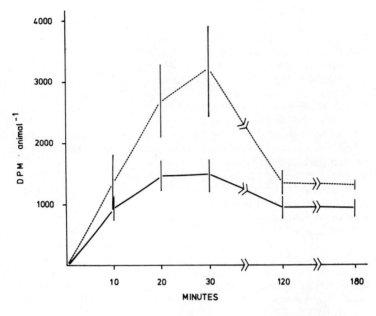

Figure 2a.

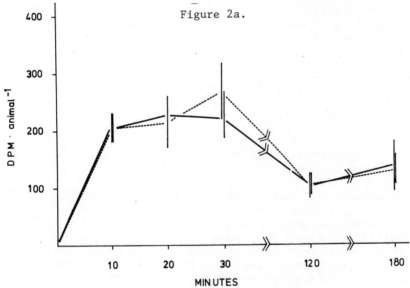

Figure 2b.

Figure 2. Results of *I. obsoleta* and *B. varium* experiments. _____ represents unsonified sediment; ---------- represents sonified sediment. Vertical bars indicate 95% confidence intervals. Animals were transferred from labelled to unlabelled sediment at 30 min, except in experiment 3 in which they were transferred at 60 min.

Experiment #2. Ingestion of [14]C-labelled material was not affected by sonification of the < 63 μm sediment or by the length of
time available for ingestion (Fig. 3). There was no difference in
amount of [14]C retained (Fig. 3) or in retention efficiency (59.8%
unsonified *vs* 84.3% sonified).

Experiment #3. When *I. obsoleta* was presented with 308-1000
μm sediment, it ingested three times more labelled material from the
sonified sand than from the unsonified sand (Fig. 4) (p < .001).
Ingestion of [14]C increased with time on both unsonified and sonified
sediment after 20 and 40 min ingestion (p < .05), but there was no
significant increase between 40 and 60 min. The animals that had
fed on sonified sand retained approximately 3 times more [14]C than
the group on unsonified sand (Fig. 4) (p < .001), but there was no
difference between animals that fed 2 h or 4 h on unlabelled sediment. There was no significant difference in retention efficiency
due to sediment sonification (74.3%, unsonified *vs* 66.4%, sonified).

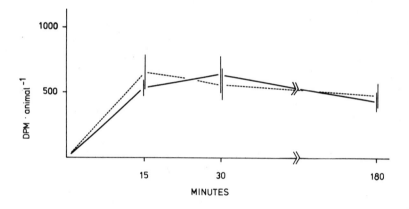

Figure 3. (For legend see Figure 2)

Animals were observed to feed actively on the surfaces of the
sand grains by scraping with the radula and by muscular pumping of
the proboscis. In the dishes with sonified sediment, there was a
greenish film floating on the water surface. The animals appeared
to be greatly attracted to this film, climbing the walls of the dish
and feeding upon the film very actively.

Experiment #4. When *B. varium* was presented with < 63 μm sediment, ingestion of [14]C-labelled material was not altered by sonification (Fig. 5a). The amount ingested increased with time allowed
for feeding (Fig. 5a) (p < .01). Snails that ate sonified sediment
retained approximately 20% more [14]C after 3 h on unlabelled sediment
than those that initially fed on unsonified sediment (Fig. 5a) (p <
.025), but retention efficiencies did not differ significantly between the two groups (33.2%, unsonified *vs* 38.1%, sonified).

When *B. varium* was presented with 63-125 μm sediment, neither
sonification or time allowed for feeding on labelled sediment affected the ingestion of [14]C (Fig. 5b). There was no difference in

the amount of $^{14}$C retained after 3 h between animals that fed on unsonified or sonified sediment. Retention efficiency was not affected by sonification (69.5% unsonified *vs* 65.0% sonified).

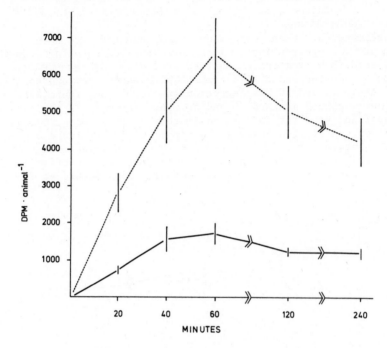

Figure 4. (For legend see Figure 2)

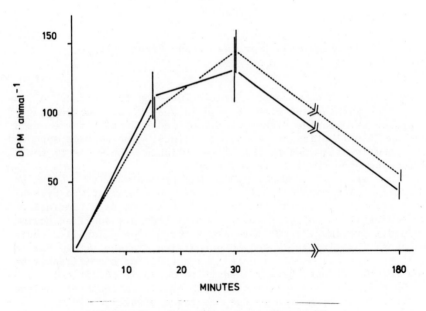

Figure 5a. (For legend see Figure 2)

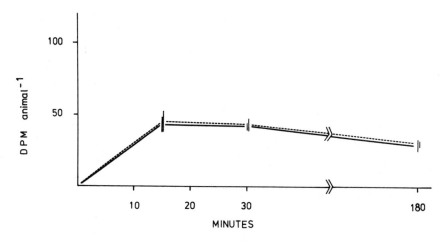

Figure 5b. (For legend see Figure 2)

## Crystalline Style Experiments

Experiment #1.  The number of algae (mostly diatoms) in the water overlying the treated sediment doubled with the addition of c.s.e. compared to the seawater control (Table 1).  There was no difference in the number of bacteria in the overlying water (Table 1).

Experiment #2.  The number of algae in the overlying water was approximately 36% greater when < 63 μm sediment was treated with c.s.e than with seawater (Table 1).  Bacteria were 25% more numerous in the seawater sample than in the c.s.e. sample (Table 1).

Experiment #3.  Approximately 32% more algae were recovered in the water overlying < 63 μm sediment that had been treated with c.s.e., compared to sediment that was mixed with seawater (Table 1). Bacteria were approximately 32% more numerous in the c.s.e-treated samples than the seawater-treated samples (Table 1).

Experiment #4.  Addition of 0.5 mg c.s.e./ml resulted in approximately twice as many algae being recovered in the overlying water, compared to samples treated with lower c.s.e. concentrations (Table 1).  Glycerol addition resulted in slight increases in algal counts compared to the seawater control (Table 1).


## DISCUSSION AND CONCLUSIONS

Even though the ingestion measurements are not comparable between experiments due to labelling methods, it is clear that *I. ob-soleta* was not nearly as active on < 63 μm sediment as on sand (Figs. 2-4).  Ingestion rates were lower with < 63 μm sediment, even though this fraction had higher weight-specific [14]C labelling than coarser sediment (Lopez, unpub.).  Direct observations confirmed the low ingestion rates measured in experiments 1b and 2.  Because sonifica-

Table 1. Results of crystalline style extract experiments. Values in parentheses are ± 1 standard deviation from mean; n = number of microscopic fields counted per sample; p = probability that s.s.e.-treated samples are not different from seawater controls.

| | expt. #1 | | expt. #2 | | expt. #3 | |
| --- | --- | --- | --- | --- | --- | --- |
| | sw | c.s.e. | sw | c.s.e. | sw | c.s.e. |
| algae x ml$^{-1}$ | 4.8 x 10$^4$ (0.0 - 11.9) | 1.0 x 10$^5$ (.23 - 1.7) | 2.3 x 10$^5$ (1.9 - 2.6) | 3.6 x 10$^5$ (2.9 - 4.3) | 1.7 x 10$^5$ (1.3 - 2.3) | 2.5 x 10$^5$ (2.0 - 3.0) |
| n | 40 | 40 | 20 | 20 | 20 | 20 |
| p | | p < .01 | | p < .001 | | p < .001 |
| bacteria x ml$^{-1}$ | 1.8 x 10$^7$ (1.5 - 2.0) | 2.0 x 10$^7$ (1.6 - 2.3) | 6.0 x 10$^7$ (4.6 - 7.5) | 4.5 x 10$^7$ (3.4 - 5.7) | 1.7 x 10$^7$ (1.1 - 2.3) | 2.5 x 10$^7$ (1.8 - 3.2) |
| n | 10 | 10 | 20 | 20 | 20 | 20 |
| p | | n.s. | | p < .01 | | p < .001 |

expt. #4
mg c.s.e. x ml$^{-1}$

| | 0 | 0.005 | 0.05 | 0.5 | glycerol |
| --- | --- | --- | --- | --- | --- |
| algae x ml$^{-1}$ | 2.0 x 10$^5$ (1.6 - 2.5) | 2.4 x 10$^5$ (1.9 - 2.8) | 2.4 x 10$^5$ (1.8 - 2.9) | 4.2 x 10$^5$ (3.6 - 4.7) | 2.5 x 10$^5$ (2.1 - 3.0) |
| n | 15 | 15 | 15 | 15 | 15 |
| | | n.s. | | | |

0.5 vs all other samples p < .001

0 vs 0.005 vs 0.05   n.s.

0 vs glycerol   p < .01

tion did not affect ingestion of labelled material from < 63 μm sediment (Figs. 2b and 3) and because of the similarity of results between the experiments using unsorted muddy sand (Fig. 2a) and that using coarse sand (Fig. 4), it appears that the ingestion of epipsammic bacteria from large sand grains by *I. obsoleta* was affected by bacterial attachment, while ingestion of bacteria associated with < 63 μm sediment was not affected by altering the attachment of bacteria. The latter result is not surprising because *I. obsoleta* feeds on fine sediment with muscular pumping of the proboscis, but the low feeding rates were not expected. The fact that: 1) *I. obsoleta* found and fed upon microbial surface films (experiment #3), 2) it is often found in nature feeding upon microbial mats (Wetzel, 1976), and 3) it did not feed actively upon < 63 μm sediment fraction while demonstrating active feeding upon sand under the same laboratory conditions, suggests that the deposit-feeding activity of juvenile *I. obsoleta* consists mainly of epistrate browsing and of feeding upon microbial films. The common name "mud snail" may be a misnomer for young *I. obsoleta*.

*B. varium* may be a facultative deposit-feeder upon the silt-clay fraction of sediment, feeding mainly upon the epiphytic growth on *Zostera* blades. The snails did not appear to feed actively upon 63-125 μm sediment, which may be near the maximum particle size that these small snails can ingest. They fed very actively upon the < 63 μm sediment fraction; the material passed through the gut in approximately 20 min. The animals produced copious amounts of very sticky mucus, and ingestion of sediment-laden mucus strands was observed. Feeding upon sediment adhering to shells, the snail's own and those of others, appeared to be rather common.

The fact that sonification did not affect retention efficiency of labelled microorganisms by *I. obsoleta* is consistent with a description of extracellular digestion in this species (Brown, 1969), although absorption of food might also occur within the tubules of the diverticula (McLean, 1971). A microorganism is equally likely to be digested and assimilated within the *I. obsoleta* stomach whether or not it is attached to a sediment particle.

It is likely that *B. varium*, like other microphagous mesogastropods, digests and absorbs its food primarily within the cells of the digestive diverticula (Owen, 1966). Microorganisms that are unattached, or that become detached, may be sorted from the mineral particles and transported to the digestive cells, while microorganisms remaining tightly attached would be passed quickly from the stomach to the intestine. Given this argument, one might have predicted higher retention efficiency of labelled microorganisms by *B. varium* from sonified sediment than from unsonified sediment. There were no statistical differences in retention efficiencies in the *B. varium* experiment (experiment 4, Figs. 5a,b), although when *B. varium* was presented with < 63 μm sediment, the amount of $^{14}C$ retained was approximately 20% higher from the sonified sediment than from the unsonified sediment. In contrast, the hydrobiid gastropod *Hydrobia ventrosa* did show increased digestion efficiency of diatoms from < 10 μm sediment after the sediment had been agitated in a blender (Lopez and Levinton, 1978). The availability of attached diatoms for *H. ventrosa* appeared to be controlled partly by the attachment strength of the diatoms on the sediment particles.

Comments on the Role of the Crystalline Style

The presence of the crystalline style in molluscs is closely
correlated with microherbivorous feeding, and a number of functions
have been attributed to it (Yonge, 1930; Graham, 1939; Morton, 1952;
Kristensen, 1972).  Bivalve crystalline styles have a powerful emul-
sifying activity, and this property may be important in detachment
of microorganisms from particles (Kristensen, 1972).  α-amylase,
which has been commonly detected in crystalline styles (Owen, 1966),
detached *Enteromorpha* zoospores from glass surfaces (Christie *et
al.*, 1970).  The present study, showing that *M. edulis* crystalline
style extract consistently detached more algae from sediment than
did seawater, supports Kristensen's (1972) suggestion that an important function of the style is to aid in microbial detachment.  The
crystalline style of *Abra tenuis*, however, appears to have little
effect on detachment of microorganisms from sand grains (Hughes,
1977).  The crystalline style certainly has other functions than the
detachment of microorganisms as a preparatory step of intracellular
digestion, as suggested by the fact that *I. obsoleta* digests microbial food extracellularly, yet has a crystalline style when deposit-
feeding.


Molluscan Deposit-Feeders:  Adaptations and Constraints

I suggest that the relation between feeding rate and digestion
efficiency among deposit-feeders may be controlled by whether digestion is primarily intra- or extra-cellular and whether the animals
have specific adaptations for detaching microorganisms from sediment
particles.  Deposit-feeders with predominantly intracellular digestion may be constrained by their ability to detach microorganisms
from sediment particles.  If this is the case, one would expect to
see adaptations of such animals for detaching cells.  Lopez and
Kofoed (1979) described a new feeding behavior in hydrobiid gastropods, in which microorganisms are scraped from the surface of sand
grains.  Sand grains small enough to be swallowed are usually
scraped and spit out rather than swallowed.
Large benthic diatoms are more effectively grazed by hydrobiids
than small diatoms (Fenchel and Kofoed, 1976).  More of the larger
diatoms are motile species which are attached to sediment only by a
mucus trail and have less robust tests than smaller, epipsammic species (Harper and Harper, 1967).  The smaller, tightly attached species are less vulnerable to grazing by hydrobiids.
Some tellinid bivalves may be adapted for detaching microorganisms from sand grains.  The tellinacean stomach is robust and contains a "straight and massive crystalline style and exceptionally
well-developed gastric shield" suggesting a triturating function
(Yonge, 1949).  The sand-ingesting *Macoma secta* removes bacteria
from sand grains in the stomach perhaps by extracellular digestion
and trituration within the stomach (Reid and Reid, 1969).  *M. balthica* has been classified as a "sand grain feeder" which abrades microorganisms from sediment particles within the stomach (Gilbert,
1977).  Trituration between particles in the stomach of *Tellina*

*tenuis* is important in detaching microorganisms and organic debris (Hughes, 1977).

Intracellular digestion combined with effective microbial detachment and preliminary extracellular digestion may be a very successful strategy of deposit-feeding because it would allow an animal to process a large amount of sediment quickly while efficiently digesting its food. Once microorganisms are detached, sediment particles can very quickly be egested and the stomach then refilled with fresh sediment. It may be that detachment of microorganisms is more rapid and simpler than digestion, but there is little evidence on this problem (see Christie *et al.*, 1970).

The adaptation of initial digestion and detachment followed by intracellular digestion might be expected to be associated with those species living in poorly-sorted sandy sediments, because the adaptation is predicated on an effective separation of detached food particles from inert mineral or detrital particles. Deposit-feeding nuculid bivalves, which feed on fine sediments (Yonge, 1939), exhibit mainly extracellular digestion (Owen, 1956, 1974). The loss of the primitive method of intracellular digestion from this otherwise primitive group is correlated with feeding mode: "Under such conditions the retention of intracellular digestion would result in considerable waste, since the cells of the digestive diverticulae would ingest relatively large quantities of inorganic material." (Owen, 1956).

Deposit-feeders with extracellular digestion have the advantage that all microorganisms, attached and free, are available as food. There are more complete, but empty, diatoms tests and protozoan skeletons within the intestine than in the stomach of nuculid bivalves (Owen, 1956). The amphipod *Corophium volutator* utilized over 70% of the microorganisms that were unavailable for digestion by *Hydrobia ventrosa* (Lopez and Levinton, 1978).

Deposit-feeders with extracellular digestion may be constrained by the length of time particles must be kept in the stomach for effective digestion (see Wieser, 1978).

ACKNOWLEDGMENTS

I gratefully acknowledge the support and stimulating discussions provided by T. Fenchel, J. Hylleberg, C. Jenner, L. Kofoed, and J. Levinton. I thank K. Tenore and two reviewers for critical comments on an earlier draft, and C. Cantrell and E. Lopez for assisting in manuscript preparation. Laboratory space was provided at the Institute of Marine Sciences, UNC, Morehead City, NC. This research was supported by the Curriculum in Marine Sciences, UNC, Chapel Hill, and by a grant from UNC University Research Council.

REFERENCES

Baker, J. H., and L. A. Bradnam. 1976. The role of bacteria in the nutrition of aquatic detritivores. Oecologia (Berl.). 24: 95-104.

Brown, S. C.  1969.  The structure and function of the digestive sys-
    tem of the mud snail *Nassarius obsoletus* (Say).  Malocologia
    9(2):  447-500.

Cammen, L. M., P. A. Rublee, and J. E. Hobbie.  1978.  The signifi-
    cance of microbial carbon in the nutrition of the polychaete
    *Nereis succinea* and other aquatic deposit feeders.  UNC Sea Grant
    Publ. 78-12.

Christie, A. O., L. V. Evans, and M. Shaw.  1970.  Studies on the
    ship-fouling alga *Enteromorpha*.  II.  The effect of certain en-
    zymes on the adhesion of zoospores.  Ann. Bot. 34:  467-482.

Dale, N. G.  1974.  Bacteria in intertidal sediments:  Factors re-
    lated to their distribution.  Limnol. Oceanogr. 19:  509-518.

Fenchel, T.  1970.  Studies on the decomposition of organic detri-
    tus from the turtle grass *Thalassia testudinum*.  Limnol. Oceanogr.
    15:  14-20.

Fenchel, T.  1975.  Factors determining the distribution patterns
    of mud snails (Hydrobiidae).  Oecologia (Berl.) 20:  1-17.

Fenchel, T., and L. H. Kofoed.  1976.  Evidence for exploitative in-
    terspecific competition in mud snails (Hydrobiidae).  Oikos
    27(3):  367-376.

Fletcher, M.  1977.  The effects of culture concentration and age,
    time, and temperature on bacterial attachment to polystyrene.
    Can. J. Microbiol. 23(1):  1-6.

Gilbert, M. A.  1977.  The behavior and functional morphology of de-
    posit feeding in *Macoma balthica* (Linne, 1958), in New England.
    J. Mollusc. Stud. 43(1):  18-27.

Graham, A.  1939.  On the structure of the alimentary canal of
    style-bearing prosobranchs.  Proc. Zool. Soc. Lond. (B) 109:  75-
    112.

Hargrave, B. T.  1970.  The utilization of benthic microflora by
    *Hyallela azteca* (Amphipoda).  J. Animal Ecol. 39:  427-437.

Harper, M. A.  1969.  Movement and migration of diatoms on sand
    grains.  Br. Phycol. Jour. 4(1):  97-103.

Harper, M. A., and J. F. Harper.  1967.  Measurements of diatom ad-
    hesion and their relationship with movement.  Br. Phycol. Bull.
    3:  195-207.

Hobbie, J. E., R. J. Daley, and S. Jaspar.  1977.  Use of nuclepore
    filters for counting bacteria by fluorescent microscopy.  App.
    Env. Microbiol. 33(5):  1225-1228.

Hobbie, J. E., and P. Rublee.  1977.  Radioisotope studies of het-
    erotrophic bacteria in aquatic ecosystems.  *In* J. Cairns, Jr.
    (ed.), Aquatic Microbial Communities, pp. 441-476.  Garland Publ.
    Co., New York.

Hughes, T. G.  1977.  The processing of food material within the
    gut of *Abra tenuis* (Bivalvia: Tellinacea).  J. Moll. Stud. 43:
    162-180.

Jannasch, H. W., and P. M. Pritchard.  1972.  The role of inert par-
    ticulate matter in the activity of aquatic microorganisms.  Mem.
    Ist. Ital. Idrobiol. 29 Suppl.:  289-308.

Jenner, C. E.  1956.  The occurrence of a crystalline style in the
    marine snail, *Nassarius obsoletus*.  Biol. Bull. 111(1):  304.

Kofoed, L. H.  1975.  The feeding biology of *Hydrobia ventrosa*

(Montague). I. The assimilation of different components of food. J. Exp. Mar.

Kristensen, J. H. 1972. Structure and function of crystalline styles of bivalves. Ophelia. 10: 91–108.

Lopez, G. R., and J. S. Levinton. 1978. The availability of micro-organisms attached to sediment particles as food for *Hydrobia ventrosa* Montague (Gastropoda: Prosobranchia). Oecologia (Berl.) 32: 263–275.

Lopez, G. R., J. S. Levinton, and L. B. Slobodkin. 1977. The effect of grazing by the detritivore *Orchestia grillus* on *Spartina* litter and its associated microbial community. Oecologia (Berl.) 30: 111–127.

Lopez, G. R. and L. H. Kofoed. 1979. Epipsammic browsing and deposit-feeding in mud snails (Hydrobiidae). Ms. submitted for publication.

Meadows, P. S., and J. G. Anderson. 1968. Microorganisms attached to marine sand grains. J. mar. biol. Ass. U.K. 48: 161–175.

Meyer-Reil, L.-A., R. Dawson, G. Liebezeit, and H. Tiedge. 1978. Fluctuations and interactions of bacterial activity in sandy beach sediments and overlying waters. Mar. Biol. 48: 161–171.

McLean, N. 1971. On the function of the digestive gland in *Nassarius* (Gastropoda: Prosobranchia). The Veliger. 13(3): 273–274.

Morton, J. E. 1952. The role of the crystalline style. Proc. Malacol. Soc. Lond. 29: 85–92.

Munro, A. L. S., and T. D. Brock. 1968. Distinction between bacterial and algal utilization of soluble substances in the sea. J. Gen. Microbiol. 51: 35–42.

Nagata, Y. 1977. Light-induced adhesion of *Spirogyra* cells to glass. Plant Physiol. 59(4): 680–683.

Neumann, D. 1961. Ernahrungsbiologie einer rhipidoglossen Kiemenschnecke. Hydrobiologia. 17: 133–151.

Newell, R. C. 1965. The role of detritus in the nutrition of two marine deposit feeders, the Prosobranch *Hydrobia ulvae* and the bivalve *Macoma balthica*. Proc. Zool. Soc. Lond. 144(1): 25–45.

Nicolaisen, W., and E. Kanneworff. 1969. On the burrowing and feeding habits of the amphipods *Bathyporeia pilosa* Lindstrom and *Bathyporeia sarsi* Watkin. Ophelia. 6: 231–250.

Nicotri, M. E. 1977. Grazing effects of four marine intertidal herbivores on the microflora. Ecology. 58: 1020–1032.

Owen, G. 1956. Observations on the stomach and digestive diverticulae of the Lamellibranchia. II. The Nuculidae. Quart. J. Microscop. Sci. 97(4): 541–567.

Owen, G. 1966. Digestion. *In* K. M. Wilbur and C. M. Yonge (ed.), Physiology of Mollusca, V. II. Academic Press, New York and London, pp. 53–96.

Owen, G. 1974. Feeding and digestion in the Bivalvia. In O. Lowenstein (ed.), Advances in Comparative Physiology and Biochemistry, V. 5. pp. 1–35. Academic Press, NY.

Paerl, H. W. 1975. Microbial attachment to particles in marine and freshwater ecosystems. Microbial Ecol. 2: 73–83.

Puleo, J. R., M. S. Favero, and N. J. Petersen. 1967. Use of ultrasonic energy in assessing microbial contamination on surfaces.

Appl. Microbiol. 15(6):  1345–1351.

Reid, R. G. B., and A. Reid.  1969.  Feeding processes of members of the genus *Macoma* (Mollusca:  Bivalvia).  Can. J. Zool. 47(4): 649–657.

Robertson, J. R.  1979.  Evidence for tidally correlated feeding rhythms in the eastern mud snail, *Ilyanassa obsoleta*.  The Nautilus. 93(1):  38–40.

Round, F. E.  1965.  The epipsammon:  a relatively unknown freshwater algal association.  Br. Phycol. Bull. 2:  456–462.

Tenore, K. R.  1977a.  Growth of the polychaete *Capitella capitata* cultured in different levels of detritus derived from various sources. Limnol. Oceanogr. 22:  936–941.

Tenore, K. R.  1977b.  Utilization of aged detritus derived from different sources by the polychaete *Capitella capitata*.  Mar. Biol. 44:  51–55.

Tenore, K. R., R. B. Hanson, B. E. Dornseif, and C. N. Weiderhold. 1979.  The effect of organic nitrogen supplement on the utilization of different sources of detritus.  Limnol. Oceanogr. 84: 350–355.

Thayer, G. W., S. M. Adams, and M. W. LaCroix.  1975.  Structural and functional aspects of a recently established *Zostera marina* community.  *In* L. E. Cronin (ed.), Estuarine Research, V. 1. pp. 518–540.  Academic Press, New York.

Wavre, M., and R. O. Brinkhurst.  1971.  Interactions between some tubificid oligochaetes and bacteria found in the sediments of Toronto Harbour, J. Fish. Res. Bd. Canada. 28:  335–341.

Weise, W., and G. Rheinheimer.  1978.  Scanning electron microscopy and epifluorescence investigations of bacterial colonization of marine sand sediments.  Microbial Ecol. 4:  175–188.

Wetzel, R. L.  1976.  Carbon resources of a benthic salt marsh invertebrate *Nassarius obsoleta* Say (Mollusca:  Nassariidae).  *In* M. Wiley (ed.), Estuarine Processes, V. II.  pp. 293–308.  Academic Press, NY.

Wieser, W.  1953.  Die Beziehung zwishcen Mundhohlengestalt Ernahrungsweise und Vorkommen bei freilebenden marinen Nematoden. Eine okologishe-morphologische Studie.  Ark. Zool. (2)4:  439–484.

Wieser, W.  1956.  Factors influencing the choice of substratum in *Cumella vulgaris* (Crustacea, Cumacea).  Limnol. Oceanogr. 1: 274–285.

Wieser, W.  1978.  Consumer strategies of terrestrial gastropods and isopods.  Oecologia (Berl.). 36:  191–201.

Wieser, W., and M. Zech.  1976.  Dehydrogenases as tools in the study of marine sediments.  Mar. Biol. 36:  113–122.

Yingst, J. Y.  1976.  The utilization of organic matter in shallow marine sediments by an epibenthic deposit-feeding holothurian. J. Exp. Mar. Biol. Ecol. 23:  55–69.

Yonge, C. M.  1930.  The crystalline style of the mollusca and a carnivorous habit cannot normally coexist.  Nature, Lond. 117: 444–445.

Yonge, C. M.  1939.  The protobranchiate mollusca:  a functional interpretation of their structure and evolution.  Phil. Trans. Roy. Soc. Lond. B230:  79–147.

Yonge, C. M.  1949.  On the structure and adaptations of the Telli-
    nacea, deposit-feeding Eulamellibranchia.  Phil. Trans. Royal
    Soc. Lond. (B) 609 (234):  29-76.
Zobell, C. E.  1938.  Studies on the bacterial flora of marine bot-
    tom sediments.  J. Sedim. Petr. 8(1):  10-18.
Zvyagintsev, D. G., and G. M. Galkina.  1967.  Ultrasonic treatment
    as a method for preparation of soils for microbiological analysis.
    Microbiology (Consultant's Bureau). 36:  910-916.

# The Role of Bioturbation in the Enhancement of Bacterial Growth Rates in Marine Sediments

Josephine Y. Yingst
Donald C. Rhoads

ABSTRACT: Bioturbation by particle reworking and
the exchange of sediment pore water with overlying
water increases the rate of nutrient mixing within sediment and accelerates the rate of flushing
of metabolites and growth inhibitors out of sediment. These changes are most common on mud bottoms dominated by equilibrium assemblages of highly mobile infaunal deposit feeders.

Sediment bacteria, an important food source
for many deposit feeders, have the potential for
increasing their metabolic activities and population growth rates under optimum conditions. Physical and chemical changes in habitat conditions
resulting from infaunal reworking could therefore
presumably enhance metabolic activities and growth
rates of microorganisms. We propose that bioturbation is a means for stimulating bacterial growth
rates. In addition, we would expect that biogenic
processes leading to the enhancement of bacterial
growth rates are best developed in soft-bottom
equilibrium communities.

We discuss the influence of life habit, trophic type, and reworking activities of macro- and

meiobenthos on standing stocks and growth rates
of bacterial populations in sediments. We also
describe our experimental approach for comparing
bacterial growth in regimes with different bioturbation rates.


## INTRODUCTION

Bioturbation is a universal phenomenon in benthic habitats
where macro- and meiofauna inhabit granular substrata. Bioturbation, in the form of both particle reworking and the exchange of
sediment pore water with overlying water by respiratory pumping, is
especially important in benthic ecosystems densely populated by infaunal deposit feeders. This biogenic reworking increases the rate
of nutrient mixing within the sediment and also accelerates the rate
of flushing of metabolites and growth inhibitors out of the sediments.

Bacteria, as well as other microorganisms, provide an important food source for many deposit feeders (Fenchel, 1970; Hargrave,
1970; Kofoed, 1975; Giere, 1975; Yingst, 1976; Lopez *et al.*, 1977;
Tenore, 1977). Sediment bacteria have the potential for increasing
their metabolic activities and population growth rates under optimum habitat conditions (Sorokin, 1964; Brock, 1971; Barsdate *et al.*,
1974; Aller, 1978; Gerlach, 1978; Aller and Yingst, 1978, in press).
Physical and chemical changes in a sedimentary habitat, resulting
from biogenic reworking, could therefore presumably enhance metabolic activities and growth rates of microorganisms. As Gerlach (1978)
pointed out, a bacteria-feeding metazoan theoretically could be relatively stationary, feeding on a very limited volume of sediment,
if it could provide (or create) conditions surrounding itself that
would stimulate bacterial growth.

We hypothesize that infaunal bioturbation is a means of stimulating bacterial growth in sediments. We also propose that at least
part of the energy expended in biogenic reworking is returned to a
bioturbating organism in the form of an enhanced food resource. Because this report represents an early stage in our research, the
above hypothesis has not been fully tested. In the following para-
graphs we shall review past research concerning the influence of the
activities of macro-, meio-, and microorganisms (other than bacteria) on bacterial population growth rates and metabolic activities
in marine sediments. We shall also describe our experimental approach for testing the hypothesis.


## BIOTURBATION AND FAUNAL SUCCESSION

Bioturbation has only recently been studied in the context of
faunal succession. Although these kinds of studies are few in number, biogenic habitat modifications that take place following a
physical disturbance of the seafloor appear to follow a pattern.
Many diagenetic changes in sediment properties can be related to the

progressive change in trophic structure and life habits of colonizing metazoans, especially macrofauna (Pearson and Rosenberg, 1976;
Rhoads *et al.*, 1977; McCall, 1977; Rhoads *et al.*, 1978).

Pioneering species tend to be tubiculous or otherwise sedentary
organisms that live near the sediment surface and feed on the surface or from the overlying water. Their influence on the sediments
through biogenic reworking may be intense, but is restricted to the
topmost centimeter or two (e.g., Myers, 1973). In contrast, high
order or equilibrium successional stages tend to be dominated by errant infaunal deposit-feeders that feed well below the interface and
intensively rework the sediment to depths of several centimeters
(Pearson and Rosenberg, 1976; Rhoads *et al.*, 1978; Fig. 1). Because
organism-sediment relations appear to be best developed in high-
order successional stages, it is likely that the biogenic mechanisms
which lead to the enhancement of microbial growth rates are probably
best developed in relatively undisturbed parts of the seafloor.

Our research deals with microbial growth rates only, not metabolic activity. The reader will understand that a complex relationship exists between microbial population growth and microbial metabolic activity. From the point of view of an organism that feeds
on bacteria, turnover rates are of paramount importance. In addition, although we are exploring the relationship between consumers
and a bacterial food source, we do not mean to imply that bacteria
are the only food utilized by deposit feeders. Cammen *et al.* (1978)
have shown that detrital particles themselves can be directly utilized by, and provide a major nutritive source for, some deposit-
feeding organisms.


ORGANISM-SEDIMENT INFLUENCES ON MICROBIAL POPULATIONS

In sediments, microbial distributions, activities, and growth
rates are influenced by a number of processes which can be related
to organism-sediment relations of both macro- and meiofauna. Bioturbation consists of two processes: biogenic manipulation of particles and pore water exchange. As a generalization, fluid transport is 10 to 100 times greater on a weight basis than particle reworking (Aller, 1977). Particle manipulation during feeding, burrow
construction, and locomotion can result in: 1) the transfer of
newly sedimented organic particles at the sediment surface to depth
in the sediment, thus providing fresh energy-rich substrates for microbial oxidation over a considerable depth interval (Aller and
Yingst, 1978; Anderson and Meadows, 1978; Aller, in press; Aller and
Yingst, in press); 2) upward transfer of reduced compounds (electron
acceptors) from below the aerobic-anaerobic boundary or redox potential discontinuity (RPD) to the aerobic zone (Hayes, 1964; Fenchel
and Riedl, 1970; Rhoads, 1974); 3) selective concentration of the
most organic-rich particles into fecal pellets (Whitlach, 1974),
which are recolonized by microorganisms following defecation
(Newell, 1965; Fenchel, 1970, 1972; Hargrave, 1972; Cullen, 1973;
Driscoll, 1975; Lopez *et al.*, 1977); and 4) breakdown of particle
aggregates, including fecal pellets, by grazing and burrowing activities, increasing the effective particle surface area for subsequent

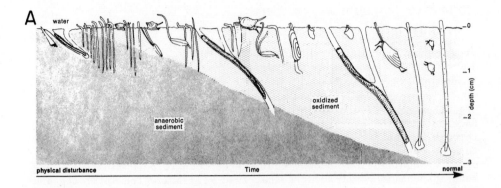

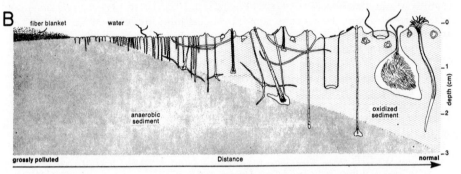

Figure 1. The trophic structure of benthic communities changes over time following major disturbances (A) and along chronic pollution gradients (B). Pioneering or pollution tolerant species (left side) tend to be tubicolous or otherwise sedentary organisms that live near the sediment surface and feed at the surface or from the water column. Bioturbation rates are low. In contrast, high order successional stages (right side) tend to be dominated by bioturbating infauna that feed at depth within the sediment. Organism-sediment relationships are most highly developed in equilibrium communities.

Our study of the relationship between bioturbation and microbial turnover rates involves comparing bacterial standing stocks and *in situ* bacterial turnover rates between pioneering and equilibrium assemblages in central Long Island Sound. This figure is from Rhoads, McCall and Yingst (1978). The pollution gradient diagram (B) is modified from Pearson and Rosenberg (1976).

microbial colonization (Cullen, 1973; Rhoads *et al*., 1977; Lopez and Levinton, 1978).

Periodic irrigation of constructed tubes or shafts and the less regular exchange of water within sediments during random burrowing by errant species can also affect microbial distributions and activities by increasing 1) the depth of the aerobic zone of the sediment (defined by the RPD boundary) and 2) the rate of transfer of dissolved nutrients into the bottom and flushing of metabolites or growth inhibitors from sediments orders of magnitude faster than molecular diffusion rates (Hargrave, 1970; Fenchel, 1970; Rhoads, 1974; Hylleberg, 1975; Aller, 1977, 1978; Rhoads *et al*., 1978).

Both the trophic or feeding type and the life habits of benthic organisms have been shown to affect particle and fluid transport within sediments (Aller, 1977, 1978, in press; Aller and Yingst, 1978). As Aller (1977) pointed out, mobile animals are more important than sedentary ones in homogeneously mixing sediment by burrowing. Sub-surface deposit-feeders are more important than surface deposit-feeders in reworking sediments vertically and in exchanging pore water with overlying water. Therefore, as indicated earlier, one might expect that in high-order successional stages dominated by highly mobile infaunal deposit-feeders would be enhanced bacterial growth and possibly metabolic activity to a greater extent than in pioneering assemblages.

The burrowing and feeding activities of individual meiofaunal taxa also affect the micro-distribution and growth of sediment bacteria (Cullen, 1973; Ott and Scheimer, 1973; Riemann and Schrage, 1978). The effect of the total microfaunal assemblage, though composed of small organisms, may be extensive.

The aerobic-anaerobic boundary or redox potential discontinuity (RPD) has been recognized for many years as the site of high microbial activity, both in the water column (Sorokin, 1964, 1965, 1972) and within sediments (Hayes, 1964; Fenchel and Riedl, 1970; Sorokin, 1978). The genus *Thiobacillus* (Vishniac and Santer, 1957) is especially abundant at this boundary. *Thiobacilli* oxidize reduced sulfur compounds which diffuse or are advected upward from the anaerobic zone. The importance of this chemical boundary to the vertical distribution of microorganisms in sediments can be seen in Fig. 2. A 5% agar medium ($ST_3$ of Tatewaki and Provasoli, 1964) was prepared using an extract of anaerobic sediment. The medium was cut into elongate strips and inserted vertically into a bioturbated sediment at station NWC, central Long Island Sound (for station location see Rhoads *et al.*, 1977). When the agar stick was removed from the sediment after 12 days, the aerobic zone just above the RPD was populated by numerous rod-shaped bacteria (Fig. 2).

In addition to increased numbers and/or activity of bacteria, high numbers of bacteria-feeding organisms have been found at the RPD. In the water column of a Volga River reservoir, Sorokin (1965) observed that high numbers of zooplankton occurred at the aerobic-anaerobic boundary and greatly reduced bacterial densities in that zone by grazing. In marine sands, Fenchel and Riedl (1970) found the maximum biomass of ciliates, nematodes and gnathostomulids to be at the RPD.

Mobile infaunal macrofauna may also feed at the RPD. Particle reworking and irrigation by the macrofaunal community at station NWC, Long Island Sound (Rhoads, 1974; Rhoads, Aller, Goldhaber, 1977) and by tubificid obligochaetes in lake sediments (Davis, 1974) influence the vertical position of the RPD layer by increasing the effective diffusion rates of oxygen into the bottom (Rhoads, Aller *et al.*, 1977). In the absence of infaunal reworking in Long Island Sound sediments, the RPD is very close to the sediment surface (1-3 mm). As in Figure 1, faunal succession proceeds from the interface downward into the bottom. The chemical effect of this "infaunalization" is an increased supply of oxygen to the sediments and a progressive depression of the RPD. This downward movement of the benthos may reflect increased competition among infaunal metazoa

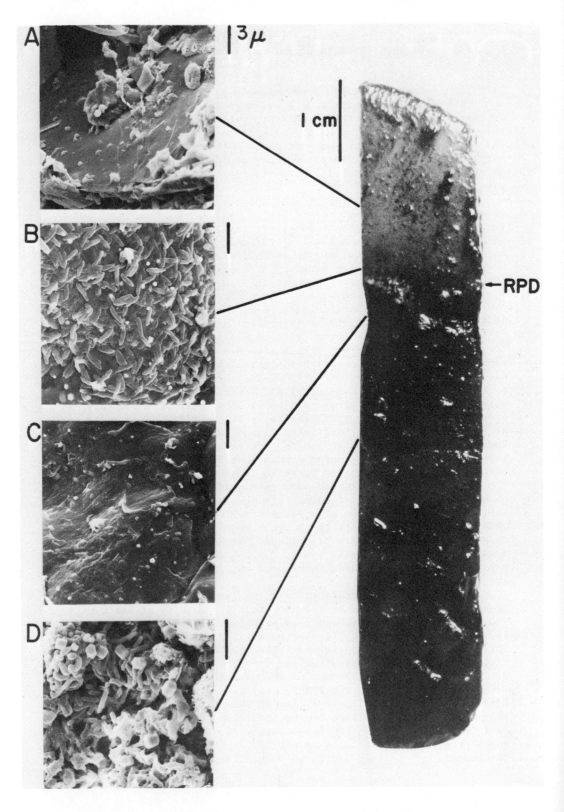

---

Explanation for Figure 2:  An agar stick placed vertically
in sediment for 12 days in central Long Island Sound.  This sediment is populated by an equilibrium community.  Agar subsamples
were examined with SEM from the aerobic zone (A) (0-2 cm), and
a zone immediately above the RPD (B) showing high numbers of
bacteria (Thiobacilli?).  The RPD zone (C) has been dissolved
away presumably by acid produced by the oxidation of sulphides
in that region.  No bacteria are associated with the dissolved
region.  The anaerobic zone (D) (dark area of the agar) is located below the RPD boundary.  High concentrations of bacteria
are seen in a narrow zone above the RPD presumably as a result
of the availability of electron acceptors (reduced substrates)
which diffuse or are advected by bioturbation across the RPD.

---

for microbial production at the RPD.  This feeding-RPD depth may
represent a dynamic equilibrium between peak grazing intensity
during summer months, microbial production, and the upward rebound
of the RPD during the winter when metazoan activities are low.  In
environments of stable climate and where detrital inputs are low,
one would expect infaunal feeding depths to be especially deep.

Nematodes also appear to stimulate bacterial growth through
mucus secretions that accumulate during feeding and burrowing.  Organic particles adhere to this mucus, a rich microflora develops,
and the nematodes subsequently ingest the mucus (Riemann and
Schrage, 1978).  Ott and Scheimer (1973) describe desmodorid nematodes (Family Stilbonematidae) that live under anaerobic conditions
and feed on bacteria growing on a mucus layer that covers their cu-
ticle.  Ciliates living under anaerobic conditions in sediments also
appear to harbor ectosymbiotic bacteria in mucus sections on their
outer surfaces.  These symbionts utilize released metabolites of the
ciliates and the ciliates, in turn, feed on the bacteria (Fenchel
*et al.*, 1977).

Macrofauna, including amphipods (Johannes, 1965; Fenchel, 1970;
Hargrave, 1970; Levinton *et al.*, 1977; Lopez and Levinton, 1978),
the shrimp *Palaemonetes* sp. (Welsh, 1975), and the gastropod *Hydrobia ulvae* (Fenchel, 1972), enhance bacterial activity and decomposition rate of organic detritus.  Meiofauna, in particular nematodes,
increase the oxidation rate of *Zostera* detritus in combination with
the polychaetes *Nephthys incisa* (Tenore *et al.*, 1977) or *Capitella
capitata* (Lee *et al.*, 1976), relative to the oxidation rate of *Zostera* in controlled systems where bacteria alone or bacteria and only
polychaetes are present.  Ciliates that graze on bacteria also appear to enhance rates of organic detritus decomposition (Johannes,
1965; Barsdate *et al.*, 1974; Fenchel, 1977).

The majority of the above investigators measured microbial activity but not changes in bacterial numbers or biomass.  Fenchel
(1977), however, measured both biomass and microbial activity and
found that grazing by protozoans of bacteria associated with hay particles kept the bacterial biomass at levels approximately half of
that found in an ungrazed system, but that the grazing strongly enhanced the decomposition rate of hay particles.

The idea that positive feedback relationships exist between feeding activities of infaunal deposit-feeders and the abundance of sediment microorganisms is not new (Newell, 1965; Levinton, 1972; Driscoll, 1975; Hylleberg, 1975; Rhoads *et al.*, 1977; Gerlach, 1978; Rhoads *et al.*, 1978). Hylleberg (1975) introduced the term "gardening" to describe the stimulatory effect of feeding of the lugworm *Abarenicola pacifica* on microbial and meiofaunal populations. Through irrigation, the worm pumps oxygen and nutrients into a feeding pocket from water overlying its burrow. Its own excretory products may also be involved in stimulating the growth of bacteria, other microorganisms, and nematodes in the pocket. These organisms are subsequently consumed by the lugworm. The actual site of enhanced microbial growth appears to be at the margins of the feeding pocket, where aerobic sediment contacts anaerobic sediment. Other species of deposit-feeders, for example the maladanid polychaete *Clymenella torquata*, also appear to feed at the aerobic-anaerobic boundary. *C. torquata*, a conveyor-belt species (Rhoads and Stanley, 1965), extends shafts of coarse-grained oxygenated sediments progressively deeper into the bottom as it selectively ingests fine sediment particles at the RPD (Fig. 3).

---

Explanation for Figure 3:   The influence of the feeding and burrowing activities of an equilibrium species, *Clymenella torquata*, on intertidal sediment from Barnstable Harbor, Cape Cod, Mass.   Note that intensive feeding areas at depth are surrounded by an RPD.   Tube irrigation produces the localized aerobic areas where intensive polychaete feeding occurs.   Mounds at the surface are formed by sediment which is defecated at the sediment-water interface by the worms.

---

Another example of the "gardening" phenomenon is found in the burrowing mud shrimp *Upogebia litoralis* that collects plant material on the sediment surface, packs it into the inner walls of its burrow, and periodically ingests the bacteria and fungi that develop on the plant detritus (Frey and Howard, 1975; Ott *et al.*, 1976). The detritus-feeding amphipod *Hyalella azteca* increases the rate of recolonization of its own fecal pellets by microorganisms through the secretion of metabolites (Hargrave, 1970).

TESTING THE HYPOTHESIS

Many bioturbating organisms ingest bacteria as they rework sediments.   Therefore standing stocks of bacteria may be significantly lower in high bioturbation rate regimes than in sediments with low bioturbation and grazing rates (Aller and Yingst, in press; and our unpublished data).   Our hypothesis states, however, that bacterial growth rates in highly bioturbated sediment should be enhanced over those in sediments with low rates of biogenic reworking.   In order to test this hypothesis, one must measure bacterial growth in sediments that experience different rates of biogenic reworking under *in situ* conditions where:   1) grazing pressure on bacteria is eliminated, 2) rates of biogenic reworking are not altered, 3) the physical and chemical conditions within the sediment, which have been induced by the bioturbating infauna, are not significantly changed, and 4) the methods of measurement do not involve the addition of growth media.

We believe that all four of the above conditions can be satisfied by a modification of the capillary tube techniques of Perfil'ev and Gabe (1969).   The capillary tubes are inserted horizontally into the sediment at specific depth intervals within the top 10 cm.   Sediment particles with their associated bacterial flora move into the ends of the tubes and serve as the initial inoculum.   With this technique, we have been able to examine the growth of bacteria on particles in the absence of grazing and under different rates of reworking activity.   The internal diameter of the capillary tubes (1.25 mm) is large enough that all particle sizes can enter the tubes yet small enough to exclude grazing meio- and macrofauna.   The tubes are short (0.5 cm) so that they present a minimal barrier to horizontal diffusion (Yingst, Rhoads, Aller, in prep.).   Microscopic

examination of sediment inside of the capillary tubes has revealed
the presence of meiofaunal organisms in only 3 instances. Ciliates
and flagellates are able to enter the tubes. However, their numbers
are low in muddy Long Island Sound sediments, and they are observed
infrequently within the capillary tubes. Most of the bacteria in
the capillary tubes are attached to particles. A small percentage,
however, are free in the interstitial water and occasional cells are
found to be attached to the inner wall of the glass tubes.

Figure 4 shows two generalized growth curves of bacteria based
on capillary tube experiments in which bacterial growth on particles
within tubes is compared between a highly bioturbated sediment and
the same sediment devoid of bioturbators. The cumulative change in
bacterial numbers/unit weight of sediment over a depth interval of up
to 10 cm is plotted against duration of emplacement of the tubes in
the sediment (see Yingst and Rhoads, in prep.; Yingst *et al.*, in
prep.).

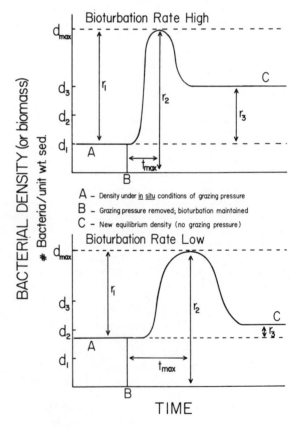

Figure 4. Generalized growth responses of bacteria to high and low rates of bioturbation. In high bioturbation rate (and high grazing rate) systems, the standing stock of bacteria (A) is low relative to low bioturbation rate (low grazing rate) systems. If grazing pressure is removed (B), but the chemical and physical effects of bioturbation are maintained, different growth responses of bacteria are observed in the two regimes:

$t_{max}$ = exponential growth period following removal of grazing effects.

$D_{max}$ = maximum density of bacterial following removal of grazing pressure (B).

$r_1$ = density difference between ambient bacterial densities and $D_{max}$.

$r_2$ = absolute abundance at $t_{max}$.

$r_3$ = difference between the ambient standing stock of bacteria (A) and the equilibrium density in the absence of grazing pressure. See text for explanation of experimental methods and discussion of growth responses of bacteria in different bioturbation rate conditions.

Initially, particles move into the tubes from the surrounding sediment with their associated bacteria. There is no significant change in the weight of sediment inside of the tubes after the first sampling period. Once inside, the bacteria on particles multiply rapidly. Tube contents are stained with acridine orange and epifluorescence microscopy reveals extensive colony formation after only 4 hours of isolation from the ambient sediment (at $22^{o}$C). This increase in bacterial numbers appears to be largely the result of the division of cells attached to particles that initially moved into the tubes and not from continued immigration of bacteria from sediment outside of the tubes. These results agree with those of Bott and Brock (1970), who demonstrated that *in situ* division of bacteria initially attached to glass slides exposed to pond water accounted for the greatest fraction of bacterial population growth rather than continued immigration and attachment of cells from the surrounding medium. In our capillary tube experiments, the growth phase of bacteria on the sediment particles continues to a maximum density of cells ($D_{max}$, Fig. 4) at $t_{max}$ (time to peak abundance). After this time, as a result of the accumulation of growth inhibitors and possible food and space limitations, cell numbers decline. Eventually a new equilibrium density is reached which is above standing stock densities in the ambient sediment outside the tubes ($r_3$).

If our hypothesis is correct, we expect that: 1) the growth response (doubling time during the exponential phase of growth) will be shorter in the highly bioturbated system; 2) $r_3$, the difference between the new equilibrium density (C) and *in situ* standing stocks in the ambient sediment (A), will be greater in the highly bioturbated system; and 3) $t_{max}$, the time to reach maximum density after tubes are filled by sediment particles (B), will be shorter in the highly bioturbated system. On the other hand, we would not necessarily expect that $D_{max}$, $r_1$ (the difference between ambient standing stock densities and $D_{max}$), or $r_2$ (the total number of bacteria at $D_{max}$) would be significantly different in either the high or low bioturbation rate system. The peak density of bacteria can be determined by factors which are only indirectly related to population growth rates, for example the number of bacteria or quality of particles initially entering the capillary tubes.

Bacterial population growth rates obtained with our capillary tube method, while useful for comparing bacterial growth responses to different rates of reworking activity in sediments, are not assumed to be representative of absolute rates of bacterial growth under natural conditions. The emplacement of glass tubes into the bottom may have a stimulatory effect. However, because the capillary tube effect is always present, we are able to compare relative rates of bacterial population growth.

With our method we are concentrating on the influence of rates of biogenic reworking on bacterial growth in sediments. One would also like to know how growth rates are related to bacterial activity. A recent attempt to differentiate metabolically active bacteria from total numbers *in situ* combines a method for detecting cytochemical dehydrogenase activity with direct counting techniques (Zimmerman *et al.*, 1978). While promising, this combination of methods does not detect all active bacteria, especially those $\leq 0.4\mu$ in length, and the effect of sampling and incubation on changing the dehydrogenase activity is unknown. In addition, the relationship between the number of respiring cells and number of dividing cells has not been determined.

It is likely that positive feedback mechanisms exist between deposit-feeders and decomposers. Enhancement of growth and metabolic activity, through the process of bioturbation, may represent one such important feedback mechanism. These feedback mechanisms must be measured in benthic systems where complex organism-sediment interactions remain relatively undisturbed by the measurement process. We believe that the capillary tube method, as outlined in this paper, may be one technique that satisfies this requirement.

## ACKNOWLEDGMENTS

The studies on which this paper is based were supported by NSF grant GA-42-838 (D. C. Rhoads, principal investigator). We thank D. Muschenheim, J. Germano, W. Ullman, R. Aller, M. Reed and M. Pimer for aid in the field. D. Gaudreau, R. Aller and D. Muschenheim helped with laboratory experiments and D. Muschenheim drafted figures for this paper. Advice and critical comments were provided by many, including: R. Aller, L. Provasoli, J. Hobbie and P. Jumars.

## REFERENCES

Aller, R. C. 1977. The influence of macrobenthos on chemical diagenesis of marine sediments, 600 pp. Ph.D. thesis. Yale University, Connecticut.

Aller, R. C. 1978. Experimental studies of changes produced by deposit feeders on pore water, sediment, and overlying water chemistry. Am. J. Sci. 278: 1185-1234.

Aller, R. C. In press. Diagenetic processes near the sediment-water interface of Long Island Sound, I: Decomposition and nutrient element geochemistry (S,N,P). *In* B. Saltzman (ed.), Estuarine Physics and Chemistry: Studies in Long Island Sound. Advances in Geophysics. Vol. 22. Academic.

Aller, R. C., and J. Y. Yingst. 1978. Biogeochemistry of tube dwellings: a study of the sedentary polychaete *Amphitrite ornata* (Leidy). J. Mar. Res. 36(2): 201-254.

Aller, R. C., and J. Y. Yingst. In press. Relationship between microbial distributions and the anaerobic decomposition of organic matter in surface sediments of Long Island Sound, U.S.A. Mar. Biol.

Anderson, J. G., and P. S. Meadows. 1978. Microenvironments in marine sediments. Proc. Roy. Soc. Edinburgh, 76B: 1-16.

Barsdate, R. J., R. Prentski, and T. Fenchel. 1974. Phosphorus cycle of model ecosystems: significance for decomposer food chains and effect of bacterial grazers. Oikos. 25: 239-251.

Bott, T. L., and T. D. Brock. 1970. Growth and metabolism of periphytic bacteria: methodology. Limnol. Oceanogr. 15: 333-342.

Brock, T. D. 1971. Microbial growth rates in nature. Bacteriol. Rev. 35: 39–58.

Cammen, L., P. Rublee, and J. Hobbie. 1978. The significance of microbial carbon in the nutrition of the polychaete *Nereis succinea* and other aquatic deposit feeders. UNC Sea Grant Publication UNC-SG-78-12, 84 pp.

Cullen, D. J. 1973. Bioturbation of superficial marine sediments by interstitial meiobenthos. Nature. 242: 323–324.

Davis, R. B. 1974. Tubificids alter profiles of redox potential and pH in profundal lake sediment. Limnol. Oceanogr. 19: 19: 342–346.

Driscoll, E. G. 1975. Sediment-animal-water interaction. Buzzard's Bay, Massachusetts. J. Mar. Res. 33(3): 275–302.

Fenchel, T. 1970. Studies on the decomposition of organic detritus derived from the turtle grass *Thalassia testudinum*. Limnol. and Oceanogr. 15(1): 14–20.

Fenchel, T. 1972. Aspects of decomposer food chains in marine benthos. Verh. Deutsch. Zool. Ges. 65 Jahresbersamml. 14: 14–22.

Fenchel, T. 1977. The significance of bactivorous protozoa in the microbial community of detrital particles, pp. 529–544. *In* J. Cairns (ed.), Aquatic Microbial Communities. Garland.

Fenchel, T., and R. J. Riedl. 1970. The sulfide system: a new biotic community underneath the oxidized layer of marine sand bottoms. Mar. Biol. 7: 255–268.

Fenchel, T., T. Perry, and A. Thane. 1977. Anaerobiosis and symbiosis with bacteria in free-living ciliates. J. Protozool 24(1): 154–163.

Frey, R. W., and J. D. Howard. 1975. Endobenthic adaptations of juvenile thalassinidean shrimp. Bull. Geol. Soc. Denmark 24: 283–297.

Gerlach, S. A. 1978. Food-chain relationships in subtidal silty sand marine sediments and the role of meiofauna in stimulating bacterial productivity. Oecologia (Berl.). 33: 55–69.

Giere, O. 1975. Population structure, food, relation and ecological role of marine oligochaetes, with special reference to meiobenthic species. Mar. Biol. 31: 139–156.

Hargrave, B. T. 1970. The effect of a deposit-feeding amphipod on the metabolism of benthic microflora. Limnol. Oceanogr. 15: 21–30.

Hargrave, B. T. 1972. Aerobic decomposition of sediment and detritus as a function of particle surface area and organic content. Limnol. Oceanogr. 17: 583–596.

Hayes, F. R. 1964. The mud-water interface, pp. 121–145. *In* H. Barnes (ed.), Oceanogr. Mar. Biol. Ann. Rev. 2. Allen and Unwin.

Hylleberg, J. 1975. Selective feeding by *Abarenicola pacifica* with notes on *Abarenicola vagabunda* and a concept of gardening in lugworms. Ophelia. 14: 113–137.

Johannes, R. E. 1965. Influence of marine protozoa on nutrient regeneration. Limnol. Oceanogr. 10: 434–442.

Kofoed, L. N. 1975. The feeding biology of *Hydrobia ventrosa* (Montagu). 1. The assimilation of different components of the food. J. Exp. Mar. Biol. Ecol. 19: 233–241.

Lee, J. J., K. R. Tenore, J. H. Tietjen, and C. Mastropaole. 1976.

An experimental approach toward understanding the role of meio-
fauna in a detritus-based marine food web, pp. 140-147. *In* C. E.
Cushing (ed.), Proc. 4th Natl. Symp. on Radioecology. Dowden,
Hutchinson and Ross.

Levinton, J. S. 1972. Stability and trophic structure in deposit-
feeding and suspension-feeding communities. Am. Nat. 106: 472-
486.

Lopez, G. R., J. S. Levinton, and L. B. Slobodkin. 1977. The effect of grazing by the detritivore *Orchestia grillus* on *Spartina*
litter and its associated microbial community. Oecologia (Berl.).
30: 111-127.

Lopez, G. R., and J. S. Levinton. 1978. The availability of microorganisms attached to sediment particles as food for *Hydrobia*
*ventrosa* Montagu (Gastropoda: Prosobranchia). Oecologia (Berl.).
32: 263-275.

McCall, P. L. 1977. Community patterns and adaptive strategies of
the infaunal benthos of Long Island Sound. J. Mar. Res. 35(2):
221-266.

Myers, A. C. 1973. Sediment reworking, tube building, and burrowing in a shallow subtidal marine bottom community: rates and effects, Ph.D. Dissertation, Univ. of Rhode Island, Kingston, R.I.,
117 pp.

Newell, R. C. 1965. The role of detritus in the nutrition of two
marine deposit-feeders, the prosobranch *Hydrobia ulvae* and the
bivalve *Macoma balthica*. Proc. Zool. Soc. Lond. 144: 25-45.

Ott, J. A., and F. Scheimer. 1973. Respiration and anaerobiosis
of free living nematodes from marine and limnic sediments.
Nethel. J. Sea. Res. 7: 233-243.

Ott, J. A., B. Fuchs, R. Fuchs, and A. Malasek. 1976. Observations
on the biology of *Callianassa stebbingi* Borradaille and *Upogebia*
*litoralis* Risso and their effect upon the sediment. Senckenberg.
Marit. 8: 61-79.

Pearson, T. H., and R. Rosenberg. 1976. A comparative study of the
effects on the marine environment of wastes from cellulose in-
dustries in Scotland and Sweden. Ambio. 5: 77-79.

Perfil'ev, B. V., and D. R. Gabe. 1969. Capillary methods of investigating microorganisms. Univ. of Toronto.

Riemann, F., and M. Schrage. 1978. The mucus-trap hypothesis on
feeding of aquatic nematodes and implications for biodegradation
and sediment texture. Oecologia (Berl.). 34: 75-88.

Rhoads, D. C. 1974. Organism-sediment relations on the muddy sea
floor. Oceanogr. Mar. Biol. Ann. Rev. 12: 263-300.

Rhoads, D. C., and D. J. Stanley. 1965. Biogenic graded bedding.
J. Sed. Petrol. 35(4): 956-963.

Rhoads, D. C., R. C. Aller, and M. Goldhaber. 1977. The influence
of colonizing benthos on physical properties and chemical diagenesis of the estuarine seafloor, pp. 113-138. *In* B. C. Coull
(ed.), Ecology of Marine Benthos, V. 6. Univ. of S. Carolina.

Rhoads, D. C., P. L. McCall, and J. Y. Yingst. 1978. Disturbance
and production on the estuarine seafloor. Am. Sci. 66(4): 577-
586.

Sorokin, Yu. I. 1964. On the trophic role of chemosynthesis in
water bodies. Int. Rev. Ges. Hydrobiol. 49(2): 307-324.

Sorokin, Yu. I. 1965. On the trophic role of chemosynthesis and bacterial biosynthesis in water bodies. Mem. Ist. Ital. Idrobiol. 18 Suppl.: 187-205.

Sorokin, Yu. I. 1972. The bacterial population and the processes of hydrogen sulphide oxidation in the Black Sea. J. Cons. Int. Explor. Mer. 34(3): 423-454.

Sorokin, Yu. I. 1978. Microbial production in the coral-reef community. Arch. Hydrobiol. 83(3): 281-323.

Tatewaki, M., and L. Provasoli. 1964. Vitamin requirements of three species of *Antithamnion*. Botanica Marina. 6: 193-203.

Tenore, K. R. 1977. Food chain pathways in detrital feeding benthic communities: a review, with new observations on sediment resuspension and detrital recycling, pp. 37-53. *In* B. C. Coull (ed.), Belle Baruch Symp. on the Ecology of Marine Benthos, V. 6. Univ. of S. Carolina.

Tenore, K. R., J. H. Tietjen, and J. J. Lee. 1977. Effect of meiofauna on incorporation of aged eelgrass *Zostera marina* detritus by the polychaete *Nephthys incisa*. J. Res. Bd. Can. 34(4): 563-567.

Vishniac, W., and M. Santer. 1957. The Thiobacilli. Bact. Rev. 21: 195-213.

Welsh, B. L. 1975. The role of grass shrimp *Palaemonetes pugio* in a tidal marsh ecosystem. Ecology. 56: 513-530.

Whitlach, R. B. 1974. Food-resource partitioning in the deposit-feeding polychaete *Pectinaria gouldii*. Biol. Bull. 147: 227-235.

Yingst, J. Y. 1976. The utilization of organic matter in shallow marine sediments by an epibenthic deposit-feeding holothurian. J. Exp. Mar. Biol. Ecol. 23: 55-69.

Yingst, J. Y. 1978. Patterns of micro- and meiofaunal abundance in marine sediments, measured with the adenosine triphosphate assay. Mar. Biol. 47: 41-54.

Zimmerman, R., R. Iturriaga, and J. Becker-Birck. 1978. Simultaneous determination of the total number of aquatic bacteria and the number thereof involved in respiration. Appl. Environ. Microbiol. 36(6): 926-936.

# Particle Feeding by Deposit-Feeders:
# Models, Data, and a Prospectus

Jeffrey S. Levinton

ABSTRACT: In this essay I summarize the mechanisms of particle feeding by deposit-feeding benthos and provide models for consideration. Particle selection by benthos is dependent upon feeding
rate, food quality, physical characteristics of the
particles and depth/mobility of the animals in the
sediments. Particle size optimization and the relationship between particle size and body size are
important considerations for successful feeding
on particles.

INTRODUCTION

Recent interest in deposit-feeding has centered around the relationship of particle size to nutrition, niche partitioning, and
the possible optimization of particle feeding by deposit feeders.
I wish to summarize what we know about particle feeding by deposit
feeders in general and the marine hydrobiid snails in particular.
I shall also point out shortcomings of our current models and research methods and suggest directions for future research.

Deposit feeding is feeding upon particles, either several at a time or in short succession, on or in the bottom by appressing or aiming a feeding organ against or towards the substratum. This definition obscures the details of size, origin and qualitative differences among particles. This is justifiable because many macrofaunal deposit feeders consume a wide variety of particles, and effective comparisons can be made between species that select different particle-types. I have avoided an exact definition of the term particle. Ingesting particles "several at a time or several in short succession" implies the exclusion of macroalgal feeders.

Benthic deposit feeders are of particular interest because it should be possible to define the "niche," i.e., the habitat, feeding rate, qualitative and size differences in particle feeding, and effects of the animal upon sediment texture, chemistry and pore water chemistry. In studies of interspecific competition and succession, we can thus define virtually all of the relevant parameters needed to estimate ecological overlap between species. This makes predictions of competitive interactions feasible and suggests experiments appropriate to overlap in relevant niche dimensions. It is also possible to define rates and processes of nutrient cycling in sediments. The effects of feeding upon particles can be studied with regard to effects upon the microflora (e.g., Hargrave, 1970; Lopez et al., 1977), changes of particle size (Fenchel, 1970) and the role of nutrient concentration in mineralization of particulate organic matter (e.g., Wetzel, 1976; Lopez et al., 1977). Burrowing behavior, fecal pellet formation and the very geometry of deposit-feeding living positions can all be related to sedimentary chemistry and nutrient cycling (Rhoads and Young, 1970; Rhoads, 1967; Jorgensen, 1977; Aller, 1980). All of these effects can be studied in the field and in microcosms.


PARAMETERS

Feeding Rate

Rate of particle ingestion is important from the standpoint of deposit-feeder nutrition and nutrient cycling in sediments. If ingestion is totally non-selective, then the rate of ingestion per unit time per animal would be a useful measure of feeding. Egestion rate is a practical way of measuring ingestion rate (e.g., Kristensen, 1973; Hargrave, 1972). However, at least two complications make egestion rates misleading as an estimate of food intake and influence on nutrient cycling:
1)  The deposit feeder feeds selectively on a particle spectrum or a range of organic particles and not on the whole sediment. Because the maldanid polychaete *Clymenella torquata* does not usually ingest particles greater than one mm in diameter, fine particles are usually sorted towards the surface, leaving a lag deposit of coarse particles near the bottom of the tube (Rhoads and Stanley, 1965). When swallowing particles, *Hydrobia* spp. ingest fine particles at more rapid rates than coarse particles (Fenchel and Kofoed, 1976; Lopez and Levinton, 1978; Levinton, 1979).

2)  Gut retention time and feeding rate might vary with changes in sediment texture or organic content (Cammen *et al.*, 1978; Taghon *et al.*, 1978). Furthermore, gut emptying rate might decrease with increased starvation time (Calow, 1975).

Despite potential variation, some generalizations can be made. Fecal production, as estimated by log surface area of pellets, is linear with log animal dry weight (Hargrave, 1972). Cammen *et al.* (1978) reevaluated published data and found that ingestion rate of organic matter is closely related to body size across a wide spectrum of deposit-feeding species (bivalves excluded). Ingestion rate of organic matter was independent of the concentration of organic matter in the food. This indicates that the feeding rate of organic matter is regulated by the nutritional requirements of the deposit-feeder. Species living in sediments typically poor in organic matter probably exert greater influence in terms of sediment reworking and nutrient cycling because they must process more sediment to obtain food. This relationship may not hold below the level of species. A given species might be adapted to a certain narrow sediment type and may not be able to greatly alter its feeding rate when transferred to sediments of differing organic content (but see Frankenburg and Smith, 1967).

Food Quality

The possible importance of food quality was investigated by Whitlatch (1974) for *Pectinaria gouldii*. This polychaete is common in New England soft-sediments and is responsible for considerable sediment turnover (Gordon, 1966). Worms showed an approximately linear relation between body length and mean particle diameter. However, positive selection was shown for fecal material, floc aggregates and mineral grains encrusted with organic matter. Through selective feeding worms concentrated an average of 32.7% possible organic matter found in the sediment to an average of 42.7% in gut contents.

A widely accepted model for deposit-feeding is the efficient assimilation of microbial biomass concomitant with a generally miniscule digestion and assimilation of non-living particulate organic detritus (Newell, 1965; Hargrave, 1970, 1972; Fenchel, 1970; Lopez *et al.*, 1977; Wetzel, 1976). Non-living organic matter is envisaged to consist of compounds too refractory for the digestive capacities of macrofaunal deposit feeders. Food availability would thus be regulated by not only the feeding rate of the deposit feeder, but by the rate at which non-living substrates can be converted into microbial biomass. As grazing often stimulates microbial activity, feeding activities can actually increase the standing stock of food for deposit feeders (Barsdate *et al.*, 1974; Lopez *et al.*, 1977).

Even if the assimilation of nutrients from non-living organic matter is low, the role of the latter in deposit feeders may be more important than is now generally supposed. Uncertainty exists as to what percent microbial carbon is of total sedimentary carbon. Estimates based upon ATP extractions suggest less than 3% living carbon in sediments (Ferguson and Murdoch, 1975; Christian *et al.*,

1975).  Estimates from muramic acid are higher by more than one
order of magnitude (e.g., Morrison *et al*., 1977).  The significance
of this methodological disagreement cannot be overestimated.  If
the carbon of sediments is 99% non-living, then even a 100% assimilation efficiency will yield only a 1% return for a strict microbial
feeder.  But a mere 1.01% assimilation efficiency on non-living
carbon will gain the same degree of nutrition!  It is safe to say
that current techniques are not accurate enough to measure such a
small percent assimilation.

To complicate matters, it is unclear as to whether microbial
carbon is sufficient to satisfy the requirements of deposit-feeding
populations.  Wetzel (1976) presents data consistent with the adequacy of microbial carbon for the mud snail *Ilyanassa obsoletus*.
However, several other studies have suggested that sedimentary microbial carbon is far too dilute to balance production and respiration (e.g., Tunnicliffe and Risk, 1977; Cammen *et al*., 1978).  Meiofauna, non-living carbon or DOC are possible sources of balance.
Structural carbohydrates may be assimilated by some deposit-feeding
species adapted to ingesting mainly organic detritus (e.g., Forster
and Gabbot, 1971), but they seem to be beyond the enzymatic capabil-
ities of many deposit feeders that ingest sediment (e.g., Calow,
1975; Kristensen, 1972; Crosby and Reid, 1971).

Nitrogen uptake seems to be more clearly linked to microbial
production and is likely to exert a regulatory role in deposit-
feeder production (e.g., Newell, 1965; Fenchel, 1972; Tenore, 1977;
Tenore *et al*., 1979).  Most particulate organic matter would be nitro-
gen-deficient without a microbial community.  As detritus ages, the
establishment of a rich microbial community results in a decrease of
the carbon:nitrogen ratio (Harrison and Mann, 1975; Mann, 1972).  Trituration of and grazing on organic detrital particles hastens colonization by microbial organisms and increases mineralization (Fenchel,
1970; Lopez *et al*., 1977).  Some forms of organic detritus are more
labile than others, suggesting that microbial colonization and detrital mineralization may vary with detritus type (e.g., Tenore *et
al*., 1979).  Given the importance of the microbial community as a
nitrogen source, models of renewable resources are appropriate in
predicting deposit-feeder standing stock (Levinton, 1979).  For the
genus *Hydrobia*, slowly renewing large diatoms may be the rate-
limiting renewable resource (Fenchel and Kofoed, 1976; Levinton,
1979).

Physical Characteristics of Particles

Several mechanical properties of particles provide potential
variation in deposit-feeder adaptations and suggest possible limitations to deposit feeding.  The most obvious property is size; we
neglect shape for the moment and simply consider the diameter of a
sphere with the same volume as the particle (spherical equivalent
diameter - SED).  Particle diameter has been related to body size
(e.g., Whitlatch, 1974; Fenchel, 1975) and the size of feeding
organs (Hughes, 1973, 1975).  Other important mechanical properties
include:  particle shape, density, and surface smoothness (Hughes,
1975; Self and Jumars, 1978).  Dense, heavy particles might simply
drop off of ciliated feeding tracts.

Depth, Orientation and Mobility Below Sediment-Water Interface

Position and feeding depth below the sediment-water interface influences sediment reworking and provides a gradient permitting microhabitat subdivision among deposit-feeding species (Rhoads, 1967; Levinton, 1977; Nicolaisen and Kanneworf, 1969; Levinton and Bambach, 1975). Interactions between species living and feeding at the same depth involve competition for space and perhaps food (Levinton, 1977; Woodin, 1974). Competition among infaunal deposit feeders and suspension feeders may result in stratified occurrence of species (Levinton and Bambach, 1975; Peterson, 1977).

## MODELS

### Particle Size Optimization

In order to investigate deposit-feeding specialization, it is useful to describe how a deposit feeder optimizes food intake. For example, the relationship of deposit-feeder body weight to rate of feeding on organic matter, independent of percent organic matter in the sediment, suggests an optimization of food intake. Can we further describe optimization patterns with regard to particle size?

Unfortunately, models optimizing food intake on the basis of particle size are plagued by the question of whether the animal feeds upon whole morsels of food (e.g., diatoms) or upon sedimentary grains coated with an attached microbial biota (e.g., bacteria or small diatoms). For example, the mud snail *Hydrobia* may derive nutrition from large mobile diatoms or bacteria and small algae attached to non-living particles (Kofoed, 1975; Fenchel and Kofoed, 1976); Lopez and Levinton, 1978). If feeding is upon edible morsels such as free-living diatoms, then it probably makes sense to feed upon the largest particle that can be ingested, as balanced by consideration of increased handling time per particle and the nature of the feeding organ (e.g., sieve-like). If non-living particles with attached epibionts are ingested, then surface/volume constraints might apply. The surface to volume ratio (SV) of a sphere is:

$$SV = \frac{6}{D} \, ,$$

where D is the diameter. Thus, if a constant volume of sediment is ingested per unit time, we might expect particle feeding rates to decrease hyperbolically with increasing particle size.

Several other parameters might be important in devising an optimal recovery model for deposit feeders:
1) gut passage time
2) assimilation efficiency per unit time particle is in the gut
3) gut volume
4) average time between feedings
5) correlations between particle diameter and nutritive value that depart from the surface/volume ratio of a sphere
6) mobility as a function of sediment properties

Taghon *et al.* (1978) investigated the interrelationship of some of these parameters via simulation of simple equations relating energy acquisition to cost functions subtracted from gain functions. As might be expected, smaller particles are preferred when particle surfaces are grazed. Under some conditions, however, larger particles might be taken as well. If the gut must be filled and gut volume is large, ingesting larger particles might decrease the cost of particle acquisition. Also, when the cost of rejecting larger particles is great, an optimal strategy might include feeding on larger particles.

The conditions for selecting larger particles rest on the assumption that the cost of collecting a large particle is the same as that of a small particle. As most deposit feeders tend to feed upon volumes of particles, it is likely that the cost of collecting particles is less for small particles than for large ones. The results of Taghon *et al.* are therefore probably not useful in these details.

A second form of optimization involves adjustments of a deposit feeder to sediments of differing particle size spectra; a deposit feeder that prefers small particles might find itself in a sediment consisting only of large grains of sand. Deposit-feeding invertebrates such as *Capitella capitata*, *Hydrobia* spp. and *Ilyanassa obsoletus* may each be found in a wide range of sediment types. Fine particles may be richer in attached microflora because of a favorable surface area/volume ratio. Such particles, however, might be rare in some sediments. Should the deposit feeder still feed preferentially on fine particles?

Alternatively deposit-feeders feeding upon diatoms might be found in sandy sediments with a wide range of diatom concentrations. Should the deposit feeder forage when diatoms are rare and sand grains are common? Consider the following simple model. Assume that the deposit feeder makes no distinction in feeding rate on the basis of size of particle—only quality. Consider two identically sized particle types; type 1 are diatoms, and type 2 are nutritionally useless sand grains. Assume that for every N particle encountered, x particles must be diatoms for the deposit-feeder energy requirements to be satisfied. If p is the proportion of diatoms in the sediment, then the probability, P, of getting the requisite x diatoms is:

$$P = \sum_{x}^{N} p^x(1-p)^{N-x}$$

Figure 1 shows a threshold below which the probability of getting the requisite diatoms is zero. Under such circumstances, should the animal continue to feed in this way or should it cease feeding or switch to another feeding mode? A similar argument could be made for an animal that can only swallow particles below a certain size. In certain sediments, the edible small particles might be too rare to sustain energetic requirements. Should the animal switch feeding modes? Cammen *et al.* (1978) present data consistent with an evolved response to ingesting a requisite amount of organic matter per unit time.

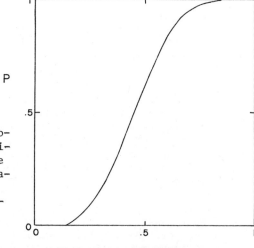

Figure 1. Example of bino-
mial model of obtaining a sufficient diet of diatoms (particle
'A') when the proportion of diatoms varies. In this example,
P is the probability of obtaining five diatoms for every ten
particles consumed.

The Particle Size-Body Size Paradigm

One of the potentially most important observations made about
deposit feeders is the positive correlation between feeding upon
larger particles and increased body size. This capacity might result in the evolution of body size differences among coexisting species to relax competition for given particle size classes (Fenchel,
1975).

Although several studies document a particle-size versus body-
size relationship (e.g., Fenchel, 1975; Whitlatch, 1974), few have
elucidated the mechanism that permits differential particle size
selection. Hughes (1973) investigated feeding by the tellinacean
bivalve *Abra tenuis* on sand grains. The diameter of the inhalent
siphon was linearly related to shell length. Correspondingly, inhalent siphon diameter was linearly related to particle diameter.
No further sorting was accomplished in the mantle cavity or on the
palps and the animals could not distinguish among grains of identical size but could among different specific gravities. A deposit-
feeding ampharetid polychaete can distinguish between particles of
differing specific gravities; the capacity for selectivity increases
with increasing body size (Self and Jumars, 1978).

Problems with the Size Paradigm

The results presented thus far suggest several problems with
the hypothesis that body size is a simple mechanism for the evolution of ecological displacement among deposit-feeding species:
1)    If the deposit feeder, at a given body size, has a spectrum of
      rates of ingestion of different particle sizes, then the parti-
      cle size spectrum will depend upon the sediment upon which the
      deposit feeder feeds (Fenchel and Kofoed, 1976). Figure 2 il-
      lustrates the relative feeding rates of three different-sized hy-

pothetical despoit feeders upon two differing sediment types.
We hypothesize an increased capacity for feeding upon large
particles as in Hughes (1973). Note that the symmetrical unimodal sediment A results in a right-skewed particle size distribution in the small animals but a virtually symmetrically
unimodal distribution in the guts of large individuals. In
the case of sediment B, the gut contents of all three body
sizes are all right-skewed and similar in appearance. Note
that when sediment A is fed upon, there is a striking degree
of difference between the predicted gut contents of small and
large-sized individuals. But with sediment B, the differences
are not pronounced. Thus, body size differences have varying
effects on niche separation depending upon the texture of the
sediment.

2)  Although body size differences may permit niche displacement
    with respect to particle size ingested, a large individual
    must pass through stages when its size matches those of smaller
    species. Similarly, a species living at depths within the
    sediment may avoid competition with species near the sediment-
    water interface. But as a juvenile, the former species (e.g.,
    a bivalve) is likely to live at the same depth as a shallow-
    burrowing species. Thus, niche relationships among continuous-
    ly growing species are probably far more complex than a simple
    comparison among full-sized adults would indicate. In the case
    of *Hydrobia*, juveniles grow rapidly into the modal size of the
    adult population (Fenchel, 1975).

3)  Switching. We have postulated strict relationships between
    body size and ingested particle size. But what if the deposit
    feeder can switch to different particle sizes in different sed-
    iments? Optimal foraging theories (e.g., Charnov, 1976) pre-
    dict switches to different food types that depend upon relative
    abundance and quality of the alternative types. If a preferred
    particle size were rare, would the deposit feeder switch to
    feeding preferentially on a less preferred but common particle
    size? Obviously, such a switch can only occur if the deposit
    feeder has morphologic and behavioral flexibility that would
    permit the switch. If such plasticity exists, then niche dif-
    ferences would vary depending upon the relative abundance of
    different alternative particle types.

Particle Size Specialization in *Hydrobia*

Hydrobiid snails are found in freshwater, brackish and marine
environments throughout the world (Davis, 1979). They occur on a
variety of substrates ranging from fine-grained sediments to rock
surfaces (e.g., Newell, 1965; Fenchel, 1975; Sanders *et al*., 1962;
Levinton *et al*., 1977; Davis, 1979). Although the group consists
of two distinct families (Hydrobiidae and Pomatiopsidae), all species share a remarkably homogeneous rostral, radular and buccal mass
structure (Davis, 1979). A few adaptive radiations in fresh water
have generated some remarkable morphological divergences, but most
Hydrobiid snails are similar enough in appearance to cause systematic problems (e.g., Davis, 1979; Muus, 1967). This suggests an
ecological homogeneity and possible severe competition among coexisting species.

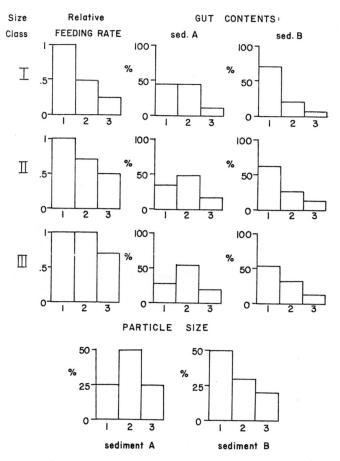

Figure 2. Predicted gut contents when different-sized deposit-feeders (size class I represents smallest animals) feed on one of two differing sediments.

Studies of feeding have been mainly limited to the marine and brackish water genus *Hydrobia*. *Hydrobia* has been regarded as a strict deposit feeder that depends upon particles within the sediment (Newell, 1965; Fenchel, 1975; Hylleberg, 1975; Kofoed, 1975). Like most deposit feeders studied thus far, *Hydrobia ventrosa* collected in Denmark is capable of efficient digestion and assimilation of diatoms and bacteria (Kofoed, 1975). It may derive most of its nutrition, however, from large diatoms (Fenchel and Kofoed, 1976).

Fenchel (1975) described a relationship between mean particle (diatom) size and body length that was similar for three species of *Hydrobia*. Thus small snails, irrespective of the particular species, fed upon small diatoms. Large individuals fed upon large diatoms. Thus, coexisting species might relax interspecific competition for food via the evolution of body size differences. Indeed, two species combinations of *Hydrobia ventrosa* and *H. ulvae* in the

Limfjord, Denmark show such character displacement.  Populations of
singly occurring *H. ulvae* have mean shell lengths not statistically
distinguishable from *H. ventrosa*, as predicted by the competitive
model.  However, a third species of *Hydrobia*, *H. neglecta*, fails to
show such obvious patterns.  Furthermore, the general pattern of
character displacement is not repeated in localities along the shore
of eastern Jutland (J. Hylleberg, pers. comm.).

   Fenchel and Kofoed (1976) measured ingestion rates on different
particle size fractions of sediment.  With these rates, they attempted to predict the relative feeding rates when different-size
particles were mixed in the same food.  When multiplied by the particle size class frequencies in natural *Hydrobia* sediments, the
feeding rates measured in this way adequately predicted ingestion
by large snails.  From gut contents and grazing experiments they
concluded that". . .there seems to be a critical minimum size of
particles which the snails can ingest; . . .this is probably in the
range of 8-16 µm for snails of the size range 3-4 mm. . ." (Fenchel
and Kofoed, 1976, p. 371).  As will be shown below, this inference
is questionable.

   The evidence gathered for Hydrobiid snails thus suggests that
deposit feeders are food-limited, as suggested several years ago by
Levinton (1972).  Available diatoms may be the principal limiting
factor.  The renewal rate of the diatom population may thus control
food availability.  Levinton (1979) suggests that it would be prof-
itable to examine various microbial resource-renewal models in order
to predict the standing stock of diatoms in equilibrium with different grazing rates.

   Deposit feeders may also be limited by particle abundance.  In-
gestible particles may be too rare to provide a sufficient source
of surface-living microbes.  This can happen in two ways.  First,
due to the hydrodynamic regime, some sediments may simply be defi-
cient in fine particles.  This probably explains the usual abundance
of deposit feeders in fine-grained sediments (Sanders, 1958).  *Hydrobia* and *Macoma* (Bivalvia:  Tellinacea) population density can be
related to the abundance of fine particles in the sediment (Newell,
1965).  Second, fine particles could be bound up in fecal pellets
and not be available for ingestion.  Ingestion of fecal pellets may
be rarer than reported in the literature (Levinton and Lopez, 1977).
Avoidance of coprophagy is consistent with the findings of Lopez and
Levinton (1978) that fecal material is largely stripped of digest-
ible microbes.  Therefore the rate of pelletization balanced by the
rate of pellet breakdown may partially determine particle availability (Levinton *et al.*, 1977; Levinton and Lopez, 1977).

   The availability of particles is partially determined by the
relative feeding rates of a deposit feeder as a function of particle
diameter.  It is therefore important to know both the feeding rate
as a function of particle size as well as the size distribution of
the sediment.  As mentioned above, it is also possible that some deposit feeders might alter their particle feeding behavior with a
change in the size spectrum of the sediment.

   The feeding rate of *Hydrobia* (3.0-3.5 mm in shell length) shows
an approximate linear decline with the logarithm of particle size
(Fenchel and Kofoed, 1976).  But particle ingestion rate does not
decrease with particles less than 8 to 16 µm, as suggested by Fenchel

and Kofoed. Rather, snails feed the fastest on this size range
(Lopez and Levinton, 1978). Both studies report egestion rates
(= ingestion rates) by snails on different sieved size fractions.
Egestion is measured as pellets $hr^{-1}$, because pellet size does not
change appreciably with increasing particle size (Fenchel and
Kofoed, 1976). One then assumes that overall feeding rate in a
mixed-size fraction sediment is the weighted average of the feeding
rates on individual size fractions.

I have investigated particle size feeding by *Hydrobia totteni*(?) on mixed fractions of particles using a Coulter Corporation
model TA II electronic particle counter. Two mixed fractions of
glass beads were presented to snails of ca 3 mm in length. One
fraction consisted of an approximately even distribution of size
classes by volume. The other was constructed such that large particles occupied the great majority of the sediment. In both cases
particles were previously cultured with bacteria. Figure 3 shows
that the relative rates of ingestion do not vary between the different size-frequency distributions. Furthermore, there is an obvious
strong preference for fine particles that does not diminish with decreasing particle size. This evidence supports the assumption of
Fenchel and Kofoed (1976) that feeding rates and sediment particle
size distribution alone are sufficient to predict particle ingestion. The data do not, however, support the assumption that *Hydrobia* can switch its particle ingestion rates in sediments of different grain size distributions. This would not support a model predicting optimal foraging.

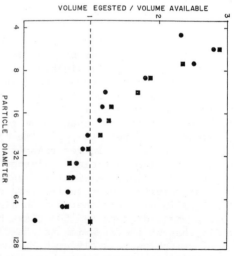

Figure 3. Relative rates
of ingestion of different size
classes of glass beads for *Hydrobia totteni* (?). Circles represent cases where all size classes
were approximately equally abundant (by volume). Squares represent a sediment where large
size classes (greater than ca.
40 microns) were enriched by a
factor of 6.

From the point of view of optimizing food intake, it is of interest that fine particles with the greatest surface area/volume
ratio are ingested at the fastest rate. This would maximize the
intake of surface-bound microbial organisms per unit volume of ingested sediment. We might ask, is the decline of feeding with increasing particle diameter consistent with the surface/volume ratio
of a sphere (6/D)? If an approximate hyperbolic relationship is
present, a log-log plot of feeding rate versus particle diameter
should yield a straight line. Figure 4 shows a probable departure

from a hyperbola. It seems likely that the particle feeding rate
may be best explained by the mechanics of the buccal apparatus of
*Hydrobia*; however, our present knowledge is insufficient to explain
the observed data.

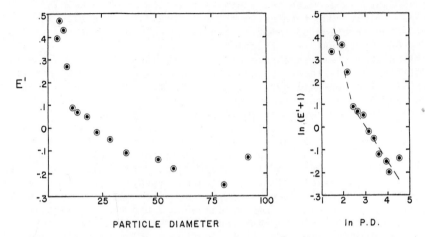

Figure 4. Comparison of relative feeding rates on
individual glass bead size classes (as measured by Ivlev's
index of preference) to a hyperbola. Left diagram seems to
depart from a hyperbola. Log-log plot on the right may be
composed of two straight lines of differing slope, but results are equivocal (hypothetical lines are dashed in).

Although neither the response to different particle size-frequency distributions nor the pattern of feeding rates on different
particle sizes is indicative of optimization, an important switch
in feeding mode occurs with coarse particles. Lopez and Levinton
(1978) measured egestion rates of *H. ventrosa* on different size
fractions of natural sediment and found a steadily decreasing rate
from sizes of less than 10 μm to about 125 μm. However, egestion
(measured as pellets·snail$^{-1}$·h$^{-1}$) increased when snails were presented particles larger than 125 μm. Pellets, however, consisted
of very fine particles – diatoms that were scraped off the surfaces
of sand grains. A similar experiment with *H. totteni* yielded similar results. When snails were presented glass beads greater than
80 μm in diameter coated with diatoms, pellets consisted of about
50% small diatoms and 50% glass beads of a size frequency similar
to that of the experimental "sediment." Thus *Hydrobia* is capable
of switching to a different feeding behavior when easily swallowed
particles are not available. This switch greatly complicates simple statements about differences in particle size ingestion vis-a-
vis niche separation (e.g., Fenchel, 1975; Whitlatch, 1974). Sediments of differing textures may differ strongly in the relative
proportions of swallowing versus scraping. Kofoed and Lopez (unpublished) have evidence to suggest that such complications exist
for *H. ventrosa*.

A final but very important point revolves around the question
of the evolutionary potential of body size differences with respect

to particle size. Fenchel (1975) demonstrates that size differences among sympatric species of *Hydrobia* are sufficient to permit substantial differences in the particle size frequency distribution of gut contents. But such differences could not evolve unless the variance in body size <u>within</u> a single year class was sufficient to permit particle size differentiation. Furthermore, there must exist sufficient genetic variability to permit natural selection to change the mean body size of a population. Thus, both detailed particle ingestion and inheritance studies are necessary. It would be of great interest to know the heritability of body size in *Hydrobia* but we do have some data on small differences within populations.

We (Levinton and DeWitt, in preparation) have examined differences between snails (*H. totteni*?) of about 2.8 mm and 3.2 mm in shell length; both size classes come from a single adult year class in Flax Pond, New York. Snails of each size group were presented with a sample of glass beads of a broad size-frequency spectrum. Using particle size classes delineated by the Coulter Counter, we calculated the volume egested/volume available for each size class; this gives a measure of preference, P. We present the ratio of P for large snails divided by the P for small snails in Figure 5. This ratio is unbounded in magnitude and biased toward preferential feeding by the larger snails on larger particles. If the ratio is less than 1, then particles are ingested in relatively greater proportions by the small snails. If it is greater than 1, then the particle size class is ingested relatively more rapidly by large snails. Small snails ingest particle sizes up to about 32 µm in diameter in greater proportions. However, above 32 µm there is a dramatic increase in the ability of large snails to ingest particles, as compared to small snails. The change is precipitous enough to suggest a threshold effect. The enrichment of small particles in the guts of small snails as compared to large snails reflects the ability of larger snails to ingest larger particles rather than a relative inability to ingest smaller particles. This result is analogous to that found by T. Hughes for particle size selectivity in the bivalve *Abra tenuis*. In summary, the essential assumption that within-year class variation must show significantly different particle size ingestion is substantiated. Furthermore, the nature of the difference is apparently a threshold effect.

PROSPECTUS

Some fundamental gaps in our understanding of deposit feeding stand in the way of adequately measuring the ecological consequences of qualitative food specialization and particle size specialization. The question of microbial versus dead organic matter in nutrition is central to the construction of deposit-feeding models (e.g., Taghon *et al.*, 1978). Furthermore, it will be necessary to quantitatively estimate for the diet of some deposit feeders the relative proportions of surface-bound and free-living microflora. This is crucial in predicting particle size preference. If surface-bound microbes are of greatest importance, then it pays to specialize on fine particles. But the reverse would be true if free-living diatoms are the primary constituent of the diet.

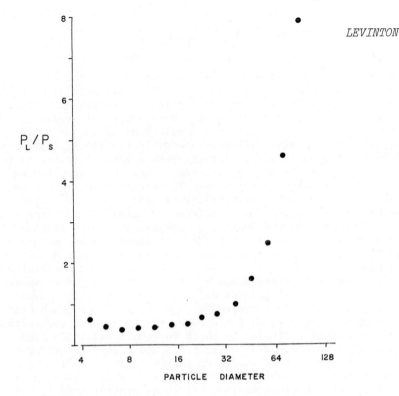

Figure 5.   Ratio of particle preference of
large (3.2 mm) snails ($P_L$) relative to small (2.8
mm) snails ($P_s$) as a function of particle diameter.


Available evidence suggests that the feeding behavior of a deposit feeder such as *Hydrobia* can be complex.  Because a switch from
swallowing to scraping is possible, it is difficult to construct
simple models of particle ingestion and microbial renewal (as in
Levinton, 1979) because the food source is not adequately understood.  Furthermore, different grazing intensities can apparently
influence the digestibility of dominant surface-bound microbes
(Lopez and Levinton, 1978).

Despite these complications, however, the swallowing mode of
feeding is well understood.  Ingestion seems to be a stochastic process with respect to particle size.  Relative feeding rates on given
particle sizes seem constant in sediments of differing particle size
frequency distributions.  Thus ingestion as a function of particle
size is most probably regulated by mechanical constraints of the
buccal apparatus.  Finally, we can observe obvious differences in
particle ingestion between different-sized snails within a single
year class.

It is my hope that this discussion points out that, to be properly understood, particle size specialization must be studied in detail with attention paid to the mechanical mechanism of particle
collection, switches in feeding behavior, qualitative effect on
feeding by different food sources, and the particle size ingestion
rates between animals of small body size differences.  A final area
for future work must be the genetic basis for body size differences
and heritability measurements in natural populations.

(Author's note: Recent experiments show that *Hydrobia* shows less preference for fine particles when coarse particles are rare in the sediments).

ACKNOWLEDGMENTS

This paper has benefitted from conversations with G. Lopez. I thank T. DeWitt and M. Reigner for assistance in the laboratory. Supported by NSF grant (Biological Oceanography) OCE78-09057. This paper is contribution number 327 to the Program in Ecology and Evolution, State University of New York at Stony Brook.

REFERENCES

Aller, R. 1980. This volume.

Barsdate, R. J., R. T. Prentki, and T. Fenchel. 1974. Phosphorus cycle of model ecosystems: significance for decomposer food chains and effect of bacterial grazers. Oikos. 25: 239-251.

Calow, P. 1975. Defecation strategies of two freshwater gastropods, *Ancylus fluviatilis* Mull. and *Planorbis contortus* Linn. (Pulmonata), with a comparison of field and laboratory estimates of food absorption rate. Oecologia (Berl.). 20: 51-63.

Cammen, L., P. Rublee, and J. Hobbie. 1978. The significance of microbial carbon in the nutrition of the polychaete *Nereis succinea* and other aquatic deposit feeders. Univ. of North Carolina Sea Grant publ. UNC-SG-78-12, pp. 1-84.

Charnov, E. L. 1976. Optimal foraging: attack strategy of a mantid. Amer. Natur. 110: 141-152.

Christian, R. R., K. Bancroft, and W. J. Wiebe. 1975. Distribution of microbial adenosine triphosphate in salt marsh sediments at Sapelo Island, Georgia. Soil Sci. 119: 89-97.

Crosby, N. D. and R. G. B. Reid. 1971. Relationships between food, phylogeny, and cellulose digestion in the Bivalvia. Can. J. Zool. 49: 617-622.

Davis, G. M. 1979. The origin and evolution of the Pomatiopsidae, with emphasis on the Mekong River hydrobiid gastropods. Monogr. no. 20, pp. 1-120, Acad. Natur. Sci. Philadelphia.

Fenchel, T. 1970. Studies on the decomposition of organic detritus derived from the turtle grass *Thalassia testudinum*. Limnol. Oceanogr. 15: 14-20.

Fenchel, T. 1972. Aspects of decomposer food chains in marine benthos. Verh. Deutsch. Zool. Ges. 65: 14-22.

Fenchel, T. 1975. Character displacement and coexistence in mud snails (Hydrobiidae). Oecologia (Berl.). 20: 19-32.

Fenchel, T. and L. H. Kofoed. 1976. Evidence for exploitative interspecific competition in mud snails (Hydrobiidae). Oikos. 27: 367-376.

Ferguson, R. L. and M. B. Murdoch. 1975. Microbial ATP and organic carbon in sediments of the Newport River estuary, North Carolina. pp. 229-250, *In* L. E. Cronin, (ed.) Estuarine Research, Vol. 1, Academic Press, New York.

Frankenberg, D. and K. L. Smith.  1967.  Coprophagy in marine ani-
    mals.  Limnol. Oceanogr. 12:  443–450.
Gordon, D. C.  1966.  The effects of the deposit-feeding polychaete
    *Pectinaria gouldii* on the intertidal sediments of Barnstable
    Harbor.  Limnol. Oceanogr. 11:  327–332.
Hargrave, B. T.  1970.  The utilization of benthic microflora by
    *Hyalella azteca* (Amphipoda).  J. Anim. Ecol. 39:  427–437.
Hargrave, B. T.  1972.  Prediction of egestion by the deposit-feed-
    ing amphipod *Hyalella azteca*.  Oikos. 23:  116–124.
Harrison, P. G. and K. H. Mann.  1975.  Detritus formation from eel-
    grass (*Zostera marina* L.):  the relative effects of fragmenta-
    tion, leaching and decay.  Limnol. Oceanogr. 20:  924–934.
Hughes, T. G.  1973.  Deposit feeding in *Abra tenuis* (Bivalvia:
    Tellinacea).  J. Zool. 171:  499–512.
Hughes, T. G.  1975.  The sorting of food particles by *Abra* sp. (Bi-
    valvia:  Tellinacea).  J. Exp. Mar. Biol. Ecol. 20:  137–156.
Jorgensen, B. G.  1977.  Bacterial sulfate reduction within reduced
    microniches of oxidized marine sediments.  Mar. Biol. 41:  7–17.
Kofoed, L. H.  1975.  The feeding biology of *Hydrobia ventrosa*
    (Montagu).  I.  The assimilation of different components of the
    food. ¨J. Exp. Mar. Biol. Ecol. 19:  233–241.
Kristensen, J. H.  1972.  Carbohydrases of some marine invertebrates
    with notes on their food and on the natural occurrence of the
    carbohydrates studied.  Mar. Biol. 14:  130–142.
Levinton, J. S.  1972.  Stability and trophic structure in deposit-
    feeding and suspension-feeding communities.  Amer. Natur. 106:
    472–486.
Levinton, J. S.  1977.  The ecology of deposit-feeding species:
    Quisset Harbor, Massachusetts. pp. 191–228.  *In* B. C. Coull (ed.)
    Ecology of Marine Benthos.  Univ. of South Carolina Press, Columbia.
Levinton, J. S.  1979.  Deposit feeders, their resources and the
    study of resource renewal. Florida State Univ. Press, Tallahassee,
    in press.

Levinton, J. S. and R. K. Bambach.  1975.  A comparative study of
    silurian and recent deposit-feeding bivalve communities.  Paleo-
    biology. 1:  97–124.
Levinton, J. S. and G. R. Lopez.  1977.  A model of renewable re-
    sources and limitation of deposit-feeding benthic populations.
    Oecologia (Berl.). 31:  177–190.
Levinton, J. S., G. R. Lopez, H. H. Lassen, and V. Rahn.  1977.
    Feedback and structure in deposit-feeding marine benthic commu-
    nities, pp. 409–416.  *In* Proc. Eleventh European Mar. Biol.
    Symp., Galway.
Lopez, G. R. and J. S. Levinton.  1978.  The availability of micro-
    organisms attached to sediment particles as food for *Hydrobia
    ventrosa* Montagu (Gastropoda:  Prosobranchia).  Oecologia
    (Berl.). 32:  263–275.
Lopez, G. R., J. S. Levinton, and L. B. Slobodkin.  1977.  The
    effect of grazing by the detritivore *Orchestia grillus* on *Spartina*
    litter and its associated microbial community.  Oecologia
    (Berl.). 30:  111–127.
Mann, K. H.  1972.  Macrophyte production and detritus food chains
    in coastal waters.  1st Ital. Idrobiol. 29(suppl.):  353–383.

Morrison, S. J., J. D. King, R. J. Bobbie, R. E. Bechtold, and D. C. White. 1977. Evidence for microfloral succession on allochthonous plant litter in Apalachicola Bay, Florida, USA. Mar. Biol. 41: 229-240.

Muus, B. J. 1967. The fauna of Danish estuaries and lagoons. Kommn. Danm. Fisk.-og Havunders. (N.S.) 5: 1-316.

Newell, R. C. 1965. The role of detritus in the nutrition of two marine deposit feeders, the prosobranch *Hydrobia ulvae* and the bivalve *Macoma balthica*. Proc. Zool. Soc. London. 144: 25-45.

Nicolaisen, W. and E. Kanneworf. 1969. On the burrowing and feeding habits of the amphipods *Bathyporeia pilosa* Lindstrom and *Bathyporeia sarsi* Watkin. Ophelia. 6: 231-250.

Peterson, C. H. 1977. Competitive organization of the soft-bottom macrobenthic communities of southern California lagoons. Mar. Biol. 43: 343-359.

Rhoads, D. C. 1967. Biogenic reworking of intertidal and subtidal sediments in Barnstable Harbor and Buzzard's Bay, Massachusetts. J. Geol. 75: 461-476.

Rhoads, D. C. and D. J. Stanley. 1965. Biogenic graded bedding. J. Sed. Petrol. 35: 956-963.

Rhoads, D. C. and D. K. Young. 1970. The influence of deposit-feeding organisms on sediment stability and community trophic structure. J. Mar. Res. 28: 159-178.

Sanders, H. L. 1958. Benthic studies in Buzzard's Bay. I. Animal-sediment relationships. Limnol Oceanogr. 3: 245-258.

Sanders, H. L., E. M. Goudsmit, E. L. Mills, and G. E. Hampson. 1962. A study of the intertidal fauna of Barnstable Harbor, Massachusetts. Limnol. Oceanogr. 7: 63-79.

Self, R. F. and P. A. Jumars. 1978. New resource axes for deposit feeders? J. Mar. Res. 627-641.

Taghon, G., R. L. Self, and P. A. Jumars. 1978. Predicting particle selection by deposit feeders: a model and its implications. Limnol. Oceanogr. 23: 752-759.

Tenore, K. R. 1977. Growth of the polychaete, *Capitella capitata*, cultured on different levels of detritus derived from various sources. Limnol. Oceanogr. 22(5): 936-941.

Tenore, K. R., R. B. Hanson, B. E. Dornseif, and C. N. Weiderhold. 1979. The effect of organic nitrogen supplement on the utilization of different sources of detritus. Limnol. Oceanogr. 24: 350-355.

Tunnicliffe, V. and M. J. Risk. 1977. Relationships between the bivalve *Macoma balthica* and bacteria in intertidal sediments: Minas Basin, Bay of Fundy. J. Mar. Res. 35: 499-507.

Wetzel, R. L. 1976. Carbon resources of a benthic salt marsh invertebrate *Nassarius obsoletus* Say (Mollusca: Nassariidae). pp. 293-308. *In* M. Wiley (ed.) Estuarine Processes, V. 2, Academic Press, New York.

Whitlatch, R. B. 1974. Food-resource partitioning in the deposit-feeding polychaete *Pectinaria gouldii*. Biol. Bull. 147: 227-235.

Woodin, S. A. 1974. Polychaete abundance patterns in a marine soft-sediment environment: the importance of biological interactions. Ecol. Monogr. 44: 171-187.

# SUMMARY

Marine benthic research has undergone a tremendous expansion, evolving from a science emphasizing mostly descriptive and structural aspects, to one that attempts to unravel the complexities of benthic processes and community interactions. The papers included in this volume not only summarize much of the knowledge concerning benthic dynamics through 1979, but suggest new and more difficult problems and point the way to further research.

## I.  SECONDARY PRODUCTION

There are inherent difficulties in measuring production of marine benthic organisms, and all too often this has resulted in inadequate production estimates. Warwick points out the need to incorporate larval macrofauna and predator cropping and subsequent regeneration of siphons, tentacles, etc. into production estimates. He recognizes that measuring production is confounded with small organisms and that P/B ratios are extremely variable. Further, he generalizes that carnivore production is always low, deposit-feeder production decreases linearly with depth, and suspension feeders have variable production in shallow water which decreases to near zero in depths exceeding 50 m. Mills also proposes a depth-stratified secondary production scheme that is based primarily on how phytoplankton are distributed. Using examples from eastern North America, he proposes that shallow water systems lead primarily to demersal fish dominance (via the benthos), whereas in deeper water, pelagic fish production dominates via a grazing food web. Intermediate

441

depths cannot be generalized. Within a habitat, Glemarec demonstrates that seasonal temperature fluctuations influence benthic
production at every trophic level. Specific physical anomalies reduce the success of macrofaunal recruitment and thus benthic secondary production is dependent upon physically-controlled, year-to-year
pulses of recruits. Pamatmat suggests that facultative anaerobiosis
is widespread among benthic animals. While the production yield
from anaerobic metabolism is low, it may actually save energy by allowing the animal to avoid shutdown of all metabolic processes under
low oxygen conditions. To accurately prepare a population energy
budget, Pamatmat argues that $O_2$ uptake is not sufficient, and one
must know the percent of time the organism spends in the aerobic and
anaerobic phase.

The information in this section cautions us to avoid generalizations on secondary production and to look at individual organisms,
habitats, and life-styles in more detail. This emphasis should certainly spawn increased research and provide the details Mills claims
are the forte of benthic ecologists.


## II. POPULATIONS STUDIES

Population dynamics is emerging as a valuable approach to investigate regulation of benthos. Dayton and Oliver review the traditional methods of field experimentation as they affect populations
and discuss many of the paradigms, i.e., functional groups, competition, disturbance and predation, used by benthic population ecologists. They argue that using an experimental design based on an
existing paradigm leads to perpetuation of preconceived ideas, the
"verifications syndrome." They urge combination lab/field experimentation with the emphasis on falsification of existing hypotheses
to determine the true value of existing paradigms. It has often
been assumed that predation pressure controls macrobenthic production, but Arntz, reporting on predation by demersal fish on macrobenthos, demonstrates that over the long-term (8 yrs), neither fish
nor benthic production is related, although he shows that demersal
fish do indeed prey heavily on benthos. Yearly differences in predation pressure change macrobenthic abundances, but in the long-term
there is no relationship.

The Alongi and Tietjen, Heip, and Bell and Coull papers deal
with various aspects of meiofauna population dynamics. Alongi and
Tietjen suggest competition between nematodes with very similar buccal morphologies. They demonstrate that two deposit-feeding nematodes grow slower when cultured together than when cultured singly.
On bacteria, both species survive, but on a chlorophyte, one is
eliminated. They found no interaction, however, when a deposit
feeder was grown with an epistrate feeder, suggesting no competitive
hierarchy when these two different feeding types coexist. Heip it-
erates the need for data at the species level when trying to partition production of benthic copepods. He found that although year-
to-year number of generations and thus, production, are highly variable, the mean values for seven years of all the meiobenthic copepods changed little: when one species declined, another took its

place. Bell and Coull report on the long overlooked interaction between recently-settled macrofauna larvae and the permanent meiofauna. They show that, in the absence of predation, permanent members of the meiofauna increase as the larval macrofauna decrease.
They present a general model and argue that juvenile macrofauna may
experience significant mortality during their temporary meiofauna
stage, which is due to competitive interactions with the permanent
meiofauna.

All the papers in this section point again to the necessity for
detail, since each species is unique. These studies suggest that,
while certain generalizations may be true, data are too scanty,
poorly obtained and/or the paradigm not tested rigorously to confirm
them. The papers illustrate that the study of population dynamics
of marine benthos is in its infancy. There is considerable room for
studies using different philosophies, approaches and species to examine population/life history tactics of marine benthos.


III. NUTRIENT CYCLING

Nutrients regenerated from the bottom are known to be important
sources for phytoplankton production. Zeitzschel outlines four se-
quential events necessary for the nutrient cycle: 1) phytoplankton
bloom, 2) sedimentation of the bloom, 3) decomposition at the sediment-water interface, and 4) return of the nutrients to the euphotic
zone by turbulent mixing. The time for each step, of course, varies
depending on a variety of factors, and disturbance of events 2, 3
and 4, Zeitzschel emphasizes, is one probable cause of observed temporal and spatial variability in phytoplankton production. While
100% of the nutrients required by phytoplankton come from benthic
regeneration in Kiel Bay (Zeitzschel), only 50% of the phytoplank-
ton-required nutrients in Narragansett Bay come from benthic regeneration (Nixon). Nixon also points out that sedimented organic matter reworked and returned to the water column is low in nitrogen
compared to phosphorus, and thus may be responsible for the characteristically low N:P ratios in shallow water. Thus nitrogen is
the limiting nutrient in such shallow marine systems. Hargrave demonstrates how settled organic material, particularly phytoplankton
and zooplankton fecal pellets, control benthic metabolism. It is
not necessarily the quantity of the organic matter which controls
the respiration, but rather the quality, i.e., nitrogen and plant
pigment content, which leads to increased secondary production and
growth.

In the Sea of Japan, Propp *et al.* found that pronounced seasonal changes in nutrient fluxes out of the bottom were correlated
with temperature, and that benthic photosynthesis was greater than
respiration only during the spring. Oxygen uptake and silicon release were positively correlated with phosphate and ammonia concentrations. They point out that microbial activity and water chemistry are an integral part of nutrient regeneration, suggesting that
the latter is not simply diffusion-limited. Aller supports this by
demonstrating that macrofaunal tubes and burrows differentially affect nutrient fluxes. For example, nitrogen fluxes change little

with increased bioturbation, but silicon dramatically increased with increased worm activity. Aller proposes a two-dimensional model to explain transport of solutes and argues that burrow size and macro-benthic density control the chemical environment.

These sediment-water interface studies all point to the importance of benthic regeneration in fueling pelagic systems. The interactions are not simple. These efforts provide us with unique and necessary information about the coupling of these two marine habitats and illustrate the need for continued research on this dynamic process.


## IV.  DETRITUS

Detritus is a major food source supporting benthic secondary production, and in recent years studies on detritus have proliferated. Sibert and Naiman discuss the evolutionary importance of "decomposer" food webs and suggest that the first Precambrian food chains were detrital-based. They point out that although the importance of detritus to benthos has long been recognized, only recently have there been quantitative studies of the nutritional value of detritus and investigations of physical and biotic factors regulating detritus availability. Tenore and Rice contrast the dynamics of detritus-based food webs versus the typical herbivore/carnivore food chain; they review the chemistry of leaching and detritus composition as well as the biochemical role of protozoa and meiofauna in detrital mineralization. Hobbie and Lee point out that while microbes may be the principal food of small detritivores such as protozoans and nematodes, they doubt if microbial biomass is sufficient for a large deposit feeder. They hypothesize that the extracellular mucopolysaccharides of microbes, which are much more abundant than the microbes themselves, are the primary food sources for many benthic animals.

Fungi are also important aspects of detrital food webs and Fell *et al.* demonstrate that carbon and dry weight loss of red mangrove leaf litter increases with increased fungal activity. Furthermore, nitrogen immobilization requires fungal presence. *Zostera* detritus is also affected by various microorganisms (Harrison & Harrison). Total nitrogen increases with aging, and the Harrisons attribute this increase to the activity of colonizing microorganisms. They suggest that bacteria and algae compete for nutrients in detrital systems, but selective grazing by copepods modified the competitive effects. Hanson warns that many techniques used to measure microbial activity of decomposing detritus do not accurately assess the total microbial activity. He suggests that if nutrient substrates are used to evaluate metabolism/growth, they should be utilized by a large proportion of the microbes and synthesized into cellular components.

Different sediment/detrital particles harbor varied microbial associations. Sand with microbes affects ingestion of the mud snail *Ilyannassa obsoleta*, but silt-clay sized particle ingestion and retention were not affected by microorganism abundance, according to Lopez. Crystalline styled extract from *Mytilus* detaches bacteria and algae from sediment particles, and Lopez concludes that the availability of sediment-associated microorganisms might depend on specific detachment mechanisms and mode of digestion. Yingst and Rhoads propose that metazoan bioturbation activity, often dependent on sediment particle size, stimulates bacterial growth, thus increasing food for the deposit-feeding benthos. They argue that such activities are most common in soft-bottom equilibrium communities, whereas Levinton reviews particle feeding by deposit feeders and points out the limitations of current models to describe this activity. He emphasizes the regulating role of particle size on the feeding activity of deposit-feeding snails.

In general, two themes recur throughout this section - i.e., 1) the importance of nitrogen as the limiting nutrient in detrital systems and 2) the biological control of detritus remineralization and particle reworking. Both research avenues serve as the basis for much future research, and delineation of the mechanisms involved are, indeed, noble, worthwhile endeavors.

Columbia
February 1980

# INDEX